T0094118

Introduction to Strong Interactions

This is a problem-oriented introduction to the main ideas, methods, and problems needed to form a basic understanding of the theory of strong interactions. Each section contains solid but concise technical foundations to key concepts of the theory, and the level of rigor is appropriate for readers with a background in physics (rather than mathematics). It begins with a foundational introduction to topics including SU(N) group, hadrons, and effective SU(3) symmetric flavor lagrangians, constituent quarks in hadrons, quarks, and gluons as fundamental fields. It then discusses quantum chromodynamics as a gauge field theory, functional integration, and Wilson lines and loops, before moving on to discuss gauge–fixing and Faddeev–Popov ghosts, Becchi-Rouet-Stora-Tyutin symmetry, and lattice methods. It concludes with a discussion on the anomalies and the strong CP problem, effective action, chiral perturbation theory, deep inelastic scattering, and derivation and solution of the Dokshitzer–Gribov–Lipatov–Altarelli–Parisi equations.

Constructed as a one-term course on strong interactions for advanced students, it will be a useful self-study guide for graduate and PhD students of high energy physics, quantum chromodynamics, and the Standard Model.

Introduction to Strong Interactions

Theory and Applications

Andrey Grabovsky

CRC Press
Taylor & Francis Group
Boca Raton London New York

CRC Press is an imprint of the
Taylor & Francis Group, an **informa** business

First edition published 2023
by CRC Press
6000 Broken Sound Parkway NW, Suite 300, Boca Raton, FL 33487-2742

and by CRC Press
4 Park Square, Milton Park, Abingdon, Oxon, OX14 4RN

CRC Press is an imprint of Taylor & Francis Group, LLC

ISBN: 978-1-032-20675-2 (hbk)
ISBN: 978-1-032-22393-3 (pbk)
ISBN: 978-1-003-27240-3 (ebk)

DOI: 10.1201/9781003272403

Typeset in Latin Modern font
by KnowledgeWorks Global Ltd.

Publisher's note: This book has been prepared from camera-ready copy provided by the authors.

Contents

Preface

This material was given to the fifth year High Energy Physics students of the Novosibirsk State University in 2017–2022. These students studied quantum electrodynamics in the preceding semester and Quantum and Analytical Mechanics, Mathematical Physics and Introduction into Field Theory in previous years. Therefore, they had knowledge of Group Theory, Lagrangian and Hamiltonian approaches, Canonical Quantization and Feynman diagrams, including one-loop integration and renormalization in Quantum Electrodynamics. However, no knowledge of Functional Integration was assumed. This is an introductory compulsory course spanning one semester of 18 weeks, 18 lectures, and 18 problem solving sessions. The material in each chapter usually took from one to three weeks. More advanced topics are covered in the follow-up optional course on QCD.

The course persues the following aims:

1. Gives the main physical ideas of the theory of strong interactions: asymptotic freedom, confinement, chiral symmetry breaking, anomalies, condensates;

2. Introduces the principal concepts and methods: $SU(N)$ group symmetry, gauge invariance, renormalisation, functional integration, lattice, chiral perturbation theory, parton distribution functions, evolution equations;

3. Links the theory and experiment discussing high energy scattering and low energy effective Lagrangians.

This course tries to give solid but as short as possible technical foundation. The material in each section is mostly taken from the sources given at the end of the section. The level of rigor is physical rather than mathematical.

I am grateful to V. S. Fadin and R. N. Lee who taught strong interactions to me when I was a student, D. Yu. Ivanov, M. G. Kozlov, A. V. Reznichenko, who taught this course with me. I thank these people and I. I. Balitsky, A. V. Bogdan, R. Boussarie, A. L. Feldman, L. V. Kardapoltsev, A. I. Milstein, L. Szymanowski, S. Wallon and many other people I met at different schools and conferences for discussions improving my understanding of the subject. I also thank the students who took this course and helped to make it better. The project is being implemented by the winner of the master's program faculty grant competition 2020/2021 of the Vladimir Potanin fellowship program.

CHAPTER 1

SU(N) Group

A KEY mathematical object we deal with in the theory of strong interactions is the $SU(3)$ group. Therefore, we have to take a close look at it and at the $SU(N)$ group in general. In this chapter we will cover the properties of these groups necessary for further discussion of the theory. We need the following facts.

1. $SU(N)$ is a **non-Abelian group**, i.e. $g_1 g_2 \neq g_2 g_1$ for $g_{1,2} \in SU(N)$, $N \geq 2$.

2. In the **fundamental (defining) representation**, $U_i^j \in SU(N)$ is a $N \times N$ matrix acting in the space of vertical vectors $\psi^i = \begin{pmatrix} \psi^1 \\ \cdots \\ \psi^N \end{pmatrix}$, $\psi^{j\prime} = U_i^j \psi^i$. These functions ψ^i are often called the contravariant components of spinors.

3. The Hermitian conjugate vectors $\phi_j = (\psi^\dagger)_j = (\psi_1^*, \ldots, \psi_N^*)$ are transformed by the Hermitian conjugate matrices $U_i^{\dagger j}$, $\phi_j' = \phi_i (U^\dagger)_j^i$, which we will call the matrices in the **antifundamental representation**. The functions ψ_j are often called the covariant components of spinors.

4. One can form **tensors** with an arbitrary number of the covariant and contravariant indices with the transformation law

$$\psi_{p_1 \ldots p_l}^{n_1 \ldots n_k}(2) = U_{i_1}^{n_1} \ldots U_{i_k}^{n_k} \psi_{j_1 \ldots j_l}^{i_1 \ldots i_k}(1) U_{p_1}^{\dagger j_1} \ldots U_{p_l}^{\dagger j_l}. \tag{1.1}$$

5. $SU(N)$ is a **unitary** group, i.e. $U \in SU(N) \implies U^\dagger = U^{-1}$, $U = e^{i\theta}$, with a Hermitian matrix θ :

$$\theta = \theta^\dagger. \tag{1.2}$$

Hence, δ_j^i is an **invariant tensor**

$$\delta_l^k = U_i^k \delta_j^i U_l^{\dagger j}. \tag{1.3}$$

DOI: 10.1201/9781003272403-1

6. $SU(N)$ is a **special** group, i.e. $U \in SU(N) \implies \det U = 1$, or via

$$\det A = e^{tr \ln A},$$ (1.4)

$e^{tr(i\theta)} = 1 \implies$

$$tr(\theta) = 0.$$ (1.5)

Hence, $\varepsilon^{i_1 \cdots i_N}$ and $\varepsilon_{i_1 \cdots i_N}$ are invariant tensors

$$\varepsilon^{j_1 \cdots j_N} = \varepsilon^{i_1 \cdots i_N} U_{i_1}^{j_1} \cdots U_{i_N}^{j_N} = \varepsilon^{j_1 \cdots j_N} \det U.$$ (1.6)

7. **Center** (the subgroup commuting with all matrices from the group) of $SU(N)$ is $Z_N \ni (U_n)_j^i = e^{\frac{2\pi i n}{N}} \delta_j^i$.

8. $dim\, SU(N) = N^2 - 1$. Indeed, the total number of real parameters in θ is $2N^2$. Hermiticity (1.2) implies

$$N_{\substack{(real,\ from \\ diagonal)}} + 2 \frac{N^2 - N}{2}_{\substack{(complex, \\ from\ the \\ corner)}} = N^2$$ (1.7)

conditions for θ and tracelessness (1.5) adds one.

9. A representation is called **real** if one can find a basis such that

$$U^* = S^{-1} U S,$$ (1.8)

i.e. real representations are equivalent to their conjugate:

$$\psi' = U\psi \implies S^{-1}\psi' = S^{-1}USS^{-1}\psi = U^*S^{-1}\psi.$$ (1.9)

In $SU(2)$ the defining representation is real.

10. **Algebra** $aSU(N)$ spans over $N^2 - 1$ generators $\mathbf{t}^a$, $\theta = \theta^a \mathbf{t}^a$, normalized so that

$$\left[\mathbf{t}^a, \mathbf{t}^b\right] = i f^{abc} \mathbf{t}^c,$$ (1.10)

where f^{abc} are **structure constants**. They are chosen totally antisymmetric.

11. The **Jacobi identity** reads

$$[\mathbf{t}^a[\mathbf{t}^b\mathbf{t}^c]] + [\mathbf{t}^b[\mathbf{t}^c\mathbf{t}^a]] + [\mathbf{t}^c[\mathbf{t}^a\mathbf{t}^b]] = 0,$$ (1.11)

$$f^{bcd} f^{ade} + f^{cad} f^{bde} + f^{abd} f^{cde} = 0.$$ (1.12)

12. It is customary to denote the generators in the **fundamental representation** t^a and normalize them so that

$$tr(t^a t^b) = \frac{1}{2} \delta^{ab}.$$ (1.13)

In the defining representation

$$\{t^a, t^b\} = \frac{1}{N}\delta^{ab} + d^{abc}t^c, \qquad (1.14)$$

where d^{abc} is a totally symmetric tensor.

13. For $SU(2)$ we have

$$t^a = \frac{\sigma^a}{2}, \quad d^{abc} = 0. \qquad (1.15)$$

For $SU(3)$ the common choice is $t^a = \frac{\lambda^a}{2}$.

$$\lambda^{1,2,3} = \begin{pmatrix} \sigma^{1,2,3} & 0 \\ 0 & 0 \end{pmatrix}, \quad \lambda^6 = \begin{pmatrix} 0 & 0 \\ 0 & \sigma^1 \end{pmatrix}, \quad \lambda^7 = \begin{pmatrix} 0 & 0 \\ 0 & \sigma^2 \end{pmatrix},$$

$$\lambda^4 = \begin{pmatrix} 0 & 0 & 1 \\ 0 & 0 & 0 \\ 1 & 0 & 0 \end{pmatrix}, \quad \lambda^5 = \begin{pmatrix} 0 & 0 & -i \\ 0 & 0 & 0 \\ i & 0 & 0 \end{pmatrix}, \quad \lambda^8 = \frac{1}{\sqrt{3}}\begin{pmatrix} 1 & 0 & 0 \\ 0 & 1 & 0 \\ 0 & 0 & -2 \end{pmatrix}$$

$$\qquad (1.16)$$

are known as the **Gell-Mann matrices**. Hence

$$f^{123} = 1, \quad f^{147} = f^{246} = f^{257} = f^{345} = f^{516} = f^{376} = \frac{1}{2},$$

$$f^{458} = f^{678} = \frac{\sqrt{3}}{2}; \qquad (1.17)$$

$$d^{118} = d^{228} = d^{338} = -d^{888} = \frac{1}{\sqrt{3}},$$

$$d^{146} = d^{157} = d^{256} = -d^{247} = d^{344} = d^{355} = -d^{366} = -d^{377} = \frac{1}{2},$$

$$d^{448} = d^{558} = d^{668} = d^{778} = \frac{-1}{2\sqrt{3}}, \qquad (1.18)$$

and d and f with other indices unrelated to these ones by permutations vanish.

14. The maximal number of mutually commuting matrices t or **rank** of $SU(N)$ is $N - 1$.

15. Any matrix from the algebra in the fundamental representation may be decomposed into t^a, which leads to the **completeness relation**

$$(t^a)^j_i (t^a)^l_k = \frac{1}{2}\delta^j_k \delta^l_i - \frac{1}{2N}\delta^j_i \delta^l_k. \qquad (1.19)$$

16. In the **adjoint representation**, U_{ab} is an $N^2 - 1 \times N^2 - 1$ matrix acting on the vectors from the algebra $aSU(N)$. So for $F = F^a t^a \in aSU(N)$

U acts in the space of vertical vectors $F^a = \begin{pmatrix} F^1 \\ \dots \\ F^{N^2-1} \end{pmatrix}$.

17. It is customary to denote the generators in the **adjoint representation** T^a. Then Jacobi identity (1.11) and definition of f (1.10) give

$$(T^b)_{ac} = i f^{abc}. \tag{1.20}$$

For $SU(2)$ we have

$$f^{abc} = \varepsilon^{abc}. \tag{1.21}$$

18. The **matrices in the adjoint and fundamental representations are related** via

$$U_{ab} t^a = U t^b U^{-1} \iff U_{ab} = 2 tr(t^a U t^b U^\dagger). \tag{1.22}$$

Therefore, the **adjoint action** of the group element U on $F = F^a t^a \in aSU(N)$ with generators in the fundamental representation reads

$$U F U^{-1} = t^a (U_{ab} F^b) \implies F'^a = U_{ab} F^b. \tag{1.23}$$

19. **Casimir operators commute with all generators.** Therefore, they are proportional to 1 on the basis vectors of the irreducible representations with a coefficient depending on the irreducible representation. **The number of the independent Casimir operators is equal to the rank of the group** (Racah theorem).

20. **Quadratic Casimir operator**

$$\hat{C}_2 = \mathbf{t}^a \mathbf{t}^a.$$

For the **fundamental** representation, it reads

$$(t^a t^a)^j_i = C_F \delta^j_i = \frac{N^2 - 1}{2N} \delta^j_i. \tag{1.24}$$

For the **adjoint** one, it reads

$$(T^a T^a)_{bc} = C_A \delta_{bc} = N \delta_{bc}. \tag{1.25}$$

21. In $SU(2)$ it is the only Casimir operator. It is known as **total spin** operator and its values label the irreducible representations. All irreducible representations of $SU(2)$ can be built as **totally symmetric spinor tensors**, i.e. all $\psi^{i_1 \cdots i_j} = \psi^{\{i_1 \cdots i_j\}}$ form a basis of the irrep corresponding to the total spin $\frac{j}{2}$. The total symmetrizator and antisymmetrizator act on tensor indices as

$$\psi_{\{i_1, \ldots, i_j\}} = \frac{1}{j!} \sum_{s(i_1 \ldots i_j)} \psi_{s(i_1, \ldots, i_j)}, \tag{1.26}$$

$$\psi_{[i_1, \ldots, i_j]} = \frac{1}{j!} \sum_{s(i_1 \ldots i_j)} sgn(s(i_1 \ldots i_j)) \psi_{s(i_1, \ldots, i_j)}, \tag{1.27}$$

where the sums go over all permutations s of $i_1, ..., i_j$. The dimension of this Hilbert space H^j or the number of linearly independent $\psi^{i_1...i_j}$ is $j+1$, and the states are labeled by the eigenvalues of $\frac{\sigma^3}{2}$, $|j, m\rangle$. The decomposition into irreducible representations reads

$$H^j \otimes H^k = \oplus_{l=|k-j|}^{k+j} H^l. \tag{1.28}$$

22. In $SU(3)$ there is also a **Cubic Casimir operator**

$$\hat{C}_3 = d^{abc} \mathbf{t}^a \mathbf{t}^b \mathbf{t}^c. \tag{1.29}$$

Its values in the **adjoint** and **defining** representations read

$$d^{abc} T^a T^b T^c = 0, \quad d^{abc} (t^a t^b t^c)_i^j = C_{3F} \delta_i^j = \frac{(N^2 - 4)(N^2 - 1)}{4N^2} \delta_i^j. \tag{1.30}$$

23. Hence in $SU(3)$ **irreps are labeled by two indices, p and q.** The Hilbert space H^{pq} of the **basis spinors** where matrices of the irrep $D(p, q)$ act consists of all **traceless tensors symmetric in both covariant and contravariant indices separately**

$$\psi_{j_1...j_q}^{i_1...i_p} = \psi_{\{j_1...j_q\}}^{\{i_1...i_p\}} : \quad \psi_{j_1...j_q}^{i_1...i_p} \delta_{i_t}^{j_l} = 0, \quad 1 \le l \le q, 1 \le t \le p. \tag{1.31}$$

Dimension of the Hilbert space H^{pq}

$$\dim H^{pq} = \frac{(p+1)(q+1)(p+q+2)}{2}. \tag{1.32}$$

Irreps are also labeled with their dimension, e.g. $D(1,0) \equiv 3$, $D(0,1) \equiv 3^* \equiv \bar{3}$. The basis states of the irrep $D(p, q)$ can be labeled by 2 eigenvalues of 2 commuting generators $\mathbf{t}^3$ and $\mathbf{t}^8$. **Decomposition into irreps can be done via symmetrization and convolution with the invariant tensors.**

24. $SU(N)$ is a **compact manifold**. Indeed, for $U \in SU(N)$ we have

$$U^\dagger U = 1 \quad \implies \quad U_{ji}^* U_{jk} = \delta_{ik}, \tag{1.33}$$

i.e. the columns of the matrix are orthonormal. Therefore, $SU(2) \sim S^3$, since

$$SU(2) \ni U_2 = \begin{pmatrix} z_1 & -z_2^* \\ z_2 & z_1^* \end{pmatrix}, \quad |z_1|^2 + |z_2|^2 = 1, \tag{1.34}$$

$$\begin{matrix} z_1 = e^{i\alpha_1} \cos \alpha_3 \\ z_2 = e^{i\alpha_2} \sin \alpha_3 \end{matrix}, \quad \begin{matrix} \alpha_{1,2} \in [0, 2\pi) \\ \alpha_3 \in [0, \pi/2] \end{matrix}. \tag{1.35}$$

For $SU(3)$ one can choose the coordinates of the $\hat{r}$ unit vector of the spherical coordinate system with arbitrary phases as $z_{1,2,3}$ and the coordinates the other unit vectors of the spherical coordinate system with the adjusted phases as the elements of the other columns:

$$\left\{ \begin{array}{l} z_1 = e^{i\alpha_4} \cos \alpha_5 \\ z_2 = e^{i\alpha_6} \sin \alpha_5 \sin \alpha_7 \\ z_3 = e^{i\alpha_8} \sin \alpha_5 \cos \alpha_7 \end{array} \right. , \qquad (1.36)$$

$$S = \begin{pmatrix} z_1 & 0 & -e^{i\alpha_4} \sin \alpha_5 \\ z_2 & e^{-i\alpha_4 - i\alpha_8} \cos \alpha_7 & e^{i\alpha_6} \cos \alpha_5 \sin \alpha_7 \\ z_3 & -e^{-i\alpha_6 - i\alpha_4} \sin \alpha_7 & e^{i\alpha_8} \cos \alpha_5 \cos \alpha_7 \end{pmatrix}, \quad S|_{\alpha_i = 0} = 1,$$

$$(1.37)$$

$$\begin{array}{l} \alpha_{4,6,8} \in [0, 2\pi) \\ \alpha_{5,7} \in [0, \pi/2] \end{array} , \quad |z_1|^2 + |z_2|^2 + |z_3|^2 = 1, \quad \det S = 1. \quad (1.38)$$

This matrix S has 5 parameters, obeys (1.33) and belongs to $SU(3)$. One builds a general $SU(3)$ matrix from it via

$$SU(3) \ni U = S \begin{pmatrix} U_2 & 0 \\ 0 & 1 \end{pmatrix} \implies SU(3) \sim S^3 \otimes S^5. \quad (1.39)$$

25. One can build an **invariant or Haar measure** on this manifold:

$$\int_{SU(N)} dV = \int_{SU(N)} d(UV) = \int_{SU(N)} d(VU), \quad \int_{SU(N)} dV = 1.$$

$$(1.40)$$

Indeed, for $U, V(\alpha), V'(\alpha') = UV \in SU(N)$, one builds **left-invariant vectors** L_i :

$$L_i = iV^{-1} \frac{\partial V}{\partial \alpha_i} = i(UV)^{-1} \frac{\partial (UV)}{\partial \alpha_i} = iV'^{-1} \frac{\partial (V')}{\partial \alpha_i}$$

$$= iV'^{-1} \frac{\partial V'}{\partial \alpha'_j} \frac{\partial \alpha'_j}{\partial \alpha_i} = L'_j \frac{\partial \alpha'_j}{\partial \alpha_i} \quad (1.41)$$

and the **metric tensor**

$$g_{ij} = tr(L_i L_j) = -tr(V^{-1} \frac{\partial V}{\partial \alpha_i} V^{-1} \frac{\partial V}{\partial \alpha_j}) = tr(\frac{\partial V^{-1}}{\partial \alpha_i} \frac{\partial V}{\partial \alpha_j}). \quad (1.42)$$

Then

$$g_{ij} = g'_{mn} \frac{\partial \alpha'_m}{\partial \alpha_i} \frac{\partial \alpha'_n}{\partial \alpha_j}, \quad g \equiv \det g_{ij} = g' \left| \frac{\partial \alpha'_n}{\partial \alpha_j} \right|^2, \quad (1.43)$$

and

$$dV = \sqrt{g} d\alpha_1 \ldots d\alpha_n = \sqrt{g'} d\alpha'_1 \ldots d\alpha'_n = d(UV) \quad (1.44)$$

is the **invariant measure**.

EXERCISES

1.1 Show that given a representation one can construct three new representations from it via transposition, conjugation, and inversion.

SOLUTION. Suppose

$$U_1 U_2 = U_3 \quad \Longrightarrow \quad \begin{array}{l} U_1^* U_2^* = U_3^* \\ U_1^{-1T} U_2^{-1T} = U_3^{-1T} \\ U_1^{-1T*} U_2^{-1T*} = U_3^{-1T*} \end{array} \quad , \qquad (P1.1)$$

i.e. U^*, U^{-1T} and U^{-1T*} are representations. In $SU(N)$ $U = U^{-1T*}$, $U^* = U^{-1T}$. Therefore for the fundamental representation U, we have only its **conjugate representation** defined by the conjugate matrices U^*. In practice, this representation is used acting on the string vectors rather than on column vectors, i.e. after transposition

$$(U^*\chi)^T \to \chi^T U^\dagger. \qquad (P1.2)$$

Hence, the common but not rigorously correct name **antifundamental representation** for the Hermitian conjugate matrices $U^\dagger$.

1.2 Show that the defining representation of $SU(2)$ is real, i.e. find the transition matrix S in (1.8).

SOLUTION. Since $\sigma_2 \vec{\sigma} \sigma_2 = -\vec{\sigma}^*$, for $U = e^{i\vec{a}\vec{\sigma}}$ with real $\vec{a}$

$$\sigma_2 U \sigma_2 = U^*, \quad \sigma_2^\dagger = \sigma_2^{-1} = \sigma_2 \quad \Longrightarrow \quad S = const\, \sigma_2. \qquad (P1.3)$$

Therefore

$$S^{-1}\psi' = S^{-1}USS^{-1}\psi = U^*S^{-1}\psi \quad \Longrightarrow$$
$$(S^{-1}\psi')^T = (S^{-1}\psi)^T U^\dagger, \quad (\psi'^T S^{-1T})_k = (\psi^T S^{-1T})_l (U^\dagger)_k^l. \qquad (P1.4)$$

Therefore, one can build a sting spinor from the column spinor via

$$\phi_k = (\psi^T)_p (S^{-1T})_k^p = (\psi^1 \quad \psi^2) \frac{-i}{const} \begin{pmatrix} 0 & -1 \\ 1 & 0 \end{pmatrix} = \frac{-i}{const} (\psi^2 \quad -\psi^1)$$

$$= \frac{-i}{const} \epsilon_{kl} \psi^l, \quad \epsilon_{12} = +1 = -\epsilon_{21}. \qquad (P1.5)$$

We choose $const = -i$.

1.3 Show that one can always choose a basis of generators such that (1.13) holds, i.e.,

$$tr(t^a t^b) = C(r)\delta^{ab}, \qquad (P1.6)$$

where we can choose different $C(r)$ for different representations r of our generators t^a. It is conventional to choose $C(r = N) = \frac{1}{2}$ for the fundamental representation.

1.4 Show that f^{abc} can be chosen real and fully skew-symmetric if one employs normalization (P1.6) of the generators t^a in any representation r.

1.5 Prove completeness relation (1.19) for the fundamental generators t^a. Is it correct in the adjoint representation?

SOLUTION. Any $N \times N$ matrix $\Gamma_i^j = a\delta_i^j + b^a(t^a)_i^j \implies tr(\Gamma) = Na$, $tr(t^a\Gamma) = \frac{1}{2}b^a$. For $\Gamma_i^j = \delta_k^j\delta_i^l$ one has $a = \frac{1}{N}\delta_k^l$, $b^a = 2(t^a)_k^l$. Hence

$$(t^a)_i^j(t^a)_k^l = \frac{1}{2}\delta_k^j\delta_i^l - \frac{1}{2N}\delta_i^j\delta_k^l. \tag{P1.7}$$

1.6 Using completeness relation (1.19) prove the identities:

$$t^a t^a = \frac{N^2 - 1}{2N} \equiv C_F \equiv C_2(N), \tag{P1.8}$$

$$t^a t^b t^a = \frac{-1}{2N}t^b, \tag{P1.9}$$

$$tr(t^a t^b t^a t^c) = -\frac{1}{4N}\delta^{bc}, \tag{P1.10}$$

$$t^a t^b t^c t^a = \frac{\delta^{bc}}{4} - \frac{t^b t^c}{2N}. \tag{P1.11}$$

1.7 Show that the structure constants f^{abc} can be expressed as

$$f^{abc} = -2i\, tr(t^a[t^b, t^c]). \tag{P1.12}$$

Representation (P1.12) and completeness relation (1.19) allow us to find any convolutions of the structure constants.

1.8 Prove the following identity:

$$f^{abc}f^{abc'} = N\delta^{cc'}, \tag{P1.13}$$

compare it with (1.25) and deduce that $C_A \equiv C_2(N^2 - 1) = N$.

1.9 Prove the following identity:

$$f^{aa'c}f^{abi}f^{a'b'i} = \frac{N}{2}f^{bb'c}. \tag{P1.14}$$

1.10 Prove that for the **fundamental generators** t^a and t^b, the following relation takes place:

$$t^a t^b = \frac{\delta^{ab}}{2N} + \frac{1}{2}\left(d^{abc} + if^{abc}\right)t^c, \tag{P1.15}$$

where the symbols

$$d^{abc} = 2tr\left(t^a\{t^b, t^c\}\right) \tag{P1.16}$$

defined in (1.14) are real ($d^{abc*} = d^{abc}$), totally symmetric, and traceless $d^{aac} = 0$.

1.11 Using representations (P1.12) and (P1.16) through the fundamental generators and completeness relation (1.19), prove the identities:

$$d^{abc}d^{abc\prime} = \frac{N^2 - 4}{N}\delta^{cc\prime}, \tag{P1.17}$$

$$f^{aa\prime c}f^{abi}d^{a\prime b\prime i} = -\frac{N}{2}d^{bb\prime c}, \tag{P1.18}$$

$$f^{aa\prime c}d^{abi}d^{a\prime b\prime i} = \frac{N^2 - 4}{2N}f^{bb\prime c}, \tag{P1.19}$$

$$d^{aa\prime c}d^{abi}d^{a\prime b\prime i} = \frac{N^2 - 12}{2N}d^{bb\prime c}. \tag{P1.20}$$

1.12 Show that structure constants (1.10) and Jacobi identity

$$[\mathbf{t}^a, [\mathbf{t}^b, \mathbf{t}^c]] + [\mathbf{t},^b [\mathbf{t}^c, \mathbf{t}^d]] + [\mathbf{t}^d, [\mathbf{t}^a, \mathbf{t}^b]] = 0. \tag{P1.21}$$

do not depend on representation.

1.13 Prove the following additional Jacobi identity:

$$[\mathbf{t}^a, \{\mathbf{t}^b, \mathbf{t}^c\}] + [\mathbf{t}^b, \{\mathbf{t}^c, \mathbf{t}^d\}] + [\mathbf{t}^d, \{\mathbf{t}^a, \mathbf{t}^b\}] = 0. \tag{P1.22}$$

The identity implies

$$f^{aij}d^{bci} + f^{bij}d^{cai} + f^{cij}d^{abi} = 0. \tag{P1.23}$$

1.14 Show that

$$[\mathbf{t}^a, [\mathbf{t}^b, \mathbf{t}^c]] = \{\mathbf{t}^c, \{\mathbf{t}^a, \mathbf{t}^b\}\} - \{\mathbf{t}^b, \{\mathbf{t}^c, \mathbf{t}^a\}\}, \tag{P1.24}$$

which implies

$$f^{had}f^{bcd} = d^{acd}d^{bdh} - d^{abd}d^{cdh} + \frac{2}{N}(\delta^{ac}\delta^{bh} - \delta^{ab}\delta^{ch}). \tag{P1.25}$$

1.15 Prove the following relations for $SU(3)$:

$$d^{a_1 a_2 b}d^{a_3 ab} + d^{a_1 a_3 b}d^{a_2 ab} + d^{a_2 a_3 b}d^{a_1 ab}$$
$$= \frac{1}{3}\left(\delta^{a_1 a_2}\delta^{a_3 a} + \delta^{a_1 a_3}\delta^{a_2 a} + \delta^{a_2 a_3}\delta^{a_1 a}\right), \tag{P1.26}$$

$$\varepsilon_{ijh}\varepsilon^{i\prime j\prime h\prime}U^i_{i\prime}U^j_{j\prime} = 2(U^\dagger)^{h\prime}_h, \quad \varepsilon_{ijh}\varepsilon^{i\prime j\prime h\prime}(U^\dagger)^i_{i\prime}(U^\dagger)^j_{j\prime} = 2U^{h\prime}_h, \tag{P1.27}$$

$$\varepsilon_{ijh}\varepsilon^{i\prime j\prime h\prime}U^i_{1i\prime}U^j_{2j\prime}U^k_{3k\prime} = \varepsilon_{ijh}\varepsilon^{i\prime j\prime h\prime}(U_1 U^\dagger_l)^i_{i\prime}(U_2 U^\dagger_l)^j_{j\prime}(U_3 U^\dagger_l)^k_{k\prime}, \tag{P1.28}$$

$$\varepsilon_{ijh}\varepsilon^{i\prime j\prime h\prime}U^i_{i\prime}U^j_{j\prime}V^k_{k\prime} = 2tr(VU^\dagger). \tag{P1.29}$$

SOLUTION. Here we follow (MSW68). For any 3×3 matrix A :

$$A = \Sigma^{-1} \begin{pmatrix} a_1 & & \\ & a_2 & \\ & & a_3 \end{pmatrix} \Sigma, \qquad \text{(P1.30)}$$

$$0 = (A - a_1)(A - a_2)(A - a_3) \qquad \text{(P1.31)}$$

$$= A^3 - A^2 tr A + \frac{1}{2} A((tr A)^2 - tr A^2) - \det A. \qquad \text{(P1.32)}$$

Taking $A = \theta^a t^a$

$$\det A = \frac{1}{3!} \varepsilon_{ijk} \varepsilon^{pqr} (t^a)^i_p (t^b)^j_q (t^c)^k_r \theta^a \theta^b \theta^c$$

$$= \frac{1}{3!} \begin{vmatrix} \delta^p_i & \delta^p_j & \delta^p_k \\ \delta^q_i & \delta^q_j & \delta^q_k \\ \delta^r_i & \delta^r_j & \delta^r_k \end{vmatrix} (t^a)^i_p (t^b)^j_q (t^c)^k_r \theta^a \theta^b \theta^c$$

$$= \frac{1}{3!} tr[t^b t^a t^c + t^a t^b t^c] \theta^a \theta^b \theta^c = \frac{1}{12} d^{abc} \theta^a \theta^b \theta^c. \qquad \text{(P1.33)}$$

Therefore (P1.32) gives

$$\theta^a \theta^b \theta^c \left[t^a t^b t^c - \frac{1}{4} t^a \delta^{bc} - \frac{1}{12} d^{abc} \right] = 0. \qquad \text{(P1.34)}$$

Since θ's are arbitrary, the expression in the brackets is zero. Symmetrizing it, multiplying by t^d, and taking the trace, one gets

$$0 = tr \left\{ \left[t^{\{a} t^b t^{c\}} - \frac{1}{12} (t^a \delta^{bc} + t^b \delta^{ac} + t^c \delta^{ba}) - \frac{1}{12} d^{abc} \right] t^d \right\}$$

$$= tr(t^{\{a} t^b t^{c\}} t^d) - \frac{1}{24} (\delta^{ad} \delta^{bc} + \delta^{bd} \delta^{ac} + \delta^{cd} \delta^{ba}). \qquad \text{(P1.35)}$$

Here via (P1.15)

$$tr(t^{\{a} t^b t^{c\}} t^d) = \frac{1}{6} tr(t^a t^b t^c t^d + t^b t^a t^c t^d + t^a t^c t^b t^d + t^c t^b t^a t^d$$

$$+ t^b t^c t^a t^d + t^c t^a t^b t^d) \qquad \text{(P1.36)}$$

$$= \frac{1}{6} tr((\frac{1}{3} \delta^{ab} + d^{abg} t^g) t^c t^d) + t^c t^b (\frac{1}{3} \delta^{ad} + d^{adg} t^g)$$

$$+ t^c t^a (\frac{1}{3} \delta^{db} + d^{dbg} t^g)) \qquad \text{(P1.37)}$$

$$= \frac{1}{36} (\delta^{ab} \delta^{cd} + \delta^{cb} \delta^{ad} + \delta^{ca} \delta^{db})$$

$$+ \frac{1}{6} tr(d^{abg} t^g t^c t^d + t^c t^b d^{adg} t^g + t^c t^a d^{dbg} t^g) \qquad \text{(P1.38)}$$

$$= \frac{1}{36}(\delta^{ab}\delta^{cd} + \delta^{cb}\delta^{ad} + \delta^{ca}\delta^{db})$$

$$+ \frac{1}{6}\frac{i}{4}(f^{cdg}d^{abg} + f^{cbg}d^{adg} + f^{cag}d^{dbg})$$

$$+ \frac{1}{6}\frac{1}{4}(d^{abg}d^{cdg} + d^{cbg}d^{adg} + d^{cag}d^{dbg}). \tag{P1.39}$$

Therefore, the imaginary part of (P1.35) reproduces (P1.23) and the real part gives (P1.26). The other equalities of the problem follow from the properties of the $SU(3)$ group, e.g.

$$\varepsilon_{ijh}\varepsilon^{i'j'h'}U^i_{i'}U^j_{j'} \times U^h_k = \det U \, \varepsilon_{i'j'k}\varepsilon^{i'j'h'} = 2\delta^{h'}_k$$

$$\implies \quad \varepsilon_{ijh}\varepsilon^{i'j'h'}U^i_{i'}U^j_{j'} = 2(U^\dagger)^{h'}_h. \tag{P1.40}$$

1.16 Show that the adjoint representation

$$(T^b)_{ac} = if^{abc} \tag{P1.41}$$

is a representation. Show that it is real.

1.17 Prove that $\hat{C}_3$ (1.29) is a Casimir operator in $SU(N)$ using relation (P1.23).

1.18 Show that in an irrep of $SU(N)$

$$f^{abc}\mathbf{t}^a\mathbf{t}^b\mathbf{t}^c = \frac{iN}{2}\hat{C}_2. \tag{P1.42}$$

1.19 Calculate the cubic Casimir values for fundamental and adjoint representations of $SU(N)$ (1.30).

1.20 Show that all $\psi^{\{i_1\cdots i_k\}}$ form a basis of the irrep of $SU(2)$ corresponding to total spin $\frac{k}{2}$.

HINT: show that they transform among one another since the transformation does not change the symmetry. Consider them as a symmetrized tensor product of fundamental spinors.

1.21 Show that a unitary matrix $V \in SU(2)$ can be written in the Cayley representation

$$V = \frac{1 + iB}{1 - iB}, \quad B^\dagger = B, \quad \mathrm{tr}\,B = 0. \tag{P1.43}$$

SOLUTION. First, in the definition of V $(1 + iB)$ and $(1 - iB)^{-1}$ commute. Then,

$$V^\dagger = \frac{1 - iB}{1 + iB} = V^{-1}. \tag{P1.44}$$

$$tr B = 0 \quad \Longrightarrow \quad B = \vec{\sigma}\vec{b} \quad \Longrightarrow$$

$$V = \frac{1 + 2iB - B^2}{1 + B^2} = \frac{1 + 2i\vec{b}\vec{\sigma} - b^2}{1 + b^2}, \quad \det V = 1. \qquad (P1.45)$$

Comparing it with the exponential representation

$$V = e^{i\frac{\theta}{2}\vec{n}\vec{\sigma}} = \cos\frac{\theta}{2} + i\vec{n}\vec{\sigma}\sin\frac{\theta}{2}, \quad \vec{n}^2 = 1, \quad 0 \le \theta \le 2\pi, \qquad (P1.46)$$

$$b = tg\frac{\theta}{4}, \quad 0 \le b \le \infty. \qquad (P1.47)$$

One can see that as b goes from 0 to ∞, V covers the whole $SU(2)$ group and V^k covers the group k times.

1.22 Identify the 3 $SU(2)$ subalgebras defined by the $SU(3)$ algebra of Gell-Mann matrices (1.16) so that they leave unchanged one of the spinor's coordinates.

SOLUTION. The first algebra can be built on the $\lambda^1, \lambda^2, \lambda^3$. These matrices are Pauli matrices acting in the space of the first two spinor coordinates. They obey

$$[\lambda^i, \lambda^j] = 2i\varepsilon_{ijk}\lambda^k, \quad [t^i, t^j] = i\varepsilon_{ijk}t^k, \quad i, j, k \in \{1, 2, 3\}. \qquad (P1.48)$$

This is the "isospin" (or T, or I) subalgebra. The second algebra is constructed on the λ^4, λ^5, and $\frac{\lambda^3 + \sqrt{3}\lambda^8}{2}$. These matrices are the Pauli matrices acting in the space of the first and third spinor coordinates. The generator $\frac{\lambda^3 + \sqrt{3}\lambda^8}{2}$ acts as the "third sigma matrix" in this space with the corresponding commutation relations, in particular

$$[\lambda^4, \lambda^5] = 2i\frac{\lambda^3 + \sqrt{3}\lambda^8}{2}, \quad [t^4, t^5] = i\frac{t^3 + \sqrt{3}t^8}{2}. \qquad (P1.49)$$

This algebra generates the so-called U subgroup. The third algebra is constructed on the λ^6, λ^7, and $\frac{-\lambda^3 + \sqrt{3}\lambda^8}{2}$, which are the Pauli matrices acting in the space of the second and third spinor coordinates. Again, $\frac{-\lambda^3 + \sqrt{3}\lambda^8}{2}$ acts as the "third sigma matrix" in this space with the corresponding commutation relations, in particular

$$[\lambda^6, \lambda^7] = 2i\frac{-\lambda^3 + \sqrt{3}\lambda^8}{2}, \quad [t^6, t^7] = i\frac{-t^3 + \sqrt{3}t^8}{2}. \qquad (P1.50)$$

This algebra generates the so-called V subgroup.

1.23 Show that in $SU(2)$ symmetric $\psi^{\{i_1 i_2\}}$ is equivalent to traceless $\psi_{i_2}^{i_1}$: $\psi_i^i = 0$.

SOLUTION. We define

$$\epsilon_{ab} = -\epsilon^{ab}: \quad \epsilon_{12} = -\epsilon_{21} = \epsilon^{21} = -\epsilon^{12} = 1$$

$$\implies \quad \epsilon_{ab}\epsilon^{bc} = \epsilon^{cb}\epsilon_{ba} = \delta_a^c, \epsilon^{ab}\frac{1}{2}\epsilon_{cd} \times \epsilon^{dc}\frac{1}{2}\epsilon_{ef} = \epsilon^{ab}\frac{1}{2}\epsilon_{ef}. \tag{P1.51}$$

Then

$$\psi_a = \epsilon_{ac}\psi^c, \quad \psi^a = \epsilon^{ac}\psi_c, \tag{P1.52}$$

$$\psi_{ab} = \epsilon_{ac}\psi_b^c, \quad \psi_b^a = \epsilon_{bc}\psi^{ca} = \epsilon^{ac}\psi_{cb}, \quad \psi^{ab} = \epsilon^{ac}\psi_c^b. \tag{P1.53}$$

Therefore

$$\psi_a^a = 0 \implies \epsilon_{ab}\psi^{ab} = \epsilon_{ab}\epsilon^{ac}\psi_c^b = -\delta_b^c\psi_c^b = 0, \quad \psi^{ab} = \psi^{\{ab\}}; \tag{P1.54}$$

$$\psi^{ab} = \psi^{ba} \implies 0 = \psi^{ab}\epsilon_{ab} = -\psi_b^b. \tag{P1.55}$$

1.24 Show decomposition into irreps in $SU(2)$

$$a) \; \frac{1}{2} \otimes \frac{1}{2} = 0 \oplus 1, \quad b) \; \frac{1}{2} \otimes \frac{1}{2} \otimes \frac{1}{2} = (0 \oplus 1) \otimes \frac{1}{2} = \frac{1}{2} \oplus \frac{1}{2} \oplus \frac{3}{2}, \tag{P1.56}$$

and build the corresponding basis tensors.
SOLUTION. a) We have

$$\psi_b^a = p^a q_b = \underbrace{p^a q_b - \frac{1}{2}\delta_b^a(p^c q_c)}_{1} + \underbrace{\frac{1}{2}\delta_b^a(p^c\, q_c)}_{0}, \quad \text{or} \tag{P1.57}$$

$$\psi^{ad} = \epsilon^{db} \times \psi_b^a = \underbrace{p^{\{a}q^{d\}}}_{1} + \underbrace{p^{[a}q^{d]}}_{0}, \tag{P1.58}$$

since

$$\epsilon^{db} \times \frac{1}{2}\delta_b^a(p^c q_c) = -\epsilon^{ad}\frac{1}{2}(p^c q_c) = -\epsilon^{ad}\frac{1}{2}\epsilon_{ce}p^c q^e = \epsilon^{ad}\frac{1}{2}\epsilon_{ec}p^{[c}q^{e]} = p^{[a}q^{d]}. \tag{P1.59}$$

b)

$$\psi^{abc} = p^a q^b r^c = \underbrace{p^{\{a}q^{b\}}r^c}_{1\otimes\frac{1}{2}} + \underbrace{p^{[a}q^{b]}r^c}_{0\otimes\frac{1}{2}}$$

$$= \underbrace{p^{\{a}q^b r^{c\}}}_{\frac{3}{2}} + \underbrace{\frac{2}{3}\delta_s^{\{a}\epsilon^{b\}c}\epsilon_{de}p^{\{s}q^{e\}}r^d}_{\frac{1}{2}} + \epsilon^{ab}\frac{1}{2}\underbrace{(\epsilon_{de}p^e q^d)}_{0}\underbrace{r^c}_{\frac{1}{2}}. \tag{P1.60}$$

1.25 The $SU(2)_I$ flavor group of light quarks is called the **isospin (I) group**. Write explicitly the quark content of the meson ψ_b^a and baryon ψ^{abc} irreducible tensors.

SOLUTION. We have the irreducible fundamental spinors with $I = \frac{1}{2}$:

$$p^a = q^a = r^a = \begin{pmatrix} u^\dagger \\ d^\dagger \end{pmatrix}, \quad \bar{p}_a = \bar{q}_a = \bar{r}_a = \begin{pmatrix} \bar{u}^\dagger & \bar{d}^\dagger \end{pmatrix} \quad \Longrightarrow \quad \text{(P1.61)}$$

$$\bar{p}^b = \epsilon^{ba}\bar{p}_a = \begin{pmatrix} -\bar{d}^\dagger \\ \bar{u}^\dagger \end{pmatrix}, \quad p_b = \epsilon_{ba}p^a = \begin{pmatrix} d^\dagger & -u^\dagger \end{pmatrix}. \quad \text{(P1.62)}$$

Here we treat $u^\dagger$ and $d^\dagger$ as quark creation operators and $\bar{u}^\dagger$ and $\bar{d}^\dagger$ as antiquark creation operators with all anticommutators vanishing except for

$$\{u, u^\dagger\} = \{d, d^\dagger\} = \{\bar{u}, \bar{u}^\dagger\} = \{\bar{d}, \bar{d}^\dagger\} = 1. \quad \text{(P1.63)}$$

Therefore

$$u^\dagger = p^1 \text{ and } (-\bar{d}^\dagger) = \bar{p}^1 \text{ have } I_3 = +\frac{1}{2},$$

$$\text{and} \quad d^\dagger = p^2 \text{ and } \bar{u}^\dagger = \bar{p}^2 \text{ have } I_3 = -\frac{1}{2}. \quad \text{(P1.64)}$$

Then for $I = 1$:

$$\psi_b^a = p^a \bar{q}_b - \frac{1}{2}\delta_b^a(p^c \bar{q}_c) = \begin{pmatrix} \frac{u^\dagger \bar{u}^\dagger - d^\dagger \bar{d}^\dagger}{\sqrt{2}} \times \frac{1}{\sqrt{2}} & u^\dagger \bar{d}^\dagger \\ d^\dagger \bar{u}^\dagger & -\frac{u^\dagger \bar{u}^\dagger - d^\dagger \bar{d}^\dagger}{\sqrt{2}} \times \frac{1}{\sqrt{2}} \end{pmatrix}, \quad \text{(P1.65)}$$

or via $\psi^{ad} = \epsilon^{db} \times \psi_b^a$.

$$I_3 = +1 : \psi^{11} = -u^\dagger \bar{d}^\dagger, \quad \text{(P1.66)}$$

$$I_3 = 0 : \psi^{12} = \frac{u^\dagger \bar{u}^\dagger - d^\dagger \bar{d}^\dagger}{\sqrt{2}} \times \frac{1}{\sqrt{2}}, \quad \text{(P1.67)}$$

$$I_3 = -1 : \psi^{22} = d^\dagger \bar{u}^\dagger. \quad \text{(P1.68)}$$

For $I = 0$:

$$I = 0 : \psi = \frac{(p^c \bar{q}_c)}{\sqrt{2}} = \frac{u^\dagger \bar{u}^\dagger + d^\dagger \bar{d}^\dagger}{\sqrt{2}}. \quad \text{(P1.69)}$$

Here we singled out the normalized to 1 states.

The baryon multiplet with $I = \frac{3}{2}$ reads

$$I_3 = +\frac{3}{2} : \psi^{111} = u^\dagger u^\dagger u^\dagger, \tag{P1.70}$$

$$I_3 = +\frac{3}{2} : \psi^{112} = \frac{1}{\sqrt{3}} \frac{u^\dagger u^\dagger d^\dagger + u^\dagger d^\dagger u^\dagger + d^\dagger u^\dagger u^\dagger}{\sqrt{3}}, \tag{P1.71}$$

$$I_3 = -\frac{1}{2} : \psi^{122} = \frac{1}{\sqrt{3}} \frac{u^\dagger d^\dagger d^\dagger + d^\dagger d^\dagger u^\dagger + d^\dagger u^\dagger d^\dagger}{\sqrt{3}}, \tag{P1.72}$$

$$I_3 = -\frac{3}{2} : \psi^{222} = d^\dagger d^\dagger d^\dagger. \tag{P1.73}$$

The baryon dublets with $I = \frac{1}{2}$ read

$$\psi_1^a = \epsilon_{de} p^{\{a} q^{e\}} r^d = -\frac{1}{2}\sqrt{\frac{3}{2}} \left(\begin{matrix} \frac{2u^\dagger u^\dagger d^\dagger - u^\dagger d^\dagger u^\dagger - d^\dagger u^\dagger u^\dagger}{\sqrt{6}} \\ \frac{d^\dagger u^\dagger d^\dagger - 2d^\dagger d^\dagger u^\dagger + u^\dagger d^\dagger d^\dagger}{\sqrt{6}} \end{matrix} \right), \tag{P1.74}$$

$$\psi_2^a = (\epsilon_{de} p^e q^d) r^a = \sqrt{2} \left(\begin{matrix} \frac{(d^\dagger u^\dagger - u^\dagger d^\dagger) u^\dagger}{\sqrt{2}} \\ \frac{(d^\dagger u^\dagger - u^\dagger d^\dagger) d^\dagger}{\sqrt{2}} \end{matrix} \right). \tag{P1.75}$$

1.26 Build light meson's wave-functions taking into account both isospin and spin symmetry. Identify these states with the mesons. Find their **electric charge** Q and **(strong) hypercharge**

$$Y = 2(Q - I_3). \tag{P1.76}$$

SOLUTION. For light quarks

$$\begin{pmatrix} Q_u \\ Q_d \end{pmatrix} = \begin{pmatrix} +\frac{2}{3} \\ -\frac{1}{3} \end{pmatrix}, \quad \begin{pmatrix} Y_u \\ Y_d \end{pmatrix} = \begin{pmatrix} \frac{1}{3} \\ \frac{1}{3} \end{pmatrix}, \quad \begin{pmatrix} Y_{\bar{d}} \\ Y_{\bar{u}} \end{pmatrix} = \begin{pmatrix} -\frac{1}{3} \\ -\frac{1}{3} \end{pmatrix}. \tag{P1.77}$$

Spin 1 and spin 0 states are given in Tables 1.1 and 1.2. Spin zero state

TABLE 1.1   Vector mesons made from light quarks

	Q	I	I_3
$\lvert\rho^+\rangle = u_\uparrow^\dagger \bar{d}_\uparrow^\dagger \lvert 0\rangle$	1	1	1
$\lvert\rho^0\rangle = \frac{u_\uparrow^\dagger \bar{u}_\uparrow^\dagger - d_\uparrow^\dagger \bar{d}_\uparrow^\dagger}{\sqrt{2}} \lvert 0\rangle$	0	1	0
$\lvert\rho^-\rangle = d_\uparrow^\dagger \bar{u}_\uparrow^\dagger \lvert 0\rangle$	-1	1	-1
$\lvert\omega^0\rangle = \frac{u_\uparrow^\dagger \bar{u}_\uparrow^\dagger + d_\uparrow^\dagger \bar{d}_\uparrow^\dagger}{\sqrt{2}} \lvert 0\rangle$	0	0	0

with $I = 0$ has the s quark admixture. Therefore, it is out of the $SU(2)_I$ classification. All these states have $Y = 0$.

TABLE 1.2   Scalar mesons made from light quarks

	Q	I	I_3
$\|\pi^+\rangle = \frac{u_\uparrow^\dagger \bar{d}_\downarrow^\dagger - u_\downarrow^\dagger \bar{d}_\uparrow^\dagger}{\sqrt{2}}\|0\rangle$	1	1	1
$\|\pi^0\rangle = \frac{u_\uparrow^\dagger \bar{u}_\downarrow^\dagger - d_\uparrow^\dagger \bar{d}_\downarrow^\dagger - u_\downarrow^\dagger \bar{u}_\uparrow^\dagger + d_\downarrow^\dagger \bar{d}_\uparrow^\dagger}{2}\|0\rangle$	0	1	0
$\|\pi^-\rangle = \frac{d_\uparrow^\dagger \bar{u}_\downarrow^\dagger - d_\downarrow^\dagger \bar{u}_\uparrow^\dagger}{\sqrt{2}}\|0\rangle$	-1	1	-1

1.27   Build light baryon's wave-functions taking into account both isospin and spin symmetry. Identify them with the baryons. Find their **electric charge** Q and **hypercharge Y.**

SOLUTION. Spin $\frac{3}{2}$ states with the projection $J_3 = \frac{3}{2}$ are given in Table 1.3. Since spin and isospin wave-functions factorize, the state with the

TABLE 1.3   Baryons made from light quarks

	Q	I	I_3
$\|\Delta^{++}\rangle = u_\uparrow^\dagger u_\uparrow^\dagger u_\uparrow^\dagger\|0\rangle$	2	$\frac{3}{2}$	$\frac{3}{2}$
$\|\Delta^+\rangle = \frac{u_\uparrow^\dagger u_\uparrow^\dagger d_\uparrow^\dagger + u_\uparrow^\dagger d_\uparrow^\dagger u_\uparrow^\dagger + d_\uparrow^\dagger u_\uparrow^\dagger u_\uparrow^\dagger}{\sqrt{3}}\|0\rangle$	1	$\frac{3}{2}$	$\frac{1}{2}$
$\|\Delta^0\rangle = \frac{u_\uparrow^\dagger d_\uparrow^\dagger d_\uparrow^\dagger + d_\uparrow^\dagger d_\uparrow^\dagger u_\uparrow^\dagger + d_\uparrow^\dagger u_\uparrow^\dagger d_\uparrow^\dagger}{\sqrt{3}}\|0\rangle$	0	$\frac{3}{2}$	$-\frac{1}{2}$
$\|\Delta^-\rangle = d_\uparrow^\dagger d_\uparrow^\dagger d_\uparrow^\dagger\|0\rangle$	-1	$\frac{3}{2}$	$-\frac{3}{2}$

spin projection $\frac{1}{2}$ reads, e.g. for $|\Delta^-\rangle$

$$|\Delta^-\rangle|_{s_3 = \frac{1}{2}} = \frac{\uparrow\uparrow\downarrow + \uparrow\downarrow\uparrow + \downarrow\uparrow\uparrow}{\sqrt{3}} \otimes d^\dagger d^\dagger d^\dagger |0\rangle$$

$$= \frac{d_\uparrow^\dagger d_\uparrow^\dagger d_\downarrow^\dagger + d_\uparrow^\dagger d_\downarrow^\dagger d_\uparrow^\dagger + d_\downarrow^\dagger d_\uparrow^\dagger d_\uparrow^\dagger}{\sqrt{3}} |0\rangle. \tag{P1.78}$$

Isospin $\frac{1}{2}$ states have mixed symmetry, i.e. ψ_1^a (P1.74) is symmetric w.r.t. permutations of first 2 quarks and ψ_2^a (P1.75) is anti-symmetric. To build the totally symmetric wave-functions, we have to multiply them by the spin wave-functions with the same symmetry and sum these products:

$$|p_\uparrow^+\rangle = \frac{1}{\sqrt{2}}\left[\frac{2u^\dagger u^\dagger d^\dagger - u^\dagger d^\dagger u^\dagger - d^\dagger u^\dagger u^\dagger}{\sqrt{6}} \otimes \frac{2\uparrow\uparrow\downarrow - \uparrow\downarrow\uparrow - \downarrow\uparrow\uparrow}{\sqrt{6}}\right.$$

$$\left. + \frac{(d^\dagger u^\dagger - u^\dagger d^\dagger)u^\dagger}{\sqrt{2}} \otimes \frac{(\downarrow\uparrow - \uparrow\downarrow)\uparrow}{\sqrt{2}}\right]|0\rangle$$

$$= \frac{1}{3\sqrt{2}}[u^\dagger u^\dagger d^\dagger \otimes (2\uparrow\uparrow\downarrow - \uparrow\downarrow\uparrow - \downarrow\uparrow\uparrow)$$

$$+ u^\dagger d^\dagger u^\dagger \otimes (2\uparrow\downarrow\uparrow - \uparrow\uparrow\downarrow - \downarrow\uparrow\uparrow)$$

$$+ d^\dagger u^\dagger u^\dagger \otimes (2\downarrow\uparrow\uparrow - \uparrow\uparrow\downarrow - \uparrow\downarrow\uparrow)]|0\rangle$$

$$= \frac{1}{3\sqrt{2}}[2u_\uparrow^\dagger u_\uparrow^\dagger d_\downarrow^\dagger - u_\uparrow^\dagger u_\downarrow^\dagger d_\uparrow^\dagger - u_\downarrow^\dagger u_\uparrow^\dagger d_\uparrow^\dagger + 2 \text{ cyclic permutations}].$$

$$(P1.79)$$

All these states have $Y = 1$.

Note, that since quarks are fermions and baryon spin-flavor wave-function is totally symmetric, the Pauli rule demands that there should be a **new quantum number** for the baryon's wave-functions to be anti-symmetric. This quantum number is called **color**, and quarks transform as fundamental spinors under the $SU(3)_{\text{color}}$ group. As a result, all baryon wave-functions are anti-symmetric singlets w.r.t. $SU(3)_{\text{color}}$, e.g.

$$|\Delta^-\rangle|_{s_3 = \frac{3}{2}} = d_\uparrow^\dagger d_\uparrow^\dagger d_\uparrow^\dagger|0\rangle = \frac{1}{\sqrt{6}}\epsilon_{ijk}d_\uparrow^{i\dagger}d_\uparrow^{j\dagger}d_\uparrow^{k\dagger}|0\rangle, \quad e.t.c., \quad (P1.80)$$

and all meson wave-functions are symmetric $SU(3)_{\text{color}}$ singlets, e.g.

$$|\rho^-\rangle|_{s_3 = 1} = d_\uparrow^\dagger \bar{u}_\uparrow^\dagger|0\rangle = \frac{1}{\sqrt{3}}\delta_i^j d_\uparrow^{i\dagger}\bar{u}_{j\uparrow}^\dagger|0\rangle = \frac{1}{\sqrt{3}}d_\uparrow^{i\dagger}\bar{u}_{i\uparrow}^\dagger|0\rangle, \quad e.t.c. \quad (P1.81)$$

Here i, j, k are the color indices.

1.28 Find the spin flavor wave-functions for $|n_\uparrow^0\rangle$, $|n_\downarrow^0\rangle$, $|p_\downarrow^+\rangle$.

HINT: use the solution to the previous problem.

1.29 Find the dimension of the irrep in $SU(3)$ dim$D(p,q)$ (1.32).

SOLUTION. Thanks to symmetry, the number of independent components of $D(p,0)$ is the number of ways to put two identical separators into the list of length p made of numbers 1,2,3:

$$\{1111...|2222...|333....\}. \quad (P1.82)$$

This number is

$$\underbrace{\frac{(p+1)p}{2}}_{\substack{nonadjacent\\separators}} + \underbrace{(p+1)}_{\substack{adjacent\\separators}} = \frac{(p+1)(p+2)}{2}. \quad (P1.83)$$

For the symmetric tensor with p upper indices and q lower indices, the number of components is

$$\frac{(p+1)(p+2)}{2} \frac{(q+1)(q+2)}{2}.$$

After convolution with δ_j^i one gets a tensor with $p-1$ upper and $q-1$ lower indices. This tensor must vanish. Therefore

$$\dim D(p,q) = \frac{(p+1)(p+2)}{2} \frac{(q+1)(q+2)}{2} - \frac{(p+1)p}{2} \frac{(q+1)q}{2}$$

$$= \frac{(p+1)(q+1)(p+q+2)}{2}. \tag{P1.84}$$

1.30 Show that all traceless tensors symmetric in both covariant and contravariant indices separately form a basis of the irrep $D(p,q)$.

HINT: show that they transform among one another since the transformation does not change the symmetry.

1.31 Show the following decompositions into irreps in $SU(3)$ and build the corresponding basis tensors:

$$3 \otimes \bar{3} = 1 \oplus 8, \tag{P1.85}$$

$$3 \otimes 3 = \bar{3} \oplus 6, \tag{P1.86}$$

$$3 \otimes 6 = 8 \oplus 10, \tag{P1.87}$$

$$3 \otimes 3 \otimes 3 = 1 \oplus 8 \oplus 8 \oplus 10. \tag{P1.88}$$

SOLUTION. We need the dimensions of low-lying irreps. Via (1.32), we get

$$D(1,0) = 3 - \text{fundamental irrep}, \tag{P1.89}$$

$$D(0,1) = \bar{3} - \text{anti-fundamental irrep}, \tag{P1.90}$$

$$D(1,1) = 8 - \text{adjoint irrep}, \tag{P1.91}$$

$$D(2,0) = 6, \tag{P1.92}$$

$$D(3,0) = 10. \tag{P1.93}$$

We define

$$\epsilon_{abc} = \epsilon^{abc}: \quad \epsilon_{123} = 1 \quad \Longrightarrow \quad \epsilon^{abc}\frac{1}{2}\epsilon_{cde} \times \epsilon^{dex}\frac{1}{2}\epsilon_{xef} = \epsilon^{abc}\frac{1}{2}\epsilon_{cef}. \tag{P1.94}$$

We have

$$\psi_b^a = p^a q_b = p^a q_b - \frac{1}{3}\delta_b^a(p^c q_c) + \frac{1}{3}\delta_b^a(p^c\, q_c), \qquad (P1.95)$$

$$\psi^{ad} = \underbrace{p^{\{a}q^{d\}}}_{6} + \underbrace{p^{[a}q^{d]}}_{3} = \underbrace{p^{\{a}q^{d\}}}_{6} + \epsilon^{adc}\frac{1}{2}\underbrace{\epsilon_{cfg}p^{[f}q^{g]}}_{3}$$

$$= \underbrace{p^{\{a}q^{d\}}}_{6} + \epsilon^{adc}\frac{1}{2}\underbrace{\epsilon_{cfg}p^{f}q^{g}}_{3}, \qquad (P1.96)$$

$$p^{\{a}q^{b\}}r^c = \underbrace{p^{\{a}q^{b}r^{c\}}}_{10} + \frac{2}{3}\delta_s^{\{a}\epsilon^{b\}cx}\underbrace{\epsilon_{xde}p^{\{s}q^{d\}}r^e}_{8}, \qquad (P1.97)$$

since $\epsilon_{xde}p^{\{a}q^{d\}}r^e \times \delta_a^x = 0$.

$$\psi^{abc} = p^a q^b r^c = \underbrace{p^{\{a}q^{b\}}r^c}_{6\otimes3} + \underbrace{p^{[a}q^{b]}r^c}_{\bar{3}\otimes3}$$

$$= \underbrace{p^{\{a}q^{b}r^{c\}}}_{10} + \frac{2}{3}\delta_s^{\{a}\epsilon^{b\}cx}\underbrace{\epsilon_{xde}p^{\{s}q^{d\}}r^e}_{8} + \epsilon^{abx}\frac{1}{2}\underbrace{\epsilon_{xfg}p^f q^g}_{3}\underbrace{r^c}_{3}$$

$$= \underbrace{p^{\{a}q^{b}r^{c\}}}_{10} + \frac{2}{3}\delta_s^{\{a}\epsilon^{b\}cx}\underbrace{\epsilon_{xde}p^{\{s}q^{d\}}r^e}_{8} + \epsilon^{abc}\frac{1}{6}\underbrace{(\epsilon_{cfg}p^f q^g r^c)}_{1}$$

$$+ \underbrace{\epsilon^{abx}\frac{1}{2}(\epsilon_{xfg}p^f q^g r^c - \frac{1}{3}\delta_x^c(\epsilon_{cfg}p^f q^g r^c))}_{8}. \qquad (P1.98)$$

1.32 Show decomposition into irreps in $SU(3)$:

$$8 \otimes 8 = 1 \oplus 8 \oplus 8 \oplus 10 \oplus \overline{10} \oplus 27, \quad D(2,2) = 27, \qquad (P1.99)$$
$$10 \otimes \overline{10} = 1 \oplus 8 \oplus 27 \oplus 64, \quad D(3,3) = 64. \qquad (P1.100)$$

1.33 Show that $(t^a)_j^i$ is an invariant tensor in $3 \otimes \bar{3} \otimes 8$.

SOLUTION. Indeed, via (1.22): $(t^a)_j^i U_k^j U_i^{\dagger l} U_{ab} = (U t^b U^{-1})_j^i U_k^j U_i^{\dagger l} = (U^\dagger U t^b U^{-1} U)_k^l = (t^b)_k^l$.

1.34 The $SU(3)_F$ flavor group is built assuming symmetry among u, d, and s quarks. Write explicitly the quark content of the ψ_b^a tensor and identify the corresponding mesons.

SOLUTION. We have the irreducible fundamental ($3 = D(1,0)$) and anti-fundamental ($\bar{3} = D(0,1)$) spinors:

$$p^a = q^a = r^a = \begin{pmatrix} u^\dagger \\ d^\dagger \\ s^\dagger \end{pmatrix}, \quad \bar{p}_a = \bar{q}_a = \bar{r}_a = \begin{pmatrix} \bar{u}^\dagger & \bar{d}^\dagger & \bar{s}^\dagger \end{pmatrix}, \qquad (P1.101)$$

where s quark has zero isospin by definition. The isospin projection I_3, the electric charge operator Q and the hypercharge Y in the **fundamental** basis read

$$I_3 = t^3 = \frac{1}{2}\begin{pmatrix} 1 & & \\ & -1 & \\ & & 0 \end{pmatrix}, \quad Q = \begin{pmatrix} \frac{2}{3} & & \\ & -\frac{1}{3} & \\ & & -\frac{1}{3} \end{pmatrix} \quad \text{(P1.102)}$$

$$\implies \quad Y = 2(Q - I_3) = \frac{1}{3}\begin{pmatrix} 1 & & \\ & 1 & \\ & & -2 \end{pmatrix} = \frac{2}{\sqrt{3}}t^8. \quad \text{(P1.103)}$$

Since t^3 and t^8 commute it is natural to classify the states according to their eigenvalues of I_3 and Y. For the quarks we can build Table 1.4 via (P1.64). Here, we introduced the baryon number $B : B = \frac{1}{3}$ for all

TABLE 1.4   Quantum numbers of the quarks

	Q	I	I_3	Y	S	B
u	$\frac{2}{3}$	$\frac{1}{2}$	$\frac{1}{2}$	$\frac{1}{3}$	0	$\frac{1}{3}$
d	$-\frac{1}{3}$	$\frac{1}{2}$	$-\frac{1}{2}$	$\frac{1}{3}$	0	$\frac{1}{3}$
s	$-\frac{1}{3}$	0	0	$-\frac{2}{3}$	-1	$\frac{1}{3}$
$\bar{u}$	$-\frac{2}{3}$	$\frac{1}{2}$	$-\frac{1}{2}$	$-\frac{1}{3}$	0	$-\frac{1}{3}$
$\bar{d}$	$\frac{1}{3}$	$\frac{1}{2}$	$\frac{1}{2}$	$-\frac{1}{3}$	0	$-\frac{1}{3}$
$\bar{s}$	$\frac{1}{3}$	0	0	$\frac{2}{3}$	1	$-\frac{1}{3}$

quarks and $B = -\frac{1}{3}$ for antiquarks, and strangeness $S : S = -1$ for s quark, $S = +1$ for $\bar{s}$, and $S = 0$ for other quarks and antiquarks. From the table one can see the relation

$$Y = S + B. \quad \text{(P1.104)}$$

Then for $8 = D(1,1)$:

$$\psi_b^a = p^a \bar{q}_b - \frac{1}{3}\delta_b^a (p^c \bar{q}_c) = \begin{pmatrix} \frac{u^\dagger \bar{u}^\dagger - d^\dagger \bar{d}^\dagger}{2} & u^\dagger \bar{d}^\dagger & u^\dagger \bar{s}^\dagger \\ d^\dagger \bar{u}^\dagger & -\frac{u^\dagger \bar{u}^\dagger - d^\dagger \bar{d}^\dagger}{2} & d^\dagger \bar{s}^\dagger \\ s^\dagger \bar{u}^\dagger & s^\dagger \bar{d}^\dagger & 0 \end{pmatrix}$$

$$- \frac{2s^\dagger \bar{s}^\dagger - u^\dagger \bar{u}^\dagger - d^\dagger \bar{d}^\dagger}{6}\begin{pmatrix} 1 & & \\ & 1 & \\ & & -2 \end{pmatrix}. \quad \text{(P1.105)}$$

The octet states read

$$I_3 = +1, Y = 0 : \psi_2^1 = u^\dagger \bar{d}^\dagger, \quad \text{(P1.106)}$$

$$I_3 = -1, Y = 0 : \psi_1^2 = d^\dagger \bar{u}^\dagger, \quad \text{(P1.107)}$$

$$I_3 = +\frac{1}{2}, Y = 1 : \psi_3^1 = u^\dagger \bar{s}^\dagger, \tag{P1.108}$$

$$I_3 = +\frac{1}{2}, Y = -1 : \psi_1^3 = s^\dagger \bar{d}^\dagger, \tag{P1.109}$$

$$I_3 = -\frac{1}{2}, Y = -1 : \psi_1^3 = s^\dagger \bar{u}^\dagger, \tag{P1.110}$$

$$I_3 = -\frac{1}{2}, Y = 1 : \psi_3^2 = d^\dagger \bar{s}^\dagger, \tag{P1.111}$$

$$I_3 = 0, Y = 0 : \psi_1^1 - \psi_2^2 = \sqrt{2} \times \frac{u^\dagger \bar{u}^\dagger - d^\dagger \bar{d}^\dagger}{\sqrt{2}}, \tag{P1.112}$$

$$I_3 = 0, Y = 0 : \psi_1^1 + \psi_2^2 = -\psi_3^3 = \sqrt{\frac{2}{3}} \times \frac{u^\dagger \bar{u}^\dagger + d^\dagger \bar{d}^\dagger - 2s^\dagger \bar{s}^\dagger}{\sqrt{6}}. \tag{P1.113}$$

The normalized singlet state $1 = D(0,0)$ reads

$$I_3 = 0, Y = 0 : \psi = \frac{1}{\sqrt{3}} p^c \bar{q}_c = \frac{s^\dagger \bar{s}^\dagger + u^\dagger \bar{u}^\dagger + d^\dagger \bar{d}^\dagger}{\sqrt{3}}. \tag{P1.114}$$

The corresponding mesons for total spin $J = 1, 0$ are given in Table 1.5. Note: the overall phases of the states differ from book to book. The states with the same $Y = I = I_3 = 0$ can mix. For $J = 1$ the observed

TABLE 1.5   Mesons built from u, d, and s quarks

$J=1$	$J=0$		Q	I	I_3	S	Y
$\lvert \rho^+ \rangle$	$\lvert \pi^+ \rangle$	$u^\dagger \bar{d}^\dagger \lvert 0 \rangle$	1	1	1	0	0
$\lvert \rho^- \rangle$	$\lvert \pi^- \rangle$	$d^\dagger \bar{u}^\dagger \lvert 0 \rangle$	-1	1	-1	0	0
$\lvert \rho^0 \rangle$	$\lvert \pi^0 \rangle$	$\frac{u^\dagger \bar{u}^\dagger - d^\dagger \bar{d}^\dagger}{\sqrt{2}} \lvert 0 \rangle$	0	1	0	0	0
$\lvert K^{*+} \rangle$	$\lvert K^+ \rangle$	$u^\dagger \bar{s}^\dagger \lvert 0 \rangle$	1	$\frac{1}{2}$	$\frac{1}{2}$	1	1
$\lvert \bar{K}^{*0} \rangle$	$\lvert \bar{K}^0 \rangle$	$s^\dagger \bar{d}^\dagger \lvert 0 \rangle$	0	$\frac{1}{2}$	$\frac{1}{2}$	-1	-1
$\lvert K^{*-} \rangle$	$\lvert K^- \rangle$	$s^\dagger \bar{u}^\dagger \lvert 0 \rangle$	-1	$\frac{1}{2}$	$-\frac{1}{2}$	-1	-1
$\lvert K^{*0} \rangle$	$\lvert K^0 \rangle$	$d^\dagger \bar{s}^\dagger \lvert 0 \rangle$	0	$\frac{1}{2}$	$-\frac{1}{2}$	1	1
$\lvert \omega_8 \rangle$	$\lvert \eta_8 \rangle$	$\frac{u^\dagger \bar{u}^\dagger + d^\dagger \bar{d}^\dagger - 2s^\dagger \bar{s}^\dagger}{\sqrt{6}} \lvert 0 \rangle$	0	0	0	0	0
$\lvert \phi_1 \rangle$	$\lvert \eta_1 \rangle$	$\frac{s^\dagger \bar{s}^\dagger + u^\dagger \bar{u}^\dagger + d^\dagger \bar{d}^\dagger}{\sqrt{3}} \lvert 0 \rangle$	0	0	0	0	0

states are

$$|\omega^0\rangle \simeq \sqrt{\frac{2}{3}}|\phi_1\rangle + \sqrt{\frac{1}{3}}|\omega_8\rangle = \frac{u^\dagger \bar{u}^\dagger + d^\dagger \bar{d}^\dagger}{\sqrt{2}}|0\rangle, \tag{P1.115}$$

$$|\phi^0\rangle \simeq \sqrt{\frac{1}{3}}|\phi_1\rangle - \sqrt{\frac{2}{3}}|\omega_8\rangle = s^\dagger \bar{s}^\dagger |0\rangle. \tag{P1.116}$$

For $J = 0$ the observed states are

$$|\eta\rangle \simeq \cos\theta|\eta_8\rangle - \sin\theta|\eta_1\rangle, \tag{P1.117}$$
$$|\eta'\rangle \simeq \sin\theta|\eta_8\rangle + \cos\theta|\eta_1\rangle, \tag{P1.118}$$

with θ between $-10°$ and $-20°$.

The spin-flavor wave-functions of the mesons are built as products of flavor and spin wave-functions:

$$|J = 0\rangle = |flavor\rangle \otimes \frac{\uparrow\downarrow - \downarrow\uparrow}{\sqrt{2}}, \quad |J = 1\rangle = |flavor\rangle \otimes \begin{pmatrix} \uparrow\uparrow \\ \frac{\uparrow\downarrow + \downarrow\uparrow}{\sqrt{2}} \\ \downarrow\downarrow \end{pmatrix}. \tag{P1.119}$$

1.35 Find η_1 and η_8 masses via η and η' masses.

SOLUTION. The physical states η, η' are mass^2 operator eigenstates, i.e.

$$m_\eta^2 = \langle\eta|\hat{M}^2|\eta\rangle, \quad m_{\eta'}^2 = \langle\eta'|\hat{M}^2|\eta'\rangle, \quad \langle\eta|\hat{M}^2|\eta'\rangle = 0. \tag{P1.120}$$

Reverting (P1.118)

$$\begin{pmatrix} \eta_8 \\ \eta_1 \end{pmatrix} = \begin{pmatrix} \cos\theta & \sin\theta \\ -\sin\theta & \cos\theta \end{pmatrix} \begin{pmatrix} \eta \\ \eta' \end{pmatrix}. \tag{P1.121}$$

Hence

$$m_{\eta_8}^2 = \langle\eta_8|\hat{M}^2|\eta_8\rangle = m_\eta^2 \cos^2\theta + m_{\eta'}^2 \sin^2\theta, \tag{P1.122}$$
$$m_{\eta_1}^2 = \langle\eta_1|\hat{M}^2|\eta_1\rangle = m_\eta^2 \sin^2\theta + m_{\eta'}^2 \cos^2\theta. \tag{P1.123}$$

1.36 Show that all particles in a multiplet have the same P-parity and spin.

SOLUTION. Generators of the flavor group $SU(2)_I$ or $SU(3)_F$ commute with the generators of the Lorentz group and with reflections. Therefore thanks to Schur's lemma, the latter operators are proportional to 1, i.e. their action on all particles in a multiplet is the same.

1.37 Is it possible to have a meson with $P = (-1)^J$ and $CP = -1$, i.e 0^{+-}, 1^{-+}, etc. as a $q\bar{q}$ state? Is it possible to have a 0^{--} as a $q\bar{q}$ state? Why?

SOLUTION. After inversion the space wave-function of a $q\bar{q}$ meson gets $(-1)^L$. Quarks and antiquarks have different internal parity. Therefore, the P parity of a meson is $(-1)^L P_q P_{\bar{q}} = (-1)^{L+1}$. The symmetry of the spin wave-function of 2 $\frac{1}{2}$ spins is $(-1)^{s+1}$, where s is the total spin. Since quarks are fermions the wave-function of a neutral meson is anti-symmetric under $q \leftrightarrow \bar{q}$, i.e. after charge conjugation, $\vec{r} \to -\vec{r}$, and $s_1 \leftrightarrow s_2$: $C(-1)^{L+s+1} = -1$, where C is the meson's C parity. Note, here $P_q P_{\bar{q}}$ are not included. Hence $C = (-1)^{L+s}$. Therefore, if a $q\bar{q}$ state has $PC = (-1)^{s+1} = -1 \implies s = 0 \implies J = L, P = (-1)^L = (-1)^{L+1}$ for the meson we study in this problem. This equation has no solutions. Such states are called **exotic** since they cannot be realized as $q\bar{q}$ mesons.

1.38 Place the mesons built from u, s, d quarks in Problem 1.34 on the plane with axes (I_3, Y). Find their parity P. For particles with zero charge find their C parity. For particles with zero strangeness find their $G = (-1)^{L+s+I}$ parity. Compare their masses and estimate the violation of $SU(2)_I$ and $SU(3)_F$ symmetries.

1.39 Assuming the $SU(3)_F$ symmetry among u, d, and s quarks write explicitly the quark content of the ψ^{abc} tensor, decompose it into the basis tensors of irreps and identify the corresponding baryons. Write explicitly the spin-flavor baryon wave-functions.

HINT. The decouplet state read

$$D^{111} = \Delta^{++}, \quad D^{112} = D^{121} = D^{211} = \frac{\Delta^+}{\sqrt{3}}, \tag{P1.124}$$

$$D^{122} = D^{221} = D^{212} = \frac{\Delta^0}{\sqrt{3}}, \quad D^{222} = \Delta^-, \tag{P1.125}$$

$$D^{113} = D^{131} = D^{311} = \frac{\Sigma^+}{\sqrt{3}}, \quad D^{223} = D^{322} = D^{232} = \frac{\Sigma^-}{\sqrt{3}}, \tag{P1.126}$$

$$D^{123} = D^{213} = D^{231} = D^{132} = D^{321} = D^{312} = \frac{\Sigma^0}{\sqrt{6}}, \tag{P1.127}$$

$$D^{133} = D^{331} = D^{313} = \frac{\Xi^0}{\sqrt{3}}, \quad D^{233} = D^{332} = D^{323} = \frac{\Xi^-}{\sqrt{3}}, \tag{P1.128}$$

$$D^{333} = \Omega^-, \tag{P1.129}$$

and $\bar{D}_{abc}$ is the decouplet of the antibaryons. It has the same structure as D^{abc} with superscripts and baryons replaced by subscripts and antibaryons. The normalization of states is such that in $\bar{D}_{abc} D^{abc}$ all

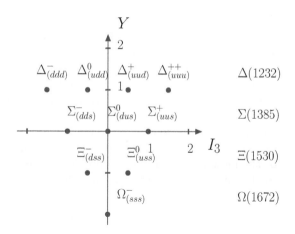

Figure 1.1 Baryon decouplet with $L = 0$, $J = \frac{3}{2}$.

baryons are normalized to 1. The octet

$$B_b^a = \begin{pmatrix} \frac{\Sigma_8^0}{\sqrt{2}} + \frac{\Lambda_8}{\sqrt{6}} & \Sigma_8^+ & p^+ \\ \Sigma_8^- & -\frac{\Sigma_8^0}{\sqrt{2}} + \frac{\Lambda_8}{\sqrt{6}} & n^0 \\ \Xi_8^- & \Xi_8^0 & -\frac{2\Lambda_8}{\sqrt{6}} \end{pmatrix}, \qquad (\text{P1.130})$$

$$\bar{B}_b^a = \begin{pmatrix} \frac{\overline{\Sigma_8^0}}{\sqrt{2}} + \frac{\overline{\Lambda_8}}{\sqrt{6}} & \overline{\Sigma_8^-} & \overline{\Xi_8^-} \\ \overline{\Sigma_8^+} & -\frac{\overline{\Sigma_8^0}}{\sqrt{2}} + \frac{\overline{\Lambda_8}}{\sqrt{6}} & \overline{\Xi_8^0} \\ \overline{p^+} & \overline{n^0} & -\frac{2\overline{\Lambda_8}}{\sqrt{6}} \end{pmatrix}, \qquad (\text{P1.131})$$

where the superscript stands for line, i.e. $B_2^1 = \Sigma_8^+$.

1.40 Place the mesons and baryons built from u, s, d quarks in Problem 1.39 on the plane with axes (I_3, Y).

ANSWER. For $L = 0$ the baryon states are shown in Figures 1.1 and 1.2.

1.41 Is it possible to have the anti-symmetric $SU(3)$ singlet from (P1.98) $\epsilon_{abc} p^a q^b r^c$ as a real particle? Why?

1.42 Is it possible to have a single flavor baryon, e.g. uuu with $L = 0$ and $J = \frac{1}{2}$? Why?

1.43 Assuming $SU(3)_F$ to $SU(2)_I$ breaking is due to larger mass of the s quark, derive mass splitting formulas for different multiplets:
a) the decouplet;
b) the baryon octet;

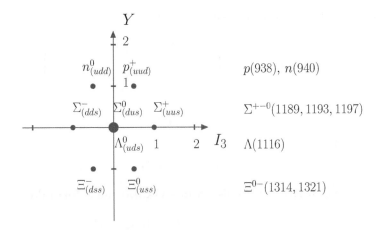

Figure 1.2 Baryon octet with $L = 0$, $J = \frac{1}{2}$.

c) the pseudoscalar meson octet;

d) the vector meson octet;

e) a multiplet transforming via an arbitrary irrep.

SOLUTION. The perturbation term should be invariant under isospin rotations and should preserve strangeness or hypercharge $Y = B + S$ (P1.104). Therefore, one can choose it as a matrix element of the hypercharge operator $Y = \frac{2}{\sqrt{3}} t^8$ (P1.103).

a) For the decouplet D^{abd}

$$-\mathcal{L}_M = m_0 \bar{D}_{abc} D^{abc} + \Delta m\, (t^8)^c_d \bar{D}_{abc} D^{abd} = a \bar{D}_{abc} D^{abc} + b\, \bar{D}_{ab3} D^{ab3}$$
$$= a(\bar{\Delta}^- \Delta^- + \bar{\Delta}^0 \Delta^0 + \bar{\Delta}^+ \Delta^+ + \bar{\Delta}^{++} \Delta^{++})$$
$$+ (a + \frac{b}{3})(\bar{\Sigma}^- \Sigma^- + \bar{\Sigma}^0 \Sigma^0 + \bar{\Sigma}^+ \Sigma^+)$$
$$+ (a + \frac{2}{3}b)(\bar{\Xi}^- \Xi^- + \bar{\Xi}^0 \Xi^0) + (a + b)\bar{\Omega}^- \Omega^-, \qquad \text{(P1.132)}$$

and one gets the **equal spacing (interval) rule**

$$\underbrace{M_\Omega - M_\Xi}_{142} = \underbrace{M_\Xi - M_\Sigma}_{145} = \underbrace{M_\Sigma - M_\Delta}_{153}. \qquad \text{(P1.133)}$$

b) For the baryon octet B^a_b

$$-\mathcal{L}_M = m_0 \bar{B}^a_b B^b_a + \Delta m_1 (t^8)^c_d \bar{B}^a_c B^d_a + \Delta m_2 (t^8)^c_d \bar{B}^d_a B^a_c. \qquad \text{(P1.134)}$$

This Lagrangian gives the **Gell-Mann – Okubo mass formula**

$$\underbrace{3M_\Lambda + M_\Sigma}_{4541} = \underbrace{2(M_N + M_\Xi)}_{4512}. \qquad \text{(P1.135)}$$

c) For the pseudoscalar meson octet P_b^a

$$P_b^a = \bar{P}_b^a = \begin{pmatrix} \frac{\pi^0}{\sqrt{2}} + \frac{\eta}{\sqrt{6}} & \pi^+ & K^+ \\ \pi^- & -\frac{\pi^0}{\sqrt{2}} + \frac{\eta}{\sqrt{6}} & K^0 \\ K^- & \overline{K^0} & -\frac{2\eta}{\sqrt{6}} \end{pmatrix}, \tag{P1.136}$$

$$C(\bar{P}_a^3 P_3^a)C^{-1} = \bar{P}_3^a P_a^3. \tag{P1.137}$$

Therefore

$$-\mathcal{L}_M = \frac{m_0^2}{2}\bar{P}_b^a P_a^b + \frac{\Delta m^2}{2}\bar{P}_3^a P_a^3, \tag{P1.138}$$

which gives the **Gell-Mann - Okubo mass formula**

$$\underbrace{4M_K^2}_{\simeq 992^2} = \underbrace{3M_\eta^2 + M_\pi^2}_{\simeq 956^2}, \tag{P1.139}$$

where we neglected $\eta_1\eta_8$ mixing.

d) For the vector meson octet V_b^a

$$V_b^a = \begin{pmatrix} \frac{\rho^0}{\sqrt{2}} + \frac{\omega_8}{\sqrt{6}} & \rho^+ & K^{*+} \\ \rho^- & -\frac{\rho^0}{\sqrt{2}} + \frac{\omega_8}{\sqrt{6}} & K^{*0} \\ K^{*-} & \overline{K^{*0}} & -\frac{2\omega_8}{\sqrt{6}} \end{pmatrix} \tag{P1.140}$$

taking into account mixing with the singlet ϕ (P1.116) and the result of Problem 1.35

$$\frac{1}{3}[4M_{K^*}^2 - M_\rho^2] = M_{\omega_8}^2 = \frac{1}{3}M_\omega^2 + \frac{2}{3}M_\phi^2. \tag{P1.141}$$

e) For a multiplet transforming via an arbitrary irrep with 8 generators $\tau_1, ...\tau_8$ normalized to obey the standard commutation rule (1.10), these generators transform among one another via the adjoint representation matrix U_{ab} (1.23): $\tau_b' = U_{ab}\tau_a$. Therefore, the matrix

$$\tau = \tau_a\lambda^a = \begin{pmatrix} \tau_3 + \frac{\tau_8}{\sqrt{3}} & \tau_1 - i\tau_2 & \tau_4 - i\tau_5 \\ \tau_1 + i\tau_2 & -\tau_3 + \frac{\tau_8}{\sqrt{3}} & \tau_6 - i\tau_7 \\ \tau_4 + i\tau_5 & \tau_6 + i\tau_7 & -2\frac{\tau_8}{\sqrt{3}} \end{pmatrix} \tag{P1.142}$$

and any function of it transforms by the adjoint action of 2 fundamental matrices (1.23). One seeks the mass splitting term transforming as hypercharge $Y = \frac{2}{\sqrt{3}}\tau_8$, i.e. as the τ_{33} component. Another combination transforming this way is e.g. the $_{33}$ element of τ^2

$$(\tau^2)_{33} = (\tau_4 + i\tau_5)(\tau_4 - i\tau_5) + (\tau_6 + i\tau_7)(\tau_6 - i\tau_7) + \frac{4}{3}\tau_8^2$$

$$= \tau_4^2 + \tau_5^2 + \tau_6^2 + \tau_7^2 + \frac{4}{3}\tau_8^2 + i[\tau_5, \tau_4] + i[\tau_7, \tau_6]$$

$$= C_2 - \tau_1^2 - \tau_2^2 - \tau_3^2 + \frac{1}{3}\tau_8^2 + \frac{\tau_3 + \sqrt{3}\tau_8}{2} + \frac{-\tau_3 + \sqrt{3}\tau_8}{2}$$

$$= C_2 - I^2 + \frac{Y^2}{4} + \frac{3}{2}Y. \tag{P1.143}$$

Here we used the commutation relations in the U and V subgroups (P1.49), (P1.50). The Casimir operator gives the same value on any member of the irrep. Therefore, it does not contribute to the mass splitting. As a result, for baryons

$$m = a + bY + c(I(I+1) - \frac{Y^2}{4}), \tag{P1.144}$$

and for mesons

$$m^2 = a + bY + c(I(I+1) - \frac{Y^2}{4}). \tag{P1.145}$$

This is the general **Gell-Mann–Okubo formula**. One can check that it reproduces the previous mass formulas (P1.133), (P1.139), and (P1.141).

1.44 Show that for multiplets turning into themselves after C transformation the coefficient of Y in Gell-Mann–Okubo formula (P1.145) is 0.

1.45 Find the flavor part of the $SU(3)_F$-invariant Lagrangian for the decay of the baryon decouplet to the baryon and pseudoscalar meson octets. Find the $SU(3)_F$ relations between the decay constants in the different channels.
SOLUTION. We follow (VH67). The flavor part of the Lagrangian has only one invariant structure

$$\mathcal{L}_F = \bar{B}_a^d \bar{P}_b^e D^{abc} \epsilon_{dec} = \bar{B}_a^d P_b^e D^{abc} \epsilon_{dec}. \tag{P1.146}$$

The values of the masses allow only decays with pions:

$$\mathcal{L}_F = \pi^- \left(\frac{\overline{\Xi_8^-}\Xi^0 - \overline{n^0}\Delta^+}{\sqrt{3}} + \frac{\overline{\Sigma_8^-}\Sigma^0 + \overline{\Sigma_8^0}\Sigma^+}{\sqrt{6}} + \frac{\overline{\Lambda_8}\Sigma^+}{\sqrt{2}} - \overline{p^+}\Delta^{++} \right)$$

$$- \pi^+ \left(\frac{\overline{\Xi_8^0}\Xi^- - \overline{p^+}\Delta^0}{\sqrt{3}} + \frac{\overline{\Sigma_8^+}\Sigma^0 - \overline{\Sigma_8^0}\Sigma^-}{\sqrt{6}} + \frac{\overline{\Lambda_8}\Sigma^-}{\sqrt{2}} - \overline{n^0}\Delta^- \right)$$

$$+ \pi^0 \left(\sqrt{\frac{2}{3}}[\overline{p^+}\Delta^+ + \overline{n^0}\Delta^0] - \frac{\overline{\Xi_8^-}\Xi^- + \overline{\Sigma_8^-}\Sigma^- + \overline{\Sigma_8^+}\Sigma^+ + \overline{\Xi_8^0}\Xi^0}{\sqrt{6}} \right.$$

$$\left. - \frac{\overline{\Lambda_8}\Sigma^0}{\sqrt{2}} \right). \tag{P1.147}$$

Therefore, the coupling constants have the following ratios, e.g.

$$g_{\Delta^{++}\to p^+\pi^+} : g_{\Sigma^+\to\Lambda_8\pi^+} : g_{\Sigma^0\to\Sigma_8^-\pi^+} : g_{\Xi^0\to\Xi_8^-\pi^+} = 1 : \frac{1}{\sqrt{2}} : \frac{1}{\sqrt{6}} : \frac{1}{\sqrt{3}}.$$
$$\text{(P1.148)}$$

1.46 Find the flavor part of the $SU(3)_F$-invariant Lagrangian for the decay of the vector mesons to 2 pseudoscalar octet mesons. Find the $SU(3)_F$ relations between the decay constants in the different channels.

SOLUTION. We follow (VH67). The flavor part of the Lagrangian has only one C-invariant structure

$$\mathcal{L}_F = V_b^a(\bar{P}_c^b(p_1)\bar{P}_a^c(p_2) - \bar{P}_c^b(p_2)\bar{P}_a^c(p_1)) \tag{P1.149}$$

since

$$CP_b^a C^{-1} = P_a^b, \quad CV_b^a C^{-1} = -V_a^b, \quad C\phi_1 C^{-1} = -\phi_1. \tag{P1.150}$$

Therefore

$$\mathcal{L}_F = K^{*0}\left(\sqrt{\frac{3}{2}}\bar{K}^0(p_1)\,\eta(p_2) - \frac{\bar{K}^0(p_1)\,\pi^0(p_2)}{\sqrt{2}} + K^-(p_1)\,\pi^+(p_2)\right)$$
$$+ K^{*+}\left(\sqrt{\frac{3}{2}}K^-(p_1)\,\eta(p_2) + \frac{K^-(p_1)\,\pi^0(p_2)}{\sqrt{2}} + \bar{K}^0(p_1)\,\pi^-(p_2)\right)$$
$$+ \rho^+\left(K^0(p_1)\,K^-(p_2) + \sqrt{2}\pi^-(p_1)\,\pi^0(p_2)\right)$$
$$+ \frac{1}{2\sqrt{2}}\rho^0\left(\bar{K}^0(p_1)\,K^0(p_2) + K^+(p_1)\,K^-(p_2) + 2\pi^+(p_1)\,\pi^-(p_2)\right)$$
$$+ \frac{1}{2}\sqrt{\frac{3}{2}}\omega_8\left(K^0(p_1)\,\bar{K}^0(p_2) + K^+(p_1)\,K^-(p_2)\right) - (p_1 \leftrightarrow p_2)$$
$$+ C\text{-conjugate}. \tag{P1.151}$$

Taking into account vector meson mixing (P1.116)

$$\omega_8 = \frac{1}{\sqrt{3}}\omega^0 - \sqrt{\frac{2}{3}}\phi^0, \tag{P1.152}$$

the coupling constants have the following ratios

$$g_{\rho^+\to\pi^+\pi^0} : g_{K^{*+}\to K^+\pi^0} : g_{\phi^0\to K^-K^+} = 1 : \frac{1}{2} : \frac{1}{\sqrt{2}}, \text{ etc.} \tag{P1.153}$$

1.47 Show that the electromagnetic current transforms as the V_1^1 component of the octet representation of $SU(3)_F$.

SOLUTION. The quark electromagnetic current reads

$$J_\mu = \frac{2}{3}\bar{u}\gamma_\mu u - \frac{1}{3}\bar{d}\gamma_\mu d - \frac{1}{3}\bar{s}\gamma_\mu s. \tag{P1.154}$$

Comparing it with the ψ_1^1 component of (P1.105) one can see that the current has the same flavor content. Under C conjugation it changes sign, i.e. it behaves as the $\frac{1}{1}$ component of the vector meson octet. Therefore

$$J_\mu = (V_\mu)_1^1. \qquad (\text{P1.155})$$

1.48 Find the $SU(3)_F$ relations for the magnetic moments in the baryon octet.

SOLUTION. We follow (VH67). The matrix element of the octet current between the 2 octet baryon states should transform as an octet. Therefore, one should build the octet flavor structure from $8 \otimes 8$. There are 2 such structures (P1.99):

$$\bar{B}_c^a B_b^c - B_c^a \bar{B}_b^c \quad | \times \delta_a^b \to 0, \qquad (\text{P1.156})$$

$$\bar{B}_c^a B_b^c + B_c^a \bar{B}_b^c - \frac{2}{3}\delta_b^a \bar{B}_c^d B_d^c \quad | \times \delta_a^b \to 0. \qquad (\text{P1.157})$$

The electromagnetic vertex has 2 form factors

$$\langle p_2 | J_\mu | p_1 \rangle = \bar{u}_{p_2}[F_1(q)\gamma_\mu + F_2(q)\frac{\sigma_{\mu\nu}q^\nu}{2m}]u_{p_2}, \quad q = p_1 - p_2, \qquad (\text{P1.158})$$

and the magnetic moment reads

$$\vec{\mu} = \mu\vec{S}, \quad \mu = \frac{e}{m}[F_1(0) + F_2(0)]. \qquad (\text{P1.159})$$

Therefore, there are 4 independent form factors

$$\langle p_2 | (V_\mu)_1^1 | p_1 \rangle = \bar{u}_{p_2}[F_1^F(q)\gamma_\mu + F_2^F(q)\frac{\sigma_{\mu\nu}q^\nu}{2m}]u_{p_2} \otimes (\bar{B}_c^1 B_1^c - B_c^1 \bar{B}_1^c)$$

$$+ \bar{u}_{p_2}[F_1^D(q)\gamma_\mu + F_2^D(q)\frac{\sigma_{\mu\nu}q^\nu}{2m}]u_{p_2}$$

$$\otimes (\bar{B}_c^1 B_1^c + B_c^1 \bar{B}_1^c - \frac{2}{3}\bar{B}_c^d B_d^c) \qquad (\text{P1.160})$$

$$= \bar{u}_{p_2}[F_1^F(q)\gamma_\mu + F_2^F(q)\frac{\sigma_{\mu\nu}q^\nu}{2m}]u_{p_2}$$

$$\otimes \left(\overline{\Xi_8^-}\Xi_8^- + \overline{\Sigma_8^-}\Sigma_8^- - \overline{p^+}p^+ - \overline{\Sigma_8^+}\Sigma_8^+\right)$$

$$+ \bar{u}_{p_2}[F_1^D(q)\gamma_\mu + F_2^D(q)\frac{\sigma_{\mu\nu}q^\nu}{2m}]u_{p_2}$$

$$\otimes \frac{1}{3}\left(\overline{\Xi_8^-}\Xi_8^- + \overline{\Sigma_8^-}\Sigma_8^- + \overline{p^+}p^+ + \sqrt{3}(\overline{\Lambda_8}\Sigma_8^0 + \overline{\Sigma_8^0}\Lambda_8)\right.$$

$$\left. + \overline{\Sigma_8^+}\Sigma_8^+ - 2\overline{n^0}n^0 - 2\overline{\Xi_8^0}\Xi_8^0 + \overline{\Sigma_8^0}\Sigma_8^0 - \overline{\Lambda_8}\Lambda_8\right). \qquad (\text{P1.161})$$

As a result, the magnetic moments are related

$$\mu_{\Sigma_8^+} = \mu_p, \quad \mu_{\Xi_8^-} = \mu_{\Sigma_8^-}, \quad \mu_n = \mu_{\Xi_8^0} = -2\mu_{\Sigma_8^0} = 2\mu_{\Lambda_8}. \qquad (\text{P1.162})$$

Moreover,

$$F_1^D(0) = Q_{\Sigma_8^0} = 0, \quad F_1^F(0) = Q_{p^+} = 1, \qquad \text{(P1.163)}$$

and the matrix element for $\Sigma_8^0 \to \Lambda_8 + \gamma$ decay has only one magnetic structure with

$$F_2^D(q) \simeq F_2^D(0) = -\frac{\sqrt{3}}{2}\mu_n \frac{m}{e}. \qquad \text{(P1.164)}$$

1.49 Calculate the $\Sigma_8^0 \to \Lambda_8 + \gamma$ decay width via the matrix element from Problem 1.48.

ANSWER. $\Gamma = \frac{3\alpha}{2m^2}\mu_n^2 \left(\dfrac{m_{\Sigma_8^0}^2 - m_{\Lambda_8}^2}{2m_{\Sigma_8^0}}\right)^3$.

1.50 Find the $SU(3)_F$ relations for the electromagnetic form-factors in the pseudoscalar meson octet and the ratio of the cross-sections of a meson pair production in the e^+e^- annihilation.

ANSWER. $F^{\pi^0}(k^2) = F^{\eta^0}(k^2) = F^{K^0}(k^2) = F^{\bar{K}^0}(k^2) = 0$,
$F^{\pi^+}(k^2) = F^{K^+}(k^2)$, $\sigma_{e^+e^- \to \eta^0\eta^0} = \sigma_{e^+e^- \to \pi^0\pi^0} = \sigma_{e^+e^- \to K^0\bar{K}^0} = 0$,
$\dfrac{\sigma_{e^+e^- \to \pi^+\pi^-}}{\sigma_{e^+e^- \to K^+K^-}} = \left(\dfrac{s-4m_{\pi^+}^2}{s-4m_{K^+}^2}\right)^{\frac{3}{2}}$.

1.51 Find the $SU(3)_F$ relations between the constants for the radiative decay of the vector mesons to the pseudoscalar meson octet. Find ratios of the corresponding decay widths, and the relations among the cross-sections of the vector and pseudoscalar meson pair production in the e^+e^- annihilation.

ANSWER. $|g_{\phi^0 \to \pi^0\gamma}|^2 + |g_{\omega^0 \to \pi^0\gamma}|^2 = 3(|g_{\phi^0 \to \eta\gamma}|^2 + |g_{\omega^0 \to \eta\gamma}|^2)$,
$g_{\rho^0,\pm \to \pi^0,\pm\gamma} : g_{\rho^0 \to \eta\gamma} : g_{K^{*\pm} \to K^{\pm}\gamma} : g_{K^{*0} \to K^0\gamma} = 1 : \sqrt{3} : 1 : -2$.

HINT: $\mathcal{L}_{V \to P\gamma} = \frac{1}{2}egP\epsilon^{\alpha\beta\mu\nu}F_{\alpha\beta}\partial_\mu V_\nu$, $\Gamma_{V \to \gamma P} = \frac{\alpha|g|^2}{96\pi}\left(\frac{m_V^2 - m_P^2}{m_V}\right)^3$,
$\sigma_{e^+e^- \to PV} = \frac{4\pi\alpha^2}{3}|g(s)|^2 \times \left(1 - 2\frac{m_V^2 + m_P^2}{s} + \frac{(m_V^2 - m_P^2)^2}{s^2}\right)^{\frac{3}{2}}$.

1.52 Find the $SU(3)_F$ relations between the constants of the neutral vector mesons to photon transition. Find the ratio of the leptonic widths of these mesons.

ANSWER. $g_{\rho^0 \to \gamma} : g_{\omega^0 \to \gamma} : g_{\phi^0 \to \gamma} = 1 : \frac{1}{\sqrt{3}}\frac{1}{\sqrt{3}} : -\sqrt{\frac{2}{3}}\frac{1}{\sqrt{3}}$, $\Gamma_{\rho^0 \to e^+e^-} :$
$\Gamma_{\omega^0 \to e^+e^-} : \Gamma_{\phi^0 \to e^+e^-} \simeq \frac{1}{m_{\rho^0}^3} : \frac{1}{9m_{\omega^0}^3} : \frac{2}{9m_{\phi^0}^3}$.

1.53 Find the $SU(3)_F$ relations between the constants of the baryon decouplet to octet radiative decay.

ANSWER. $g_{\Xi^- \to \Xi_8^-\gamma} = g_{\Sigma^- \to \Sigma_8^-\gamma} = 0$, $g_{\Delta^+ \to p^+\gamma} : g_{\Delta^0 \to n^0\gamma} : g_{\Xi^0 \to \Xi_8^0\gamma} :$
$g_{\Sigma^+ \to \Sigma_8^+\gamma} : g_{\Sigma^0 \to \Lambda_8^0\gamma} : g_{\Sigma^0 \to \Sigma_8^0\gamma} = 1 : 1 : -1 : -1 : -\frac{\sqrt{3}}{2} : \frac{1}{2}$.

1.54 Assuming $SU(3)_F$ symmetry estimate baryon mass splitting due to electromagnetic interaction.

SOLUTION. We follow (VH67). Since the electromagnetic current transforms as the $\frac{1}{1}$ component of the vector meson octet (P1.155), the mass correction must transform as a direct product of 2 currents: one for the photon emission and one for its absorption. Therefore, it transforms via $8 \otimes 8 = 1 \oplus 8 \oplus 8 \oplus 10 \oplus \overline{10} \oplus 27$ (P1.99). Here the singlet $V_{1i}^j V_{2j}^i$ does not split masses and the decouplets $\sim \epsilon^{ij\{k} V_{1i}^m V_{2j}^{n\}}$ and $\epsilon_{ij\{k} V_{1n}^i V_{2m\}}^j$ do not contain 2 electromagnetic currents. The octets and the 27 read

$$V_{1k}^i V_{2j}^k - \frac{1}{3}\delta_j^i V_{1k}^l V_{2l}^k, \quad \text{and} \quad (V_2 \leftrightarrow V_1), \tag{P1.165}$$

$$(V_1)_{\{k}^{\{i}(V_2)_{j\}}^{l\}} - \frac{1}{5}[(V_2)_{\{j}^s \delta_{k\}}^{\{i}(V_1)_s^{l\}} + (V_1)_{\{j}^s \delta_{k\}}^{\{i}(V_2)_s^{l\}}] + \frac{1}{20}\delta_{\{k}^{\{i}\delta_{j\}}^{l\}} V_{1s}^a V_{2a}^s. \tag{P1.166}$$

The subtraction terms in the octets do not split masses, while subtraction terms in the 27-plet have the same structure as octets or do not split masses either. Therefore, one may consider only the leading terms.

The product of 2 baryon octets in the mass term transforms under $8 \otimes 8$ as well. Therefore, to build an $SU(3)_F$-invariant contribution one takes the similar 4 structures

$$-\mathcal{L}_m = a\underbrace{\bar{B}_\beta^\alpha B_\alpha^\beta}_{1} + b\underbrace{\bar{B}_1^1 B_1^\beta}_{8} + c\underbrace{\bar{B}_1^\alpha B_\alpha^1}_{8} + d\underbrace{\bar{B}_1^1 B_1^1}_{27} \tag{P1.167}$$

$$= (a+c)\left(\overline{\Xi_8^-}\Xi_8^- + \overline{\Sigma_8^-}\Sigma_8^-\right) + (a+b)\left(\overline{p^+}p^+ + \overline{\Sigma_8^+}\Sigma_8^+\right)$$

$$+ a\left(\overline{n^0}n^0 + \overline{\Xi_8^0}\Xi_8^0\right) + \frac{6a+b+c+d}{6}\Lambda_8\Lambda_8$$

$$+ \frac{2a+b+c+d}{2}\overline{\Sigma_8^0}\Sigma_8^0 + \frac{b+c+d}{2\sqrt{3}}\left(\overline{\Lambda_8}\Sigma_8^0 + \overline{\Sigma_8^0}\Lambda_8\right). \tag{P1.168}$$

Hence, mass differences in $SU(2)_I$ multiplets have the relation

$$M_p - M_n = M_{\Sigma_8^+} - M_{\Sigma_8^-} + M_{\Xi_8^-} - M_{\Xi_8^0}. \tag{P1.169}$$

For the decouplet one has $10 \otimes \overline{10} = 1 \oplus 8 \oplus 27 \oplus 64$ (P1.100). Therefore, the mass splitting part contains only 2 terms:

$$-\mathcal{L}_m = a\underbrace{\overline{D}_{\alpha\beta\gamma}D^{\alpha\beta\gamma}}_{1} + b\underbrace{\overline{D}_{1\alpha\beta}D^{1\alpha\beta}}_{8} + c\underbrace{\overline{D}_{11\alpha}D^{11\alpha}}_{27} \tag{P1.170}$$

$$= a\left(\overline{\Delta^-}\Delta^- + \overline{\Xi^-}\Xi^- + \overline{\Sigma^-}\Sigma^- + \overline{\Omega^-}\Omega^-\right)$$

$$+ \frac{3a+2b+c}{3}\left(\overline{\Delta^+}\Delta^+ + \overline{\Sigma^+}\Sigma^+\right) + (a+b+c)\overline{\Delta^{++}}\Delta^{++}$$

$$+ \left(a + \frac{b}{3}\right)\left(\overline{\Delta^0}\Delta^0 + \overline{\Xi^0}\Xi^0 + \overline{\Sigma^0}\Sigma^0\right). \tag{P1.171}$$

Therefore, e.g.

$$3(M_{\Delta^+} - M_{\Delta^0}) = M_{\Delta^{++}} - M_{\Delta^-}, \tag{P1.172}$$
$$M_{\Delta^+} - M_{\Delta^0} = M_{\Sigma^+} - M_{\Sigma^0}, \tag{P1.173}$$
$$M_{\Delta^-} - M_{\Delta^0} = M_{\Sigma^-} - M_{\Sigma^0} = M_{\Xi^+} - M_{\Xi^0}, ... \tag{P1.174}$$

Compare with the data. Which of these equalities is a consequence of the $SU(2)_I$ symmetry?

1.55 Estimate **constituent** quark masses for u, d, s quarks from

a. $L = 0$ meson masses assuming that $J = 0$ and $J = 1$ mesons have different masses due to spin-spin interaction;

b. $L = 0$ baryon masses assuming that $J = \frac{1}{2}$ and $J = \frac{3}{2}$ baryons have different masses due to spin-spin interaction;

c. $L = 0$ baryon magnetic moments assuming that they are the matrix elements of the sum of the constituent quark's magnetic moments.

SOLUTION.

a. Suppose the mass of a meson comes from the constituent quark masses and spin-spin interaction

$$M_{q_1 \bar{q}_2} = m_1 + m_2 + A \frac{(s_1 s_2)}{m_1 m_2} \implies \tag{P1.175}$$

$$M_{J=1} = m_1 + m_2 + \frac{A}{4 m_1 m_2}, \tag{P1.176}$$

$$M_{J=0} = m_1 + m_2 - \frac{3A}{4 m_1 m_2}. \tag{P1.177}$$

Assuming $m_u \simeq m_d = m$,

$$m = \frac{3}{8}\left(M_\rho + \frac{1}{3}M_\pi\right) \simeq 306\,MeV, \tag{P1.178}$$

$$A \simeq m^2(M_\rho - M_\pi) \simeq 0.06\,GeV^3, \tag{P1.179}$$

$$m_s = m\frac{M_\rho - M_\pi}{M_{K^*} - M_K} \simeq 486\,MeV \implies \tag{P1.180}$$

$$M_\phi \simeq 1034\,MeV \; vs \; exp. \; 1019\,MeV. \tag{P1.181}$$

b. Suppose the mass of a baryon comes from the constituent quark masses and spin-spin interaction

$$M_{q_1 q_2 q_3} = m_1 + m_2 + m_3 + B\left(\frac{(s_1 s_2)}{m_1 m_2} + \frac{(s_1 s_3)}{m_1 m_3} + \frac{(s_3 s_2)}{m_3 m_2}\right). \tag{P1.182}$$

For the proton and delta states (P1.79) and (1.3)

$$M_p = \langle p | M_{q_1 q_2 q_3} | p \rangle = 2m_u + m_d + \frac{B}{4m_u^2} - \frac{B}{m_u m_d}, \qquad \text{(P1.183)}$$

$$M_\Delta = \langle \Delta | M_{q_1 q_2 q_3} | \Delta \rangle = 2m_u + m_d + \frac{B}{4m_u^2} + \frac{B}{2m_u m_d}. \qquad \text{(P1.184)}$$

Assuming $m_u \simeq m_d = m$,

$$M_\Delta = 3m + \frac{3B}{4m^2}, \quad M_p = 3m - \frac{3B}{4m^2} \quad \Longrightarrow \qquad \text{(P1.185)}$$

$$m = \frac{M_\Delta + M_p}{6} = 363 \, MeV, \qquad \text{(P1.186)}$$

$$B = \frac{2m_u^2}{3}(M_\Delta - M_p) = 0.027 \, GeV^3. \qquad \text{(P1.187)}$$

For uus states Σ_8^+ and Σ^+ the wave-functions are similar to the proton and Δ's with the substitution of $d \to s$. Therefore

$$M_{\Sigma_8^+} = 2m_u + m_s + \frac{B}{4m_u^2} - \frac{B}{m_u m_s}, \qquad \text{(P1.188)}$$

$$M_{\Sigma^+} = 2m_u + m_s + \frac{B}{4m_u^2} + \frac{B}{2m_u m_s} \quad \Longrightarrow \qquad \text{(P1.189)}$$

$$m_s = \frac{3B}{2(M_{\Sigma^+} - M_{\Sigma_8^+})} = 583 \, MeV \quad \Longrightarrow \qquad \text{(P1.190)}$$

$$M_\Omega = 1749 \, MeV \; vs \; exp. \; 1672 \, MeV. \qquad \text{(P1.191)}$$

c. The magnetic moment of a baryon is the average of the following operator

$$\mu = \frac{e_1}{m_1} s_1 + \frac{e_2}{m_2} s_2 + \frac{e_3}{m_3} s_3, \quad s|q_\uparrow\rangle = \frac{\hbar}{2}|q_\uparrow\rangle. \qquad \text{(P1.192)}$$

For the proton state (P1.79) assuming $m_u \simeq m_d = m$,

$$\mu_p = \langle p | \mu | p \rangle = \frac{4}{3}\mu_u - \frac{1}{3}\mu_d$$

$$= \frac{e\hbar}{2M_p}\left(\frac{8}{9}\frac{M_p}{m_u} + \frac{1}{9}\frac{M_p}{m_d}\right) \simeq \mu_N \frac{M_p}{m} \quad \Longrightarrow \qquad \text{(P1.193)}$$

$$m = \frac{\mu_N}{\mu_p} M_p = \frac{1}{2.79} M_p = 336 \, MeV. \qquad \text{(P1.194)}$$

$$\mu_\Omega = \langle \Omega | \mu | \Omega \rangle = 3\mu_s = -\frac{e\hbar}{2M_p}\frac{M_p}{m_s} \quad \Longrightarrow \qquad \text{(P1.195)}$$

$$m_s \simeq -\frac{\mu_N}{\mu_\Omega} M_p = \frac{1}{2.02} M_p = 464 \, MeV. \qquad \text{(P1.196)}$$

1.56 Estimate the ratio of the proton and neutron's magnetic moments. Compare to the data.

SOLUTION. One gets the neutron state by $u \leftrightarrow d$ substitution in the proton state (P1.79)

$$\mu_n = \langle n|\mu|n \rangle = \frac{4}{3}\mu_d - \frac{1}{3}\mu_u = -\frac{e\hbar}{2M_p}\left(\frac{4}{9}\frac{M_p}{m_d} + \frac{2}{9}\frac{M_p}{m_u}\right)$$

$$\simeq -\mu_N \frac{2}{3}\frac{M_p}{m} = -\frac{2}{3}\mu_p \ vs \ exp. \ -0.685\mu_p. \tag{P1.197}$$

1.57 Express the magnetic moments of all $L = 0$ baryons from the octet and decouplet via the quark magnetic moments. Estimate their values in nuclear magnetons and compare them to the experimental data.

HINT. Use the result of problem 1.55c.

1.58 Show that $SU(N) \sim S^3 \otimes S^5 \otimes \cdots \otimes S^{2N-1}$.

1.59 Show that $S \in SU(3)$, where S is defined in (1.36). Can one use the following matrix as S in decomposition (1.39)

$$\begin{pmatrix} e^{i\alpha_8}\sin\alpha_5\cos\alpha_7 & e^{i\alpha_8}\cos\alpha_5\cos\alpha_7 & -e^{-i\alpha_6-i\alpha_4}\sin\alpha_7 \\ e^{i\alpha_6}\sin\alpha_5\sin\alpha_7 & e^{i\alpha_6}\cos\alpha_5\sin\alpha_7 & e^{-i\alpha_4-i\alpha_8}\cos\alpha_7 \\ e^{i\alpha_4}\cos\alpha_5 & -e^{i\alpha_4}\sin\alpha_5 & 0 \end{pmatrix} ? \tag{P1.198}$$

1.60 Show that in $SU(N)$ the measure built via right invariant vectors $R_i = i\frac{\partial V}{\partial\alpha_i}V^{-1}$ coincides with the measure built via left invariant vectors L_i (1.44) and as such has both left and right invariance $dU = d(UV) = d(VU)$.

1.61 Show that for $SU(2)$ the (unnormalized) Haar measure (1.40) may be defined as

$$d\mu = i\varepsilon_{ijk}tr(L_iL_jL_k)d\alpha_1 d\alpha_2 d\alpha_3$$

$$= \varepsilon_{ijk}tr(V^{-1}\frac{\partial V}{\partial\alpha_i}V^{-1}\frac{\partial V}{\partial\alpha_j}V^{-1}\frac{\partial V}{\partial\alpha_k})d\alpha_1 d\alpha_2 d\alpha_3. \tag{P1.199}$$

SOLUTION. For $L_i = a_i + b_i^t\sigma^t$

$$i\varepsilon_{ijk}tr(L_iL_jL_k) = -2\varepsilon^{tqr}\varepsilon_{ijk}b_i^t b_j^q b_k^r = -12\det b_k^t, \tag{P1.200}$$

$$\det tr(L_iL_j) = 8\varepsilon_{ijk}(a_ia_1 + b_i^t b_1^t)(a_ja_2 + b_j^q b_2^q)(a_ka_3 + b_k^r b_3^r),$$

$$= 8\varepsilon_{ijk}b_i^t b_1^t b_j^q b_2^q b_k^r b_3^r = 8\det b_i^t b_k^t = 8(\det b_k^t)^2. \tag{P1.201}$$

Therefore

$$\det tr(L_iL_j) = 18[i\varepsilon_{ijk}tr(L_iL_jL_k)]^2. \tag{P1.202}$$

1.62 Construct the invariant measure on a) $SU(2)$, b) $SU(3)$.

ANSWER. For the parametrization given in the lecture

a)

$$\int_{SU(2)} dU = \frac{1}{4\pi^2} \int_0^{\frac{\pi}{2}} \sin(2\alpha_3) d\alpha_3 \int_0^{2\pi} \int_0^{2\pi} d\alpha_1 d\alpha_2; \qquad (P1.203)$$

b)

$$\int_{SU(3)} dU = \int_{SU(2)} dU \int_{S^5} dS, \quad \int_{S^5} dS = \frac{1}{2\pi^3} \int_0^{\frac{\pi}{2}} \sin(2\alpha_7) d\alpha_7$$
$$\times \int_0^{\frac{\pi}{2}} \cos^3 \alpha_5 \sin \alpha_5 d\alpha_5 \int_0^{2\pi} \int_0^{2\pi} \int_0^{2\pi} d\alpha_4 d\alpha_6 d\alpha_8.$$
$$(P1.204)$$

1.63 Using Haar measure definition (1.40) shows that

$$\int_{SU(N)} dU U_j^i = \int_{SU(N)} dU U_j^{\dagger i} = 0, \qquad (P1.205)$$

$$\int_{SU(N)} dU U_j^i U_l^k = \int_{SU(N)} dU U_j^{\dagger i} U_l^{\dagger k} = 0, \qquad (P1.206)$$

$$\int_{SU(N)} dU U_j^i U_l^{\dagger k} = \frac{1}{N} \delta_l^i \delta_j^k, \qquad (P1.207)$$

$$\int_{SU(N)} dU tr(UV^\dagger) tr(WU^\dagger) = \frac{1}{N} tr(WV^\dagger), \qquad (P1.208)$$

$$\int dU_1 dU_2 dU_3 dU_4 tr(U_1 U_2 U_3 U_4) tr(U_4^\dagger U_3^\dagger U_2^\dagger U_1^\dagger) = 1, \qquad (P1.209)$$

$$\int_{SU(3)} dU U_j^i U_l^k U_o^m = \frac{1}{6} \varepsilon^{ikm} \varepsilon_{jlo}. \quad (P1.210)$$

$$\int_{SU(N)} dU U_j^i U_l^{\dagger k} U_n^m U_q^{\dagger p} = \frac{-1}{N(N^2-1)} (\delta_l^i \delta_q^m \delta_n^k \delta_j^p + \delta_q^i \delta_l^m \delta_j^k \delta_n^p)$$
$$+ \frac{1}{N^2-1} (\delta_l^i \delta_q^m \delta_j^k \delta_n^p + \delta_q^i \delta_l^m \delta_n^k \delta_j^p). \quad (P1.211)$$

SOLUTION. One can consider the integral

$$\int_{SU(N)} dU U_{j_1}^{i_1} ... U_{j_p}^{i_p} U_{l_1}^{\dagger k_1} ... U_{l_n}^{\dagger k_n} \qquad (P1.212)$$

as a projector of the integrand to the scalar representation. Indeed, the integrand transforms as a direct product of $p+n$ fundamental and antifundamental reps. However, the integral does not depend on U, i.e. it should be a set of invariant tensors with number coefficients. It means that if there is no scalar rep in the direct product, the integral is 0. If, however, there are scalar reps, one can often find the coefficients of the invariant tensors contracting the indices and using the Haar measure definition. Integral (P1.207)

$$I^{ik}_{jl} = \int_{SU(N)} dU\, U^i_j U^{\dagger k}_l = a\delta^i_l\delta^k_j + b\delta^k_l\delta^i_j \quad |\times V^j_m V^{\dagger n}_k \quad V \in SU(N)$$

$$\implies \int_{SU(N)} d(UV)(UV)^i_m (UV)^{\dagger n}_l = I^{in}_{ml} = a\delta^i_l V^{\dagger n}_k V^k_m + bV^i_m V^{\dagger n}_l$$

$$= a\delta^i_l\delta^n_m + bV^i_m V^{\dagger n}_l \implies b = 0,$$

$$\int_{SU(N)} dU\, U^i_j U^{\dagger k}_l = a\delta^i_l\delta^k_j \quad |\times \delta^l_i \implies$$

$$\int_{SU(N)} dU\, \delta^k_j = 1\delta^k_j = aN\delta^k_j \implies a = \frac{1}{N}. \tag{P1.213}$$

Integral (P1.211) retains its form after convolution with $V^a_i V^{\dagger l}_b V^c_m V^{\dagger q}_d$ if it is proportional to $\delta^i_l\delta^m_q$ and $\delta^i_q\delta^m_l$. Likewise, invariance after convolution with $V^j_a V^{\dagger b}_k V^n_c V^{\dagger d}_p$ is ensured by $\delta^k_j\delta^p_n$ and $\delta^k_n\delta^p_j$. Symmetry w.r.t. $U^i_j \leftrightarrow U^m_n$ and $U^{\dagger k}_l \leftrightarrow U^{\dagger p}_q$ lead to 2 invariant combinations

$$\int_{SU(N)} dU\, U^i_j U^{\dagger k}_l U^m_n U^{\dagger p}_q = a(\delta^i_l\delta^m_q\delta^k_j\delta^p_n + \delta^i_q\delta^m_l\delta^k_n\delta^p_j)$$

$$+ b(\delta^i_l\delta^m_q\delta^k_n\delta^p_j + \delta^i_q\delta^m_l\delta^k_j\delta^p_n). \tag{P1.214}$$

Convolution with δ^l_i transforms this integral to the previous one giving us equations for a and b.

The material of this chapter is discussed in more details in the references for further reading below. General introduction to non-abelian groups is given in (PS95), and the basic group relations are derived in (MSW68). A list of advanced group relations useful for color algebra one can find in the Appendix of (FF05). Introduction to Group integration and its application to Lattice QCD is presented in (Cre85), while constituent quark model is described in (Per82). Applications of flavor symmetry and a great number of group relations between different amplitudes, form factors, and cross-sections are derived in (VH67). General introduction to the quark model is given in (Ric12) and (dS63).

FURTHER READING

[Cre85] Michael Creutz. *Quarks, gluons and lattices*, chapter 8, Group integration. Cambridge Monographs on Mathematical Physics. Cambridge University Press, Cambridge, UK, 6, 1985.

[dS63] J. J. de Swart. The Octet model and its Clebsch-Gordan coeffi-cients. *Rev. Mod. Phys.*, 35:916–939, 1963. [Erratum: *Rev. Mod. Phys.* 37, 326–326 (1965)].

[FF05] V. S. Fadin and R. Fiore. Non-forward NLO BFKL kernel. *Phys. Rev. D*, 72:014018, 2005. Appendix A.

[MSW68] A. J. MacFarlane, Anthony Sudbery, and P. H. Weisz. On Gell-Mann's lambda-matrices, d- and f-tensors, octets, and parametriza-tions of SU(3). *Commun. Math. Phys.*, 11:77–90, 1968.

[Per82] D. H. Perkins. *Introduction to high energy physics*, chapter 4, Quarks in hadrons. 1982.

[PS95] Michael E. Peskin and Daniel V. Schroeder. *An introduction to quantum field theory*, chapter 15.4. Addison-Wesley, Reading, USA, 1995.

[Ric12] Jean-Marc Richard. An introduction to the quark model. In *Ferrara International School Niccolo Cabeo 2012: Hadronic spectroscopy*, 5, 2012.

[VH67] Nguyen Van Hieu. *Lectures on the theory of unitary symmetry of elementary particles*. Atomizdat, 1967. In Russian. Лекции по теории унитарной симметрии элементарных частиц, Нгуен Ван Хьеу, Атомиздат, Москва, 1967 год.

CHAPTER 2

SU(3) Color Gauge Invariance

T HE MAIN concept we build the theory of strong interactions on is color gauge invariance. In this chapter we use it to construct the Lagrangian of quantum chromodynamics (QCD), Wilson lines, inter-quark potential, and derive the basic equations of the theory.

1. A **classical QCD Lagrangian** for one quark field and gluon fields reads

$$\mathcal{L}_{color} = \mathcal{L}_{YM} + \bar{\psi}(i\hat{D} - m)\psi, \quad \mathcal{L}_{YM} = -\frac{1}{4}F^{a\mu\nu}F^a_{\mu\nu}, \qquad (2.1)$$

where $\mathcal{L}_{YM}$ is called the **Yang-Mills Lagrangian**. The QCD Lagrangian is built **invariant under an internal local group** $SU(N_c)$, $N_c = 3$, called the **color group.** Here

$$\psi = \psi^i = \begin{pmatrix} \psi^1 \\ \psi^2 \\ \psi^3 \end{pmatrix}, \quad \bar{\psi}_i = (\psi^\dagger)_i \gamma^0 \qquad (2.2)$$

are the **quark** and **antiquark** fields, which transform as the **fundamental and anti-fundamental** spinors under the color $SU(3)$ group

$$\psi \to \psi^\theta : (\psi^\theta)^i = (U^\theta)^i_j \psi^j \qquad (2.3)$$

$$\bar{\psi} \to \bar{\psi}^\theta : \bar{\psi}^\theta_i = \bar{\psi}_j (U^{\theta\dagger})^j_i. \qquad (2.4)$$

The mass term is diagonal in color in $\mathcal{L}_{color}$

$$m = m\delta^i_j, \qquad (2.5)$$

and the **covariant derivative** is

$$D_\mu = \partial_\mu - igA_\mu, \qquad (2.6)$$

DOI: 10.1201/9781003272403-2

where A is the **gauge field associated with color transformations** called the **gluon** field. The corresponding charge g is called **color** or **strong** charge. To have **gauge (or SU(3) color) invariance** in $\mathcal{L}_{color}$, D must transform in the adjoint representation

$$D_\mu \to D_\mu^\theta : (D_\mu^\theta)_j^i = (U^\theta)_k^i (D_\mu)_l^k (U^{\theta\dagger})_j^l \tag{2.7}$$

and A must take values in the algebra spanning the fundamental generators t^a

$$(D_\mu)_j^i = \partial_\mu \delta_j^i - ig A_\mu^a (t^a)_j^i, \quad (A_\mu)_j^i = A_\mu^a (t^a)_j^i. \tag{2.8}$$

Hence, A transforms as

$$A_\mu \to A_\mu^\theta : (A_\mu^\theta)_j^i = (U^\theta)_k^i ((A_\mu)_l^k - \delta_l^k \frac{\partial_\mu}{ig})(U^{\theta\dagger})_j^l, \tag{2.9}$$

or using

$$A_\mu^\theta \simeq (1 + i\theta^a t^a)(A_\mu^b t^b - \frac{\partial_\mu}{ig})(1 - i\theta^c t^c), \tag{2.10}$$

in the infinitesimal form

$$A_\mu^{\theta a} \simeq A_\mu^a + \frac{\partial_\mu \theta^a}{g} + f^{abc} A_\mu^b \theta^c = A_\mu^a + \frac{1}{g}(\partial_\mu \theta^a + g f^{abc} A_\mu^b \theta^c)$$

$$= A_\mu^a + \frac{1}{g}(\partial_\mu \delta^{ac} - ig T_{ac}^b A_\mu^b)\theta^c = A_\mu^a + \frac{1}{g} D^{ac} \theta^c, \tag{2.11}$$

where the covariant derivative D^{ac} is built from the gluon fields spanning the adjoint generators T^a (P1.41). In the matrix form

$$A_\mu^\theta = A_\mu + \frac{1}{g}(\partial_\mu - ig[A_\mu, \theta]) = A_\mu + \frac{1}{g} D_\mu \theta. \tag{2.12}$$

The gluon **strength tensor**

$$F_{\mu\nu} = F_{\mu\nu}^a t^a = \frac{-1}{ig}[D_\mu, D_\nu] = \partial_\mu A_\nu - \partial_\nu A_\mu - ig[A_\mu A_\nu]$$

$$= \partial_\mu A_\nu - \partial_\nu A_\mu + g f^{abc} A_\mu^b A_\nu^c t^a \tag{2.13}$$

also transforms by the adjoint representation

$$F_{\mu\nu} \to F_{\mu\nu}^\theta = U^\theta F_{\mu\nu} U^{\theta\dagger}. \tag{2.14}$$

Hence, $\mathcal{L}_{YM}$ is **gauge invariant**

$$\mathcal{L}_{YM} = -\frac{1}{2}tr(F_{\mu\nu} F^{\mu\nu}) = -\frac{1}{2}tr(U^\theta F_{\mu\nu} U^{\theta\dagger} U^\theta F^{\mu\nu} U^{\theta\dagger}) = \mathcal{L}_{YM}^\theta. \tag{2.15}$$

2. A **Wilson line** is a path ordered exponential

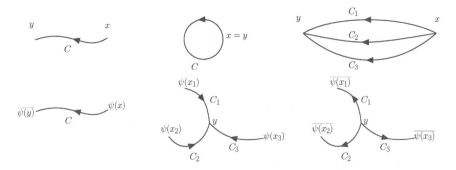

Figure 2.1   The Wilson line, loop, baryon loop, and gauge invariant objects made of Wilson lines and quark fields.

$$U_C(y,x) = \mathcal{P}_C e^{ig \int_x^y dz^\mu A_\mu(z)} = 1 + ig \int_x^y dz^\mu A_\mu(z)$$

$$+ \frac{(ig)^2}{2!} \int_x^y dz_1^{\mu_1} \int_x^y dz_2^{\mu_2} \mathcal{P}_C (A_{\mu_1}(z_1) A_{\mu_2}(z_2)) + \dots \quad (2.16)$$

$$= 1 + ig \int_x^y dz^\mu A_\mu(z) + (ig)^2 \int_x^y dz_1^{\mu_1} \int_x^{z_1} dz_2^{\mu_2} A_{\mu_1}(z_1) A_{\mu_2}(z_2) + \dots$$

$$= 1 + ig \int_0^1 dt z^\mu(t) A_\mu(z(t))$$

$$+ (ig)^2 \int_0^1 dt_1 \int_0^{t_1} dt_2 z^{\mu_1}(t_1) A_{\mu_1}(z(t_1)) z^{\mu_2}(t_2) A_{\mu_2}(z(t_2)) + \dots \quad (2.17)$$

Here C — is a contour from x to y, $z(t)$ is the parametrization of this contour such that $z(0) = x$ and $z(1) = y$, and $\mathcal{P}_C$ stands for the ordering of matrices $A = A^a t^a$ along the contour. By construction the Wilson line obeys the equation

$$\frac{\partial}{\partial y^\alpha} U_C(y,x) = ig A_\alpha(y) U_C(y,x), \quad D_\alpha(y) U_C(y,x) = 0 \quad (2.18)$$

with the initial condition

$$U_{C=\{x\}}(x,x) = 1. \quad (2.19)$$

Hence, after an $SU(3)$ gauge rotation V^θ

$$V_y^\theta D_\alpha(y) V_y^{\theta\dagger} V_y^\theta U_C(y,x) V_x^{\theta\dagger} = D_\alpha^\theta(y) (V_y^\theta U_C(y,x) V_x^{\theta\dagger}) = 0 \quad (2.20)$$

and

$$U_C(y,x) \to U_C^\theta(y,x) = V_y^\theta U_C(y,x) V_x^{\theta\dagger}. \qquad (2.21)$$

Multiplication of $V_x^{\theta\dagger}$ from the left is required to satisfy the boundary condition at $x = y$. Therefore the **Wilson loop**

$$trU_C(x,x) = tr(\mathcal{P}_C e^{ig \oint_C dz^\mu A_\mu(z)}), \qquad (2.22)$$

3-quark or baryon Wilson loop

$$\varepsilon^{ijk} \varepsilon_{lmn} U_{C_1}(y,x)_i^l U_{C_2}(y,x)_j^m U_{C_3}(y,x)_k^n, \qquad (2.23)$$

and the non-local objects

$$\varepsilon_{lmn} U_{C_1}(y,x_1)_i^l U_{C_2}(y,x_2)_j^m U_{C_3}(y,x_3)_k^n \, \psi(x_1)^i \psi(x_2)^j \psi(x_3)^k, \qquad (2.24)$$

$$\bar{\psi}(y) U_C(y,x) \psi(x), \quad F^a(y) U_C(y,x)^{ab} F^b(x) \qquad (2.25)$$

are **gauge invariant**. In the latter equation

$$U_C(y,x)^{ab} = (\mathcal{P}_C e^{ig \int_x^y dz^\mu A_\mu^c(z) T^c})^{ab} \qquad (2.26)$$

is a **Wilson line in the adjoint rep**.

3. **Ground state energy of 2 static quarks.** Consider a $q\bar{q}$ pair of heavy $(m \to \infty)$ quarks located at the distance R. The interaction energy (inter-quark potential) of this system can be calculated as a Wilson loop average

$$\frac{1}{-iT} \ln\langle \frac{1}{N_c} trU_{\square T} \rangle \underset{T\to\infty(1-i\varepsilon)}{\longrightarrow} V(R). \qquad (2.27)$$

Indeed, to get the static quark propagator one can do a **Foldy-Wouthuysen transformation** in the QCD Lagrangian with a unitary operator

$$e^{iS}, \quad S = \frac{-i}{2m}(i\vec{\partial} + g\vec{A})\vec{\gamma}, \qquad (2.28)$$

$$\begin{aligned}
\mathcal{L}_{quark} &= \bar{\psi}(i\partial_0\gamma^0 + gA_0\gamma^0 - (i\vec{\partial} + g\vec{A})\vec{\gamma} - m)\psi = |\psi = e^{-iS}\psi'| \\
&= \psi^{\dagger'} e^{iS}(i\partial_0 + gA_0 - (i\vec{\partial} + g\vec{A})\gamma^0\vec{\gamma} - \gamma^0 m)e^{-iS}\psi' \\
&= \psi^{\dagger'}(i\partial_0 + gA_0 - (i\vec{\partial} + g\vec{A})\gamma^0\vec{\gamma} - \gamma^0 m - i[S,\gamma^0]m + O(\frac{1}{m}))\psi' \\
&= \psi^{\dagger'}(i\partial_0 + gA_0 - \gamma^0 m + O(\frac{1}{m}))\psi'. \qquad (2.29)
\end{aligned}$$

Neglecting the $O(\frac{1}{m})$ correction one gets the equation for the **static quark propagator**

$$((i\partial_0 + gA_0)\gamma^0 - m)G(\vec{x},t;\vec{x}',t') = i\delta(\vec{x} - \vec{x}')\delta(t - t') \qquad (2.30)$$

with a solution

$$G(\vec{x}, t; \vec{x}', t') = \delta(\vec{x} - \vec{x}') \mathcal{T} e^{ig \int_{t'}^{t} d\tau A_0(\vec{x}, \tau)}$$
$$\times \frac{1}{2}((1 + \gamma^0)\theta(t - t')e^{-im(t-t')} + (1 - \gamma^0)\theta(t' - t)e^{im(t-t')}).$$
$$(2.31)$$

Next, consider a gauge invariant $q\bar{q}$ state with inter-quark separation R at $t = 0$

$$\Psi(0)|0\rangle = \bar{\psi}'(R, 0)U([R, 0], [0, 0])\Gamma\psi'(0, 0)|0\rangle \qquad (2.32)$$

and find its projection on itself at $t = T$

$$\langle 0|\mathcal{T}(\Psi(T)^\dagger \Psi(0))|0\rangle = \sum_n \langle 0|\Psi(T)^\dagger|n\rangle\langle n|\Psi(0)|0\rangle. \qquad (2.33)$$

Here Γ is a γ-matrix indicating the properties of the state w.r.t. space-time transformations, e. g. γ^μ for vector state, 1 for the scalar state, e.t.c. Rewriting time dependence as time evolution by the Hamiltonian

$$\Psi(T) = e^{iHT}\Psi(0)e^{-iHT}, \qquad (2.34)$$

we get

$$\sum_n \langle 0|e^{iE_0 T}\Psi^\dagger(0)e^{-iE_n T}|n\rangle\langle n|\Psi(0)|0\rangle = \sum_n |\langle n|\Psi(0)|0\rangle|^2 e^{-i(E_n - E_0)T}. \qquad (2.35)$$

Therefore

$$\langle 0|\mathcal{T}(\Psi(T)^\dagger \Psi(0))|0\rangle \xrightarrow[T \to \infty(1-i\varepsilon)]{} |\langle 0|\Psi(0)_j^i|1\rangle|^2 e^{-i(E_1 - E_0)T}$$
$$= const\, e^{-iE_{Ground\,state}T} \qquad (2.36)$$

since the contribution of higher exited states with E_2, E_3, ... dye out. Using the **static quark propagator** one can calculate this matrix element explicitly

$$\langle 0|\mathcal{T}(\Psi(T)^\dagger \Psi(0))|0\rangle = \langle 0|\mathcal{T}(\bar{\psi}'(0, T)U([0, T], [R, T])\Gamma^\dagger\psi'(R, T)$$
$$\times \bar{\psi}'(R, 0)U([R, 0], [0, 0])\Gamma\psi'(0, 0))|0\rangle \qquad (2.37)$$

$$= -\langle 0|\mathcal{T}(G(0, 0; 0, T)U([0, T], [R, T])\Gamma^\dagger$$
$$\times G(R, T; R, 0)U([R, 0], [0, 0])\Gamma)|0\rangle \qquad (2.38)$$
$$= const\, e^{-2imT} tr\langle 0|\mathcal{T}(U([0, 0], [0, T])U([0, T], [R, T])$$
$$\times U([R, T], [R, 0])U([R, 0], [0, 0]))|0\rangle \qquad (2.39)$$
$$= const\, e^{-2imT} \langle \frac{1}{N_c} tr U_{\square T}\rangle. \qquad (2.40)$$

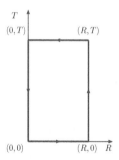

Figure 2.2 The Wilson loop $\frac{1}{N_c}trU_{\square T}$.

Here

$$const \simeq N_c tr(\frac{(1+\gamma^0)}{2}\Gamma\frac{(1-\gamma^0)}{2}\Gamma^\dagger)\delta(0)^2 \qquad (2.41)$$

does not depend on T. To measure **interaction energy** one subtracts $2m$ and gets

$$\frac{1}{-iT}\ln\langle\frac{1}{N_c}trU_{\square T}\rangle \underset{T\to\infty(1-i\varepsilon)}{\to} V(R). \qquad (2.42)$$

EXERCISES

2.1 Show that $F^{a\mu\nu}F^a_{\mu\nu}$ and $\varepsilon_{\mu\nu\alpha\beta}F^{a\mu\nu}F^{a\alpha\beta}$ are C-even.
SOLUTION. Here we follow (Pec99). By the C-transformation the current

$$\bar\psi\gamma^\mu t^a\psi \overset{C}{\to} -\bar\psi\gamma^\mu t^{aT}\psi = -\eta(a)\bar\psi\gamma^\mu t^a\psi. \qquad (P2.1)$$

The Gell-Mann matrices have the following property

$$t^{aT} = \begin{cases} t^a, a = 1,3,4,6,8 \\ -t^a, a = 2,5,7. \end{cases} \implies \eta(a) = \begin{cases} 1, a = 1,3,4,6,8 \\ -1, a = 2,5,7. \end{cases} \qquad (P2.2)$$

Strong interactions conserve C parity. Therefore the gluon field should have the same transformation law under the charge conjugation

$$A^a_\mu \overset{C}{\to} -\eta(a)A^a_\mu \implies \bar\psi\hat A\psi \overset{C}{\to} \bar\psi\hat A\psi. \qquad (P2.3)$$

Then the stress tensor

$$F^a_{\mu\nu} = \partial_\mu A^a_\nu - \partial_\mu A^a_\nu + gf^{abc}A^b_\mu A^c_\nu \overset{C}{\to} -\eta(a)F^a_{\mu\nu}, \qquad (P2.4)$$

since

$$f^{abc} \neq 0 \quad \text{for} \quad abc = \{123, 147, 156, 246, 257, 345, 367, 458, 678\},$$
$$(P2.5)$$

where 2, 5, and 7 always come in the odd numbers. Therefore both $\varepsilon_{\mu\nu\alpha\beta}F^{a\mu\nu}F^{a\alpha\beta}$ and $F^{a\mu\nu}F^a_{\mu\nu}$ are C-even.

2.2 Jacobi identity for $F_{\mu\nu}$. Show that

$$\varepsilon^{\mu\nu\alpha\beta}(D_\mu{}^{ac})F^c_{\alpha\beta} = 0. \tag{P2.6}$$

SOLUTION.

$$
\begin{aligned}
\varepsilon^{\mu\nu\alpha\beta}(D_\mu{}^{ac})F^c_{\alpha\beta} &= \varepsilon^{\mu\nu\alpha\beta}(\partial_\mu\delta^{ac} + gf^{abc}A^b_\mu) \\
&\quad \times ((\partial_\alpha A^c_\beta) - (\partial_\beta A^c_\alpha) + gf^{cde}A^d_\alpha A^e_\beta) \\
&= \varepsilon^{\mu\nu\alpha\beta}(\partial_\mu\delta^{ac})(gf^{cde}A^d_\alpha A^e_\beta) \\
&\quad + \varepsilon^{\mu\nu\alpha\beta}(gf^{abc}A^b_\mu)(2(\partial_\alpha A^c_\beta) + gf^{cde}A^d_\alpha A^e_\beta)
\end{aligned}
$$

$$
\begin{aligned}
&= \varepsilon^{\mu\nu\alpha\beta}gf^{ade}\partial_\mu(A^d_\alpha)A^e_\beta + \varepsilon^{\mu\nu\alpha\beta}gf^{ade}A^d_\alpha\partial_\mu(A^e_\beta) \\
&\quad + \varepsilon^{\mu\nu\alpha\beta}(gf^{abc}A^b_\mu)(2(\partial_\alpha A^c_\beta) + gf^{cde}A^d_\alpha A^e_\beta) \\
&= \varepsilon^{\mu\nu\alpha\beta}gf^{ade}\partial_\alpha(A^d_\beta)A^e_\mu - \varepsilon^{\mu\nu\alpha\beta}gf^{ade}A^d_\mu\partial_\alpha(A^e_\beta) \\
&\quad + \varepsilon^{\mu\nu\alpha\beta}(gf^{abc}A^b_\mu)(2(\partial_\alpha A^c_\beta) + gf^{cde}A^d_\alpha A^e_\beta) \\
&= \varepsilon^{\mu\nu\alpha\beta}gf^{ade}A^e_\mu\partial_\alpha(A^d_\beta) - \varepsilon^{\mu\nu\alpha\beta}gf^{aed}A^e_\mu\partial_\alpha(A^d_\beta) \\
&\quad + \varepsilon^{\mu\nu\alpha\beta}(gf^{abc}A^b_\mu)(2(\partial_\alpha A^c_\beta) + gf^{cde}A^d_\alpha A^e_\beta)
\end{aligned}
$$

$$
\begin{aligned}
&= 2\varepsilon^{\mu\nu\alpha\beta}gf^{acb}A^b_\mu\partial_\alpha(A^c_\beta) + \varepsilon^{\mu\nu\alpha\beta}(gf^{abc}A^b_\mu)(2(\partial_\alpha A^c_\beta) + gf^{cde}A^d_\alpha A^e_\beta) \\
&= \varepsilon^{\mu\nu\alpha\beta}g^2 f^{abc}f^{cde}A^b_\mu A^d_\alpha A^e_\beta = -\varepsilon^{\mu\nu\alpha\beta}g^2 f^{abd}f^{cde}A^b_\mu A^c_\alpha A^e_\beta \\
&= -\varepsilon^{\mu\nu\alpha\beta}g^2 \frac{1}{3}\left[f^{bcd}f^{ade} + f^{cad}f^{bde} + f^{abd}f^{cde}\right]A^b_\mu A^c_\alpha A^e_\beta = 0. \quad (P2.7)
\end{aligned}
$$

Here we used Jacobi identity (1.12) and anti-symmetry of f and ε

$$
\begin{aligned}
\varepsilon^{\mu\nu\alpha\beta}\left[f^{cad}f^{bde}\right]A^b_\mu A^c_\alpha A^e_\beta &= -\varepsilon^{\mu\nu\alpha\beta}\left[f^{bad}f^{cde}\right]A^c_\alpha A^b_\mu A^e_\beta \\
&= \varepsilon^{\mu\nu\alpha\beta}\left[f^{abd}f^{cde}\right]A^c_\alpha A^b_\mu A^e_\beta, \tag{P2.8} \\
\varepsilon^{\mu\nu\alpha\beta}\left[f^{bcd}f^{ade}\right]A^b_\mu A^c_\alpha A^e_\beta &= -\varepsilon^{\mu\nu\alpha\beta}\left[f^{ecd}f^{adb}\right]A^e_\beta A^c_\alpha A^b_\mu \\
&= \varepsilon^{\mu\nu\alpha\beta}\left[f^{cde}f^{abd}\right]A^e_\beta A^c_\alpha A^b_\mu. \tag{P2.9}
\end{aligned}
$$

2.3 Show how $F_{\mu\nu}$ transforms by a direct calculation

SOLUTION.

$$F_{\mu\nu} \to \partial_\mu \{U^\theta((A_\nu) - \frac{\partial_\nu}{ig})U^{\theta\dagger}\} - \partial_\nu \{U^\theta((A_\mu) - \frac{\partial_\mu}{ig})U^{\theta\dagger}\}$$

$$- ig[\{U^\theta((A_\mu) - \frac{\partial_\mu}{ig})U^{\theta\dagger}\}, \{U^\theta((A_\nu) - \frac{\partial_\nu}{ig})U^{\theta\dagger}\}]$$

$$= \{(\partial_\mu U^\theta)((A_\nu) - \frac{\partial_\nu}{ig})U^{\theta\dagger}\} + \{U^\theta((\partial_\mu A_\nu) - \frac{\partial_\mu \partial_\nu}{ig})U^{\theta\dagger}\}$$

$$+ \{U^\theta A_\nu (\partial_\mu U^{\theta\dagger})\} - (\mu \leftrightarrow \nu)$$

$$- ig[\{U^\theta((A_\mu))U^{\theta\dagger}\}, \{U^\theta((A_\nu))U^{\theta\dagger}\}]$$

$$- ig[\{U^\theta(-\frac{\partial_\mu}{ig})U^{\theta\dagger}\}, \{U^\theta((A_\nu))U^{\theta\dagger}\}]$$

$$- ig[\{U^\theta((A_\mu))U^{\theta\dagger}\}, \{U^\theta(-\frac{\partial_\nu}{ig})U^{\theta\dagger}\}]$$

$$- ig[\{U^\theta(-\frac{\partial_\mu}{ig})U^{\theta\dagger}\}, \{U^\theta(-\frac{\partial_\nu}{ig})U^{\theta\dagger}\}]$$

$$= -\frac{1}{ig}(\partial_\mu U^\theta)(\partial_\nu U^{\theta\dagger}) + \partial_\mu \{U^\theta A_\nu U^{\theta\dagger}\} - (\mu \leftrightarrow \nu)$$

$$- igU^\theta[A_\mu, A_\nu]U^{\theta\dagger} - igU^\theta[(-\frac{\partial_\mu U^{\theta\dagger}}{ig})U^\theta, A_\nu]U^{\theta\dagger}$$

$$- igU^\theta[A_\mu, (-\frac{\partial_\nu U^{\theta\dagger}}{ig})U^\theta]U^{\theta\dagger}$$

$$- igU^\theta[(-\frac{\partial_\mu U^{\theta\dagger}}{ig})U^\theta, (-\frac{\partial_\nu U^{\theta\dagger}}{ig})U^\theta]U^{\theta\dagger}$$

$$= U^\theta(\partial_\mu A_\nu - \partial_\nu A_\mu - ig[A_\mu, A_\nu])U^{\theta\dagger}, \tag{P2.10}$$

since

$$-igU^\theta[(-\frac{\partial_\mu U^{\theta\dagger}}{ig})U^\theta, (-\frac{\partial_\nu U^{\theta\dagger}}{ig})U^\theta]U^{\theta\dagger}$$

$$= -igU^\theta(-\frac{\partial_\mu U^{\theta\dagger}}{ig})U^\theta(-\frac{\partial_\nu U^{\theta\dagger}}{ig})U^\theta U^{\theta\dagger}$$

$$+ igU^\theta(-\frac{\partial_\nu U^{\theta\dagger}}{ig})U^\theta(-\frac{\partial_\mu U^{\theta\dagger}}{ig})U^\theta U^{\theta\dagger}$$

$$= -igU^\theta(-\frac{\partial_\mu U^{\theta\dagger}}{ig})U^\theta(-\frac{\partial_\nu U^{\theta\dagger}}{ig}) + igU^\theta(-\frac{\partial_\nu U^{\theta\dagger}}{ig})U^\theta(-\frac{\partial_\mu U^{\theta\dagger}}{ig})$$

$$= -ig(\frac{\partial_\mu U^\theta}{ig})U^{\theta\dagger}U^\theta(-\frac{\partial_\nu U^{\theta\dagger}}{ig}) + ig(\frac{\partial_\nu U^\theta}{ig}U^{\theta\dagger})U^\theta(-\frac{\partial_\mu U^{\theta\dagger}}{ig})$$

$$= (\frac{\partial_\mu U^\theta}{ig})(\partial_\nu U^{\theta\dagger}) + (\partial_\nu U^\theta)(-\frac{\partial_\mu U^{\theta\dagger}}{ig}), \tag{P2.11}$$

and

$$-igU^\theta[(-\frac{\partial_\mu U^{\theta\dagger}}{ig})U^\theta, A_\nu]U^{\theta\dagger} = U^\theta(\partial_\mu U^{\theta\dagger})U^\theta A_\nu U^{\theta\dagger} - U^\theta A_\nu(\partial_\mu U^{\theta\dagger})$$
$$= -(\partial_\mu U^\theta)U^{\theta\dagger}U^\theta A_\nu U^{\theta\dagger} - U^\theta A_\nu(\partial_\mu U^{\theta\dagger})$$
$$= -(\partial_\mu U^\theta)A_\nu U^{\theta\dagger} - U^\theta A_\nu(\partial_\mu U^{\theta\dagger}).$$
$$(P2.12)$$

2.4 Show that $\varepsilon_{\alpha\beta\gamma\delta}F^{a\alpha\beta}F^{a\gamma\delta}$ is gauge invariant.

2.5 Show that

$$\varepsilon^{\mu\nu\alpha\beta}F^a_{\mu\nu}F^a_{\alpha\beta} = \partial_\mu K^\mu, \quad K^\mu = 2\varepsilon^{\mu\nu\alpha\beta}(A^a_\nu F^a_{\alpha\beta} - \frac{1}{3}gf^{abc}A^a_\nu A^b_\alpha A^c_\beta).$$
$$(P2.13)$$

SOLUTION.

$$\partial_\mu K^\mu = \partial_\mu 2\varepsilon^{\mu\nu\alpha\beta}(A^a_\nu F^a_{\alpha\mu} - \frac{1}{3}gf^{abc}A^a_\nu A^b_\alpha A^c_\beta)$$
$$= 2\varepsilon^{\mu\nu\alpha\beta}(\partial_\mu(A^a_\nu F^a_{\alpha\beta}) - \frac{1}{3}gf^{abc}\partial_\mu(A^a_\nu A^b_\alpha A^c_\beta))$$
$$= 2\varepsilon^{\mu\nu\alpha\beta}((\partial_\mu A^a_\nu)F^a_{\alpha\beta} + A^a_\nu(\partial_\mu F^a_{\alpha\beta}) - gf^{abc}(\partial_\mu A^a_\nu)A^b_\alpha A^c_\beta)$$
$$= \varepsilon^{\mu\nu\alpha\beta}F^a_{\mu\nu}F^a_{\alpha\beta} - \varepsilon^{\mu\nu\alpha\beta}gf^{abc}A^b_\mu A^c_\nu F^a_{\alpha\beta}$$
$$+ 2\varepsilon^{\mu\nu\alpha\beta}(A^a_\nu(\partial_\mu gf^{abc}A^b_\alpha A^c_\beta) - gf^{abc}(\partial_\mu A^a_\nu)A^b_\alpha A^c_\beta)$$
$$= \varepsilon^{\mu\nu\alpha\beta}F^a_{\mu\nu}F^a_{\alpha\beta} - \varepsilon^{\mu\nu\alpha\beta}gf^{abc}A^b_\mu A^c_\nu F^a_{\alpha\beta}$$
$$+ 2\varepsilon^{\mu\nu\alpha\beta}gf^{abc}(\partial_\alpha A^a_\beta)A^b_\mu A^c_\nu \qquad (P2.14)$$

$$= \varepsilon^{\mu\nu\alpha\beta}F^a_{\mu\nu}F^a_{\alpha\beta} + \varepsilon^{\mu\nu\alpha\beta}gf^{abc}\left[-F^a_{\alpha\beta} + (\partial_\alpha A^a_\beta) - (\partial_\beta A^a_\alpha)\right]A^b_\mu A^c_\nu$$
$$= \varepsilon^{\mu\nu\alpha\beta}F^a_{\mu\nu}F^a_{\alpha\beta} + \varepsilon^{\mu\nu\alpha\beta}gf^{bcd}\left[gf^{ade}A^a_\alpha A^e_\beta\right]A^b_\mu A^c_\nu$$
$$= \varepsilon^{\mu\nu\alpha\beta}F^a_{\mu\nu}F^a_{\alpha\beta}$$
$$+ \varepsilon^{\mu\nu\alpha\beta}g^2\frac{1}{3}\left[f^{bcd}f^{ade} + f^{cad}f^{bde} + f^{abd}f^{cde}\right]A^a_\alpha A^e_\beta A^b_\mu A^c_\nu$$
$$= \varepsilon^{\mu\nu\alpha\beta}F^a_{\mu\nu}F^a_{\alpha\beta}. \qquad (P2.15)$$

Here we also used Jacobi identity (1.12) and anti-symmetry of f and ε

$$\varepsilon^{\mu\nu\alpha\beta}\left[f^{cad}f^{bde}\right]A^a_\alpha A^e_\beta A^b_\mu A^c_\nu = -\varepsilon^{\mu\nu\alpha\beta}\left[f^{cbd}f^{ade}\right]A^b_\mu A^e_\beta A^a_\alpha A^c_\nu$$
$$= \varepsilon^{\mu\nu\alpha\beta}\left[f^{bcd}f^{ade}\right]A^b_\mu A^e_\beta A^a_\alpha A^c_\nu, \qquad (P2.16)$$
$$\varepsilon^{\mu\nu\alpha\beta}\left[f^{abd}f^{cde}\right]A^a_\alpha A^e_\beta A^b_\mu A^c_\nu = -\varepsilon^{\mu\nu\alpha\beta}\left[f^{cbd}f^{ade}\right]A^c_\nu A^e_\beta A^b_\mu A^a_\alpha$$
$$= \varepsilon^{\mu\nu\alpha\beta}\left[f^{bcd}f^{ade}\right]A^c_\nu A^e_\beta A^b_\mu A^a_\alpha. \qquad (P2.17)$$

The other way round,

$$
\begin{aligned}
\varepsilon^{\mu\nu\alpha\beta} F^a_{\mu\nu} F^a_{\alpha\beta} &= \varepsilon^{\mu\nu\alpha\beta}((\partial_\mu A^a_\nu) - (\partial_\nu A^a_\mu) + g f^{abc} A^b_\mu A^c_\nu) F^a_{\alpha\beta} \\
&= 2\varepsilon^{\mu\nu\alpha\beta}(\partial_\mu A^a_\nu) F^a_{\alpha\beta} + \varepsilon^{\mu\nu\alpha\beta} g f^{abc} A^b_\mu A^c_\nu F^a_{\alpha\beta} \\
&= 2\varepsilon^{\mu\nu\alpha\beta}(\partial_\mu A^a_\nu F^a_{\alpha\beta}) - 2\varepsilon^{\mu\nu\alpha\beta}(A^a_\nu \partial_\mu - \frac{1}{2} g f^{abc} A^b_\mu A^c_\nu) F^a_{\alpha\beta} \\
&= 2\varepsilon^{\mu\nu\alpha\beta}(\partial_\mu A^a_\nu F^a_{\alpha\beta}) - \varepsilon^{\mu\nu\alpha\beta} A^a_\nu (\partial_\mu \delta^{ac} + g f^{abc} A^b_\mu) F^c_{\alpha\beta} \\
&\quad - \varepsilon^{\mu\nu\alpha\beta} A^a_\nu \partial_\mu F^a_{\alpha\beta} \\
&= 2\varepsilon^{\mu\nu\alpha\beta}(\partial_\mu A^a_\nu F^a_{\alpha\beta}) - \varepsilon^{\mu\nu\alpha\beta} A^a_\nu (D_\mu{}^{ac}) F^c_{\alpha\beta} \\
&\quad - \varepsilon^{\mu\nu\alpha\beta} g f^{abc} A^a_\nu \partial_\mu (A^b_\alpha A^c_\beta) \\
&= 2\varepsilon^{\mu\nu\alpha\beta}(\partial_\mu A^a_\nu F^a_{\alpha\beta}) - \frac{2}{3}\varepsilon^{\mu\nu\alpha\beta} g f^{abc} \partial_\mu (A^a_\nu A^b_\alpha A^c_\beta).
\end{aligned}
$$

$$(P2.18)$$

2.6 Prove the following representations for K (P2.13)

$$
K^\mu = 4\varepsilon^{\mu\nu\alpha\beta}(A^a_\nu(\partial_\alpha A^a_\beta) + \frac{1}{3} g f^{abc} A^a_\nu A^b_\alpha A^c_\beta) \tag{P2.19}
$$

$$
= 8\varepsilon^{\mu\nu\alpha\beta}(tr[A_\nu(D_\alpha A_\beta)] + \frac{ig}{3} tr[A_\nu A_\alpha A_\beta]). \tag{P2.20}
$$

SOLUTION. Using $D_\mu = \partial_\mu - ig A_\mu$ and expressing f^{abc} via the trace (P1.12), one gets

$$
\begin{aligned}
K^\mu &= 2\varepsilon^{\mu\nu\alpha\beta}(A^a_\nu F^a_{\alpha\beta} - \frac{1}{3} g f^{abc} A^a_\nu A^b_\alpha A^c_\beta) \\
&= 2\varepsilon^{\mu\nu\alpha\beta}(A^a_\nu((\partial_\alpha A^a_\beta) - (\partial_\beta A^a_\alpha) + g f^{ade} A^d_\alpha A^e_\beta) - \frac{1}{3} g f^{abc} A^a_\nu A^b_\alpha A^c_\beta) \\
&= 4\varepsilon^{\mu\nu\alpha\beta}(2tr[A_\nu(\partial_\alpha A_\beta)] - \frac{4i}{3} g tr[A_\nu A_\alpha A_\beta]) \tag{P2.21} \\
&= 4\varepsilon^{\mu\nu\alpha\beta}(2tr[A_\nu(D_\alpha A_\beta)] + \frac{2i}{3} g tr[A_\nu A_\alpha A_\beta]). \tag{P2.22}
\end{aligned}
$$

2.7 Show that the gauge transformation $A_\mu \to A'_\mu = U(A_\mu - \frac{\partial_\mu}{ig})U^\dagger$ modifies K^μ (P2.13) in the following way

$$
K^\mu \to K^\mu + \delta K^\mu, \tag{P2.23}
$$

$$
\delta K^\mu = \varepsilon^{\mu\nu\alpha\beta} \frac{8}{3g^2} tr[U\{\partial_\nu U^\dagger\} U\{\partial_\alpha U^\dagger\} U\{\partial_\beta U^\dagger\}]
$$

$$
- \frac{8}{ig}\varepsilon^{\mu\nu\alpha\beta}\partial_\alpha tr[\{\partial_\nu U^\dagger\} U A_\beta]. \tag{P2.24}
$$

SOLUTION.

$$K^\mu \to K'^\mu = 2\varepsilon^{\mu\nu\alpha\beta}(A^{a\prime}_\nu F^{a\prime}_{\alpha\beta} - \frac{1}{3}gf^{abc}A^{a\prime}_\nu A^{b\prime}_\alpha A^{c\prime}_\beta)$$

$$= 2\varepsilon^{\mu\nu\alpha\beta}(2tr[A'_\nu F'_{\alpha\beta}] - \frac{1}{3}g(-4i)tr[A'_\nu A'_\alpha A'_\beta])$$

$$= 4\varepsilon^{\mu\nu\alpha\beta}tr[\{U(A_\nu - \frac{\partial_\nu}{ig})U^\dagger\}UF_{\alpha\beta}U^\dagger]$$

$$+ \varepsilon^{\mu\nu\alpha\beta}\frac{8ig}{3}\,tr[\{U(A_\nu - \frac{\partial_\nu}{ig})U^\dagger\}$$

$$\times \{U(A_\alpha - \frac{\partial_\alpha}{ig})U^\dagger\}\{U(A_\beta - \frac{\partial_\beta}{ig})U^\dagger\}] \qquad (\text{P2.25})$$

$$= K^\mu + \varepsilon^{\mu\nu\alpha\beta}\frac{8}{3g^2}\,tr[U\{\partial_\nu U^\dagger\}U\{\partial_\alpha U^\dagger\}U\{\partial_\beta U^\dagger\}]$$

$$- \varepsilon^{\mu\nu\alpha\beta}8\,tr[UA_\nu A_\alpha\{\partial_\beta U^\dagger\}] + \varepsilon^{\mu\nu\alpha\beta}\frac{8}{ig}\,tr[UA_\nu\{\partial_\alpha U^\dagger\}\{U\partial_\beta U^\dagger\}]$$

$$- \frac{4}{ig}\varepsilon^{\mu\nu\alpha\beta}tr[\{\partial_\nu U^\dagger\}U(\partial_\alpha A_\beta - \partial_\beta A_\alpha - ig[A_\alpha, A_\beta])] \qquad (\text{P2.26})$$

$$= K^\mu + \varepsilon^{\mu\nu\alpha\beta}\frac{8}{3g^2}\,tr[U\{\partial_\nu U^\dagger\}U\{\partial_\alpha U^\dagger\}U\{\partial_\beta U^\dagger\}]$$

$$- \varepsilon^{\mu\nu\alpha\beta}\frac{8}{ig}\,tr[A_\nu\{\partial_\alpha U^\dagger\}\{\partial_\beta U\}] + 8\varepsilon^{\mu\nu\alpha\beta}tr[A_\alpha A_\beta\{\partial_\nu U^\dagger\}U]$$

$$- \varepsilon^{\mu\nu\alpha\beta}8\,tr[A_\nu A_\alpha\{\partial_\beta U^\dagger\}U] - \frac{8}{ig}\varepsilon^{\mu\nu\alpha\beta}tr[\{\partial_\nu U^\dagger\}U(\partial_\alpha A_\beta)]$$

$$\qquad (\text{P2.27})$$

$$= K^\mu + \varepsilon^{\mu\nu\alpha\beta}\frac{8}{3g^2}\,tr[U\{\partial_\nu U^\dagger\}U\{\partial_\alpha U^\dagger\}U\{\partial_\beta U^\dagger\}]$$

$$- \varepsilon^{\mu\nu\alpha\beta}\frac{8}{ig}\,tr[A_\nu\{\partial_\alpha U^\dagger\}\{\partial_\beta U\}] - \frac{8}{ig}\varepsilon^{\mu\nu\alpha\beta}\partial_\alpha tr[\{\partial_\nu U^\dagger\}UA_\beta]$$

$$+ \frac{8}{ig}\varepsilon^{\mu\nu\alpha\beta}tr[\{\partial_\nu U^\dagger\}\{\partial_\alpha U\}A_\beta] \qquad (\text{P2.28})$$

$$= K^\mu + \varepsilon^{\mu\nu\alpha\beta}\frac{8}{3g^2}\,tr[U\{\partial_\nu U^\dagger\}U\{\partial_\alpha U^\dagger\}U\{\partial_\beta U^\dagger\}]$$

$$- \frac{8}{ig}\varepsilon^{\mu\nu\alpha\beta}\partial_\alpha tr[\{\partial_\nu U^\dagger\}UA_\beta]. \qquad (\text{P2.29})$$

2.8 Derive the conjugate momenta and the classic equations of motion for the quark and gluon fields. Choose the **temporal gauge** $A_0 = 0$.

SOLUTION. For the symmetric classical QCD action

$$S = \int d^4x \left(-\frac{1}{4} F^{\mu\nu a} F^a_{\mu\nu} + \bar{\psi}(i\overleftrightarrow{D} - m)\psi \right), \tag{P2.30}$$

$$\overleftrightarrow{D} = \frac{(\vec{\partial} - igA) + (-\overleftarrow{\partial} - igA)}{2} = \overleftrightarrow{\partial} - igA = \frac{\vec{\partial} - \overleftarrow{\partial}}{2} - igA, \tag{P2.31}$$

the conjugate momenta to the fields $\psi, \bar{\psi}$, and $A^\mu = (A^0, \vec{A})$ read

$$\Pi^a_\nu = \frac{\partial \mathcal{L}}{\partial \partial_0 A^{a\nu}} = -F^{a0}{}_\nu, \tag{P2.32}$$

$$\Pi_{\psi i} = \frac{\partial \mathcal{L}}{\partial \partial_0 \psi^i} = \frac{i}{2}\psi^\dagger_i, \quad \Pi^i_{\psi^\dagger} = \frac{\partial \mathcal{L}}{\partial \partial_0 \psi^\dagger_i} = -\frac{i}{2}\psi^i. \tag{P2.33}$$

The corresponding equations of motion read

$$(i\hat{D} - m)\psi = 0 \quad or \quad (i\hat{\partial} - m)\psi = -g\hat{A}\psi, \tag{P2.34}$$

$$\bar{\psi}(-i\overleftarrow{\partial} + g\hat{A} - m) = 0, \quad D^{ac}_\mu F^{c\mu\nu} = -g\bar{\psi}t^a\gamma^\nu\psi. \tag{P2.35}$$

Note that the conjugate momentum for A^a_0 is $-F^{a00} = 0$ and the conjugate momenta for A^{ai} are

$$\Pi^a_i = -F^{a0}{}_i = F^{a0i} = F^a_{i0} = \frac{\partial A^a_0}{\partial x^i} - \frac{\partial A^a_i}{\partial x^0} + gf^{abc}A^b_0 A^c_i = E^a_i = -E^{ai}, \tag{P2.36}$$

where E^{ai} is the chromoelectric field. One can choose the temporal gauge $A^a_0 = 0$ and A^{ai} as independent canonical variables. Then

$$\Pi^a_i = -\frac{\partial A^a_i}{\partial x^0} = \frac{\partial A^{ai}}{\partial x^0} = \dot{A}^{ai} = E^a_i = -E^{ai}, \tag{P2.37}$$

and the equations of motion read

$$G^a \equiv D^{ac}_i E^{ci} + g\bar{\psi}t^a\gamma^0\psi = 0, \tag{P2.38}$$

$$-\partial_0 E^{ak} + D^{ac}_i F^{cik} = -g\bar{\psi}t^a\gamma^k\psi. \tag{P2.39}$$

One can see that the first equation here does not contain time derivatives. Therefore it is a **constraint equation** often called the **Gauss law**.

2.9 Show that in the temporal gauge $A_0 = 0$ Gauss law is inconsistent with the canonical commutation relations.

SOLUTION. The canonical commutation relations read

$$[A^{ai}(x), \dot{A}^{bj}(y)]_{x^0=y^0} = i\delta^{ij}\delta^{ab}\delta^{(3)}(x-y), \quad or \tag{P2.40}$$

$$[A^{ai}(x), \Pi^b_j(y)]_{x^0=y^0} = [A^{ai}(x), E^b_j(y)]_{x^0=y^0} = ig^i_j\delta^{ab}\delta^{(3)}(x-y), \tag{P2.41}$$

$$[A(x), A(y)]_{x^0=y^0} = 0, \tag{P2.42}$$

$$\{\psi(x)^i, \psi(y)^\dagger_j\}_{x^0=y^0} = \delta^i_j\delta^{(3)}(x-y), \tag{P2.43}$$

$$\{\psi(x), \psi(y)\}_{x^0=y^0} = \{\psi(x)^\dagger, \psi(y)^\dagger\}_{x^0=y^0} = 0. \tag{P2.44}$$

Then

$$\begin{aligned}
[A^{ak}(x), G^b(y)]_{x^0=y^0} &= [A^{ak}(x), D^{bc}_i E^{ci}(y) + g\bar{\psi}(y)t^b\gamma^0\psi(y)]_{x^0=y^0}\\
&= D^{bc}_i(y)[A^{ak}(x), E^{ci}(y)]_{x^0=y^0}\\
&= ig^{ik}D^{ba}_i(y)\delta^{(3)}(x-y) \neq 0. \tag{P2.45}
\end{aligned}$$

Therefore Gauss law cannot be implemented as an operator identity. However, it can be imposed as a constraint on the **physical sates,** i.e.

$$G(x)|\Psi\rangle_{phys} = 0. \tag{P2.46}$$

2.10 Show that in the gauge $A_0 = 0$ the Gauss law is the generator for the small time-independent gauge transformations. Show that the physical states are invariant under these transformations.

SOLUTION. For $\theta(\vec{x}) \to 0$ as $\vec{x} \to \infty$ the integral of the Gauss law can be taken by parts and the total divergence may be neglected

$$\begin{aligned}
G_\theta &= \int d^3x\, \theta^c(\vec{x})\, G^c(x) = \int d^3x\, \theta^c(\vec{x})\, \left[D^{ac}_i E^{ic}(x) + g\psi(x)^\dagger t^c\psi(x)\right]\\
&= \int d^3x\, \left[-(D^{ac}_i\theta^c(\vec{x}))E^{ia}(x) + g\psi(x)^\dagger t^c\theta^c(\vec{x})\psi(x)\right]. \tag{P2.47}
\end{aligned}$$

The action of this operator on the fields reads

$$\begin{aligned}
G_\theta A^b_j(y) &= A^b_j(y)G_\theta - \int d^3x\, (D^{ac}_i\theta^c(\vec{x}))\left[E^{ia}(x)A^b_j(y)\right]_{x^0=y^0}\\
&= A^b_j(y)G_\theta + i\int d^3x\, (D^{bc}_j\theta^c(\vec{x}))\delta^{(3)}(x-y)\\
&= A^b_j(y)G_\theta + iD^{bc}_j\theta^c(\vec{y}), \tag{P2.48}
\end{aligned}$$

$$\begin{aligned}
G_\theta\psi^j(y) &= \psi^j(y)G_\theta + \int d^3x\, \left[g\psi(x)^\dagger t^c\theta^c(\vec{x})\psi(x), \psi^j(y)\right]_{x^0=y^0}\\
&= \psi^j(y)G_\theta - g\int d^3x\, \theta^c(\vec{x})\delta(\vec{x}-\vec{y})(t^c\psi(x))^j\\
&= \psi^j(y)G_\theta - g\theta^c(\vec{y})(t^c\psi(y))^j, \tag{P2.49}
\end{aligned}$$

$$G_\theta \psi(y)_j^\dagger = \psi(y)_j^\dagger G_\theta + \int d^3x \left[g\psi(x)^\dagger t^c \theta^c(\vec{x})\psi(x), \psi(y)_j^\dagger \right]_{x^0=y^0}$$

$$= \psi(y)_j^\dagger G_\theta + g \int d^3x\, \theta^c(\vec{x})\delta(\vec{x}-\vec{y})(\psi(x)^\dagger t^c)_j$$

$$= \psi(y)_j^\dagger G_\theta + g\theta^c(\vec{y})(\psi(y)^\dagger t^c)_j. \tag{P2.50}$$

Here one uses the equal time canonical commutation relations since the Gauss law is valid for any time, which gives

$$[\psi(x)^\dagger t^a \psi(x), \psi^j(y)]_{x^0=y^0} = -\delta_k^j \delta(\vec{x}-\vec{y})(t^a\psi(x))^k$$

$$= -\delta(\vec{x}-\vec{y})(t^a\psi(x))^j, \tag{P2.51}$$

$$[\psi(x)^\dagger t^a \psi(x), \psi_j^\dagger(y)]_{x^0=y^0} = \delta_k^j \delta(\vec{x}-\vec{y})(\psi(x)^\dagger t^a)_j$$

$$= \delta(\vec{x}-\vec{y})(\psi(x)^\dagger t^a)_j. \tag{P2.52}$$

We see that the action of $-\frac{i}{g}G_\theta$ on the fields gives exactly the infinitesimal gauge transformations (2.3), (2.4), and (2.11). Such time-independent gauge transformations with $\theta(\vec{x}) \to 0$ as $\vec{x} \to \infty$ are called **small**, while the transformations which do not have this property are called **large**. As a result, one can write any small gauge transformation as

$$e^{-\frac{i}{g}G_\theta} \implies e^{-\frac{i}{g}G_\theta}|\Psi\rangle_{phys} = |\Psi\rangle_{phys}, \tag{P2.53}$$

since the Gauss law vanishes on the physical states. Therefore the physical states are invariant under the small gauge transformations. However, large gauge transformations cannot be reduced to the Gauss law and change the states.

2.11 Show that action of $e^{-\frac{i}{g}G_\theta}$ on the fields give finite gauge transformations (2.3, 2.4, 2.9).

2.12 Here we follow (JR76). Consider a **large gauge transformation** generated by

$$\Lambda_1(\vec{x}) = \frac{\vec{x}^2 - d^2 + 2id(\vec{\tau}\vec{x})}{\vec{x}^2 + d^2}, \tag{P2.54}$$

where $\vec{x}$ is a 3-vector and τ are the Pauli matrices in any $SU(2)$ subgroup of $SU(3)$. Show that $\Lambda_1 \in SU(2)$, apply this transformation to $A_\mu = 0$, calculate the **winding number** n of the field $A_i^{(1)}(x)$ you get, where

$$n = \frac{ig^3}{24\pi^2} \int d^3x\, tr[A_i^{(1)}(x) A_j^{(1)}(x) A_k^{(1)}(x)]\varepsilon^{ijk}. \tag{P2.55}$$

SOLUTION. Λ_1 is a matrix from $SU(2)$ written in the Cayley representation with $\vec{b} = \frac{\vec{x}}{d}$. The proof that it belongs $SU(2)$ and its exponential representation are given in problem 1.21 in Chapter 1. This transformation is large since in the exponential representation, it reads

$$\Lambda_1 = -e^{i2(\vec{n}\vec{\tau})arctg\frac{x}{d}} \xrightarrow[x\to\infty]{} e^{i\pi((\vec{n}\vec{\tau})+1)} = 1 \tag{P2.56}$$

and the index of the exponent does not vanish as $x \to \infty$. This transformation changes the 0 potential into

$$0 = A_\mu \to A_\mu^{(1)} = \Lambda_1(A_\mu - \frac{\partial_\mu}{ig})\Lambda_1^\dagger = \Lambda_1(-\frac{\partial_\mu}{ig})\Lambda_1^\dagger$$

$$= \frac{\vec{x}^{\,2} - d^2 + 2id(\vec{\tau}\vec{x})}{\vec{x}^{\,2} + d^2} \frac{i\partial_\mu}{g} \frac{\vec{x}^{\,2} - d^2 - 2id(\vec{\tau}\vec{x})}{\vec{x}^{\,2} + d^2} \qquad \Longrightarrow \qquad A_0^{(1)} = 0.$$

$$\text{(P2.57)}$$

$$-A_j^{(1)} = A^{(1)j} = \frac{\vec{x}^{\,2} - d^2 + 2id(\vec{\tau}\vec{x})}{\vec{x}^{\,2} + d^2} \frac{i}{g}$$

$$\times \left(\frac{2x^j - 2id\tau^j}{\vec{x}^{\,2} + d^2} - 2x^j \frac{\vec{x}^{\,2} - d^2 - 2id(\vec{\tau}\vec{x})}{(\vec{x}^{\,2} + d^2)^2} \right)$$

$$= \frac{\vec{x}^{\,2} - d^2 + 2id(\vec{\tau}\vec{x})}{\vec{x}^{\,2} + d^2} \frac{i}{g}$$

$$\times \left(\frac{(2x^j - 2id\tau^j)(\vec{x}^{\,2} + d^2) - 2x^j(\vec{x}^{\,2} - d^2 - 2id(\vec{\tau}\vec{x}))}{(\vec{x}^{\,2} + d^2)^2} \right)$$

$$= \frac{\vec{x}^{\,2} - d^2 + 2id(\vec{\tau}\vec{x})}{\vec{x}^{\,2} + d^2} \frac{2di}{g} \left(\frac{2x^j d - i\tau^j \vec{x}^{\,2} - id\tau^j d + 2ix^j(\vec{\tau}\vec{x})}{(\vec{x}^{\,2} + d^2)^2} \right)$$

$$= \frac{2d}{g(\vec{x}^{\,2} + d^2)^2} \left[\tau^j(\vec{x}^{\,2} - d^2) - 2d\varepsilon^{ijk}x^i\tau^k - 2x^j(\vec{\tau}\vec{x}) \right].$$

$$\text{(P2.58)}$$

Via

$$tr\left[\tau^i\tau^j\tau^k\right] = 2i\varepsilon^{ijk} \tag{P2.59}$$

the winding number for this gauge potential reads

$$n = \frac{ig^3}{24\pi^2} \int d^3x \, tr[A_i^{(1)}(x) A_j^{(1)}(x) A_k^{(1)}(x)]\varepsilon^{ijk} \tag{P2.60}$$

$$= \frac{2}{3\pi^2} \int d^3x \frac{d^3}{(d^2 + \vec{x}^{\,2})^3} = 1. \tag{P2.61}$$

Comparing (P2.60) with the solution of problem 1.61 from the first chapter, one can see that this integral is exactly the integral of the Haar measure on the $SU(2)$ group with the coefficient set to normalize it to 1. As x spans the whole 3 dimensional space, $\Lambda_1(x)$ (P2.54) covers the whole $SU(2)$ group, and Λ_1^k covers the group k times as one can see from the solution of problem 1.21 in Chapter 1. Therefore the winding number for the field

$$A_\mu^{(k)} = \Lambda_1^k(-\frac{\partial_\mu}{ig})\Lambda_1^{k\dagger} \tag{P2.62}$$

is k. Moreover,

$$\Lambda_1^{k\dagger}(-\frac{\partial_\mu}{ig})\Lambda_1^k = -\Lambda_1^k(-\frac{\partial_\mu}{ig})\Lambda_1^{k\dagger}. \tag{P2.63}$$

Therefore the winding number for

$$A_\mu^{(-k)} = \Lambda_1^{k\dagger}(-\frac{\partial_\mu}{ig})\Lambda_1^k \qquad (P2.64)$$

is $-k$. Note that the winding number for $A_\mu = 0$ is 0. Note also that the winding number does not depend on d and the substitution $\vec{x} \to \vec{x} + \vec{a}$ in (P2.54).

One studies **local effects**. Therefore as in this problem one considers the gauge transformations going to identity at spacial infinity

$$U \underset{x \to \infty}{\to} 1. \qquad (P2.65)$$

The 3-dimensional Euclidean space with all directions at infinity identified is equivalent to the sphere S^3. Hence, the gauge transformations map

$$S_{phys}^3 \to S_{SU(2)}^3, \qquad (P2.66)$$

since the $SU(2)$ group is the S^3 sphere itself. Mathematics teaches us that such mappings fall into homotopy classes, i.e. the transformations from different classes cannot be continuously transformed into one another. The winding number says how many times the gauge transformation covers the $SU(2)$ group.

2.13 Show that $F_{\mu\nu} = 0$ for the field A_μ gauge equivalent to 0.

SOLUTION. For $A_\mu = U(-\frac{\partial_\mu}{ig})U^\dagger$

$$\begin{aligned}
gF_{\mu\nu} &= g(\partial_\mu A_\nu - \partial_\nu A_\mu - ig[A_\mu, A_\nu]) \\
&= i\partial_\mu(U\partial_\nu U^\dagger) - i\partial_\nu(U\partial_\mu U^\dagger) + i[U(\partial_\mu U^\dagger), U(\partial_\nu U^\dagger)] \\
&= i\partial_\mu(U\partial_\nu U^\dagger) - i\partial_\nu(U\partial_\mu U^\dagger) \\
&\quad + iU(\partial_\mu U^\dagger)U(\partial_\nu U^\dagger) - iU(\partial_\nu U^\dagger)U(\partial_\mu U^\dagger) \\
&= i\partial_\mu(U\partial_\nu U^\dagger) - i\partial_\nu(U\partial_\mu U^\dagger) + iU\partial_\mu\partial_\nu U^\dagger - i\partial_\mu(U\partial_\nu U^\dagger) \\
&\quad - iU\partial_\nu\partial_\mu U^\dagger + i\partial_\nu(U\partial_\mu U^\dagger) = 0. \qquad (P2.67)
\end{aligned}$$

2.14 Consider a process starting at $t = -\infty$ with the pure gauge gluon field $A^{(i)}$ with the winding number $n_{t=-\infty} = i$ and finishing at $t = +\infty$ with the pure gauge gluon field $A^{(f)}$ with the winding number $n_{t=+\infty} = f$. Show that then the following integral called **the topological charge**

$$\nu[A] = \frac{g^2}{64\pi^2} \int \varepsilon_{\mu\nu\alpha\beta} F^{a\mu\nu} F^{a\alpha\beta} d^4x \qquad (P2.68)$$

gives the difference in the winding numbers

$$\nu[A] = n_{t=+\infty} - n_{t=-\infty} = f - i. \qquad (P2.69)$$

SOLUTION.

$$\nu[A] = \frac{g^2}{64\pi^2} \int \varepsilon_{\mu\nu\alpha\beta} F^{a\mu\nu} F^{a\alpha\beta} d^4x = \frac{g^2}{64\pi^2} \int \partial_\mu K^\mu d^4x$$

$$= \frac{g^2}{64\pi^2} \int K^0 d^3x |_{t=-\infty}^{t=+\infty} = \frac{2g}{3} \frac{g^2}{64\pi^2} f^{abc} \varepsilon^{ijk} \int A_i^a A_j^b A_k^c d^3x |_{t=-\infty}^{t=+\infty}$$

$$= \frac{2g}{3} \frac{g^2}{64\pi^2} (-4i) \varepsilon^{ijk} \int tr[A_i^{(f)} A_j^{(f)} A_k^{(f)} - A_i^{(i)} A_j^{(i)} A_k^{(i)}] d^3x = f - i.$$

(P2.70)

Here we used representation (P2.13), then (P1.12) to introduce the trace, assumed that the gluon fields vanish fast enough to drop the surface integral at spatial infinity and dropped $F_{\mu\nu}|_{t=\pm\infty}$ since it vanishes for pure gauge fields, as is shown in problem 2.13.

2.15 Show that the winding number for the vacuum field is invariant under small gauge transformations.

SOLUTION. For the pure gauge field A $F = 0$ and via (P2.13) one has

$$n = \frac{ig^3}{24\pi^2} \int d^3x\, tr[A_i(x) A_j(x) A_k(x)] \varepsilon^{ijk} = \frac{g^2}{64\pi^2} \int d^3x K^0. \quad (P2.71)$$

Under the infinitesimal gauge transformations K^0 changes via (P2.24) as

$$n \to n + \frac{8}{g} \frac{g^2}{64\pi^2} \int d^3x \varepsilon^{ijk} \partial_j tr[\{\partial_i \theta\} A_k]. \quad (P2.72)$$

Therefore for the small gauge transformations, i.e. for $U = e^{i\theta}$, $\theta \to 0$, as $x \to \infty$, the total divergence vanishes.

2.16 In the $A^0 = 0$ gauge show that for the vacuum field A and a local gauge transformation U

$$n[U(A_i - \frac{\partial_i}{ig})U^\dagger] = n[A] + n[U(-\frac{\partial_i}{ig})U^\dagger]. \quad (P2.73)$$

SOLUTION. Using (P2.24), one has for the finite gauge transformations

$$n[U(A_i - \frac{\partial_i}{ig})U^\dagger] = n[A] + \frac{\varepsilon^{ijk}}{24\pi^2} \int d^3x tr[U\{\partial_i U^\dagger\}U\{\partial_j U^\dagger\}U\{\partial_k U^\dagger\}]$$

$$+ \frac{ig}{8\pi^2} \int d^3x\, \varepsilon^{ijk} \partial_j tr[\{\partial_i U^\dagger\}U A_k]. \quad (P2.74)$$

The total divergence vanishes since for local gauge transformations $U \to 1$ as $x \to \infty$.

2.17 Calculate the winding numbers for the following gauge transformations:

$$a)\; V = \frac{1 + i\frac{\vec{\tau}\vec{x}}{x^2}}{1 - i\frac{\vec{\tau}\vec{x}}{x^2}} \;\Longrightarrow\; n\left[V(-\frac{\partial}{ig})V^\dagger\right] = -1, \tag{P2.75}$$

$$b)\; V = e^{\,i\frac{\pi\vec{\tau}\vec{x}}{\sqrt{x^2+a^2}}} \;\Longrightarrow\; n = -1, \tag{P2.76}$$

$$c)\; V = (-1)^N e^{-i\frac{\vec{\tau}\vec{x}}{x}f(x)},\; f(x) \to \begin{cases}\pi N,\; x \to \infty \\ 0,\; x \to 0\end{cases} \;\Longrightarrow\; n = N. \tag{P2.77}$$

SOLUTION. c) Here we follow (Bjo79). First, one can check that V is local

$$\lim_{x\to\infty} V = (-1)^N e^{-i\frac{\vec{\tau}\vec{x}}{x}\pi N} = 1. \tag{P2.78}$$

Next since the winding number is invariant under small gauge transformations, one can change f by such transformations to make it concentrated at $x \sim R$ as in Figure 2.3. Then one can calculate the gluon field in the small solid angle near $\vec{x} = (0,0,R)$:

$$-igA_1 = V(-1)^N i\tau_1 \frac{\sin f(x_3)}{x_3} + O(x_1, x_2), \tag{P2.79}$$

$$-igA_2 = V(-1)^N i\tau_2 \frac{\sin f(x_3)}{x_3} + O(x_1, x_2), \tag{P2.80}$$

$$-igA_3 = i\tau_3 f'(x_3) + O(x_1, x_2). \tag{P2.81}$$

Finally,

$$g^3 tr[A_1 A_2 A_3] = -tr[(\cos f - i\tau_3 \sin f)\tau_1(\cos f - i\tau_3 \sin f)\tau_2\tau_3]$$
$$\times f'\frac{\sin^2 f(x_3)}{x_3^2} + O(x_1, x_2)$$
$$= -2if'(x_3)\frac{\sin^2 f(x_3)}{x_3^2} + O(x_1, x_2), \tag{P2.82}$$

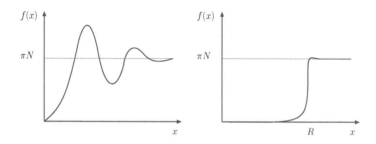

Figure 2.3 Left: A generic $f(x)$. Right: $f(x)$ after a small gauge transformation. $f(x) = 0$, $x \lesssim R$, $f(x) = \pi N$, $x \gtrsim R$.

and

$$dn = \frac{i}{24\pi^2} \times 6 \times (-2i) \int_0^{+\infty} dx_3 \int_{x_3^2 d\Omega_{12}} dx_1 dx_2 f'(x_3) \frac{\sin^2 f(x_3)}{x_3^2}$$

$$= \int_0^{+\infty} dx_3 \frac{x_3^2 d\Omega_{12}}{2\pi^2} f'(x_3) \frac{\sin^2 f(x_3)}{x_3^2}$$

$$= \frac{d\Omega_{12}}{2\pi^2} \int_0^{\pi N} df \sin^2 f = N \frac{d\Omega_{12}}{4\pi}. \tag{P2.83}$$

Since $\varepsilon^{ijk} A_i A_j A_k$ is invariant under 3-d space rotations,

$$n = \int N \frac{d\Omega_{12}}{4\pi} = N. \tag{P2.84}$$

2.18 Show that Wilson lines have the following properties: causality

$$U_{C_1}(y,x) U_{C_2}(x,z) = U_{C_1 \cup C_2}(y,z), \tag{P2.85}$$

"hermiticity"

$$U_C(y,x)^\dagger = U_{-C}(x,y), \tag{P2.86}$$

unitarity

$$U_C(y,x)^\dagger U_C(y,x) = 1. \tag{P2.87}$$

2.19 Derive how the Wilson line changes under gauge transformations.
SOLUTION. The Wilson line may be defined as

$$U_C(y,x) = \mathcal{P}_C e^{ig \int_C A^\mu(z) dz_\mu}$$

$$= \mathcal{P}_{C_n} e^{ig \int_{C_n} A^\mu(z) dz_\mu} \dots \mathcal{P}_{C_2} e^{ig \int_{C_2} A^\mu(z) dz_\mu} \mathcal{P}_{C_1} e^{ig \int_{C_1} A^\mu(z) dz_\mu}$$

$$= \lim_{n \to \infty} (1 + ig\Delta z_n^{\mu_n} A_{\mu_n}(\tilde{z}_n)) \dots$$

$$\times (1 + ig\Delta z_2^{\mu_2} A_{\mu_2}(\tilde{z}_2))(1 + ig\Delta z_1^{\mu_1} A_{\mu_1}(\tilde{z}_1)). \tag{P2.88}$$

Here the contour C is divided into n small contours C_i connecting $n+1$ points z_i on the initial contour

$$C = C_n \cup \dots \cup C_2 \cup C_1, \quad C_i = \{z_i, z_{i-1}\}, \tag{P2.89}$$

$$\Delta z_i = z_i - z_{i-1}, \quad \tilde{z}_i = \frac{z_i + z_{i-1}}{2}, \quad z_0 = x, \, z_n = y. \tag{P2.90}$$

Therefore for the gauge transformation V_x :

$$A_\mu(z) \to A'_\mu(z) = V_z (A_\mu(z) - \frac{\partial_\mu}{ig}) V_z^\dagger, \tag{P2.91}$$

one gets

$$V_{\tilde{z}_i}(1 + ig\Delta z_i^\mu A_\mu(\tilde{z}_i))V_{\tilde{z}_{i-1}}^\dagger = (V_{\tilde{z}_i} + \frac{\Delta z_i^\rho}{2}\partial_\rho V_{\tilde{z}_i})$$
$$\times (1 + ig\Delta z_i^\mu A_\mu(\tilde{z}_i))(V_{\tilde{z}_i}^\dagger - \frac{\Delta z_i^\nu}{2}\partial_\nu V_{\tilde{z}_i}^\dagger) + O(\Delta z_i^2)$$
$$= 1 + ig\Delta z_i^\mu (V_{\tilde{z}_i} A_\mu(\tilde{z}_i)V_{\tilde{z}_i}^\dagger + \frac{1}{2ig}(\partial_\mu V_{\tilde{z}_i})V_{\tilde{z}_i}^\dagger - \frac{1}{2ig}V_{\tilde{z}_i}\partial_\mu V_{\tilde{z}_i}^\dagger) + O(\Delta z_i^2)$$
$$= 1 + ig\Delta z_i^\mu V_{\tilde{z}_i}(A_\mu(\tilde{z}_i) - \frac{\partial_\mu}{ig})V_{\tilde{z}_i}^\dagger + O(\Delta z_i^2)$$
$$= 1 + ig\Delta z_i^\mu A'_\mu(\tilde{z}_i) + O(\Delta z_i^2). \tag{P2.92}$$

Hence

$$U_C(y,x) \to U_C(y,x)' = \lim_{n\to\infty} V_{z_n}(1 + ig\Delta z_n^{\mu_n} A_{\mu_n}(\tilde{z}_n))V_{z_{n-1}}^\dagger \cdots$$
$$\times V_{z_2}(1 + ig\Delta z_2^{\mu_2} A_{\mu_2}(\tilde{z}_2))V_{z_1}^\dagger V_{z_1}(1 + ig\Delta z_1^{\mu_1} A_{\mu_1}(\tilde{z}_1))V_{z_0}^\dagger$$
$$= V_y U_C(y,x)V_x^\dagger. \tag{P2.93}$$

2.20 Show that

$$F^a(y)U_C(y,x)^{ab}F^b(x) \tag{P2.94}$$

is gauge invariant. Here $F^a(y)$ is the gluon strength tensor and $U_C(y,x)^{ab}$ is the Wilson line in the adjoint rep.

HINT. Since the adjoint rep is real

$$F_1^T U F_2 = F_1^T V^{\theta T} V^\theta U V^{\theta T} V^\theta F_2 = F_1^{\theta T} U^\theta F_2^\theta. \tag{P2.95}$$

2.21 Show that

$$e^{iS}, \quad S = \frac{-i}{2m}(i\vec{\partial} + g\vec{A})\vec{\gamma} \tag{P2.96}$$

is a unitary operator.

2.22 Consider a small square Wilson line lying in the $x^1 x^2$ plane. Let a be the size of the edge. Show that

$$\frac{1}{N_c}tr(U_\square) = 1 - \frac{a^4}{4N_c}g^2 F_{12}^a F_{12}^a + o(a^4). \tag{P2.97}$$

SOLUTION.

$$U_\square = (1 - iag A_\mu(x + \frac{a}{2}n_2)n_2^\mu + \frac{(-iag)^2}{2}A_2^2(x))$$

$$\times (1 - iag A_\mu(x + an_2 + \frac{a}{2}n_1)n_1^\mu + \frac{(-iag)^2}{2}A_1^2(x))$$

$$\times (1 + iag A_\mu(x + an_1 + \frac{a}{2}n_2)n_2^\mu + \frac{(iag)^2}{2}A_2^2(x))$$

$$\times (1 + iag A_\mu(x + \frac{a}{2}n_1)n_1^\mu + \frac{(iag)^2}{2}A_1^2(x))) \qquad \text{(P2.98)}$$

$$= (1 + iga^2(-\frac{1}{2}\partial_2 A_2 - \partial_2 A_1 - \frac{1}{2}\partial_1 A_1 + \partial_1 A_2 + \frac{1}{2}\partial_2 A_2 + \frac{1}{2}\partial_1 A_1$$

$$+ ig(A_2 A_1 - A_2 A_2 - A_2 A_1 - A_1 A_2 - A_1 A_1 + A_2 A_1)$$

$$+ ig(A_1^2 + A_2^2)) = (1 + iga^2(\partial_1 A_2 - \partial_2 A_1 - ig[A_1 A_2])$$

$$= 1 + iga^2 F_{12} + O(a^3) = 1 + iga^2 F_{12}^a t^a + O(a^3). \qquad \text{(P2.99)}$$

On the other hand in the ε-vicinity of 1

$$U = e^{i\varepsilon\theta^a t^a} \simeq 1 + i\varepsilon\theta^a t^a - \frac{\varepsilon^2\theta^a\theta^b}{2}t^a t^b + O(\varepsilon^3). \qquad \text{(P2.100)}$$

Therefore taking $\varepsilon = a^2$, $\theta^a = gF_{12}^a$

$$\frac{1}{N_c}tr(U_\square) = 1 - \frac{a^4}{4N_c}g^2 F_{12}^a F_{12}^a + o(a^4). \qquad \text{(P2.101)}$$

2.23 Fourier transform the gluon propagator in $d = 4 - 2\varepsilon$

$$D_{\mu\nu}(x) = -ig_{\mu\nu}\int \frac{\mu^{2\varepsilon}d^dp}{(2\pi)^d}\frac{e^{-ipx}}{p^2 + i0} = -\frac{\Gamma(1 - \varepsilon)}{4\pi^{2-\varepsilon}}\frac{\mu^{2\varepsilon}\delta^{ab}g_{\mu\nu}}{(-x^2 + i0)^{1-\varepsilon}}. \qquad \text{(P2.102)}$$

SOLUTION.

$$\int \frac{d^d p}{(2\pi)^d} \frac{ie^{-ipx}}{p^2 + i0} = \int_0^{+\infty} d\alpha \int \frac{d^d p}{(2\pi)^d} e^{-ipx + i\alpha(p^2 + i0)}$$

$$= \int_0^{+\infty} d\alpha \int \frac{d^d p}{(2\pi)^d} e^{i\alpha((p - \frac{x}{2\alpha})^2 - \frac{x^2}{4\alpha^2} + i0)}$$

$$= \int_0^{+\infty} \frac{d\alpha}{(2\pi)^d} \left(\frac{\pi}{i\alpha}\right)^{\frac{d-1}{2}} \left(\frac{\pi}{-i\alpha}\right)^{\frac{1}{2}} e^{i\alpha(-\frac{x^2}{4\alpha^2} + i0)}$$

$$= i \int_0^{+\infty} \frac{d\alpha}{(4\pi)^{\frac{d}{2}}} (i\alpha)^{-\frac{d}{2}} e^{\frac{i}{4\alpha}(-x^2 + i0)} \qquad (P2.103)$$

$$= \left| \beta = -\frac{i(-x^2 + i0)}{4\alpha}, \; d\alpha = \frac{d\beta}{4\beta^2} i(-x^2 + i0) \right|$$

$$= -\frac{(-x^2 + i0)^{1 - \frac{d}{2}}}{4^{1 - \frac{d}{2}}} \int_{-i\infty(-x^2 + i0)}^{0} \frac{d\beta}{(4\pi)^{\frac{d}{2}}} \beta^{\frac{d}{2} - 2} e^{-\beta}$$

$$= \frac{(-x^2 + i0)^{1 - \frac{d}{2}}}{4\pi^{\frac{d}{2}}} \int_0^{+\infty} d\beta \beta^{\frac{d}{2} - 2} e^{-\beta}$$

$$= \frac{\Gamma(\frac{d}{2} - 1)}{4\pi^{\frac{d}{2}} (-x^2 + i0)^{\frac{d}{2} - 1}}. \qquad (P2.104)$$

2.24 Calculate the divergent part of the Wilson loop over a contour without self-intersections and light-cone parts up to terms $\sim O(g^2)$ in the a) cutoff and b) dimensional regularizations. Consider a smooth contour and a contour with one cusp with angle ϕ.

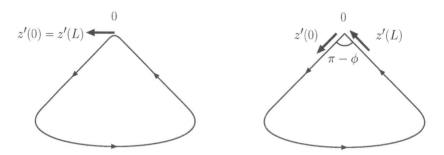

Figure 2.4 A smooth Wilson loop and a Wilson loop with one cusp.

SOLUTION.

$$\frac{1}{N_c} \langle trU_C \rangle \simeq 1 - g^2 \frac{tr(t^a t^b)}{2N_c} \oint_C dx^\mu \oint_C dy^\nu D_{\mu\nu}^{ab}(x - y). \qquad (P2.105)$$

Here the $\mathcal{P}_C$ ordering is not important since the t matrices in the gluon fields commute because of the trace.

a) After the Wick rotation and cutoff regularization

$$x^0 \to -ix_E^0, \quad \frac{g_{\mu\nu}}{(x-y)^2} \to \frac{-\delta_{\mu\nu}}{-(x-y)_E^2} \to \frac{\delta_{\mu\nu}}{(x-y)_E^2 + a^2} \quad \text{(P2.106)}$$

we use the invariant proper length parametrization of the contour

$$z(s): \quad (z')^2 = \left(\frac{\partial z^\mu}{\partial s}\right)^2 = 1, \quad (z'z'') = 0. \quad \text{(P2.107)}$$

We have

$$\frac{1}{N_c}\langle trU_C\rangle \simeq 1 - g^2 \frac{C_F}{2 \times 4\pi^2} \int_0^L ds \int_0^L dt \frac{z'(s)z'(t)}{(z(s)-z(t))_E^2 + a^2}. \quad \text{(P2.108)}$$

The only divergence in the integral comes from the points where $z(s) = z(t)$. To get it we expand the integrand in s near t :

$$\frac{\langle trU_C\rangle_1}{N_c} = 1 - \frac{g^2 C_F}{2 \times 4\pi^2} \int_0^L ds \int_0^L dt \frac{z'(t)^2 + z'(t)z''(t)(s-t)}{z'(t)^2(s-t)^2 + a^2} + O(a^0)$$

$$= 1 - g^2 \frac{C_F}{4\pi^2} \int_0^L dt \int_0^t \frac{ds}{(s-t)^2 + a^2} + O(a^0) = |s' = t - s|$$

$$= 1 - g^2 \frac{C_F}{4\pi^2} \int_0^L dt \int_0^t \frac{ds'}{s'^2 + a^2} + O(a^0)$$

$$= 1 - g^2 \frac{C_F}{4\pi^2} \int_0^L \frac{dt}{a} arctg\frac{t}{a} + O(a^0)$$

$$= 1 - g^2 \frac{C_F}{4\pi^2}(\frac{\pi L}{2a} - \ln\frac{L}{a}) + O(a^0). \quad \text{(P2.109)}$$

This contribution is in fact the contribution for the traced Wilson line of length L rather than for the loop, since we did not take into account the contribution of the point where the beginning and the end of the line meet, i.e. $z(0) = z(L)$. At this point we also have a singularity. One gets its contribution via

$$\frac{1}{N_c}\langle trU_C\rangle_2 = 2 \times (-g^2)\frac{C_F}{2 \times 4\pi^2} \int_0^R ds \int_{L-R}^L dt$$

$$\times \frac{z'(0)z'(L)}{(z'(0)s - z'(L)(t-L))^2 + a^2} + O(a^0), \quad \text{(P2.110)}$$

where $a \ll R \lesssim \frac{L}{2}$. If the loop is smooth $z'(0) = z'(L)$ than

$$\frac{1}{N_c}\langle trU_C\rangle_2 = -g^2\frac{C_F}{4\pi^2}\int_0^R ds \int_{L-R}^L dt\frac{1}{(s-t+L)^2+a^2} + O(a^0)$$

$$= |t' = L - t|$$

$$= -g^2\frac{C_F}{4\pi^2}\int_0^R ds \int_0^R dt'\frac{1}{(s+t')^2+a^2} + O(a^0)$$

$$= |t' = \rho\tau, s = \rho(1-\tau)|$$

$$= -g^2\frac{C_F}{4\pi^2}\int_0^1 d\tau \int_0^R \frac{\rho d\rho}{\rho^2+a^2} + O(a^0)$$

$$= -g^2\frac{C_F}{4\pi^2}\ln\frac{R}{a} + O(a^0). \tag{P2.111}$$

Therefore the total contribution to the loop reads

$$\frac{1}{N_c}\langle trU_C\rangle = 1 - g^2\frac{C_F}{4\pi}\frac{L}{2a} + O(a^0). \tag{P2.112}$$

The linear dependence of the loop average on the length of the contour is often called the **perimeter law**. It reveals in all orders of perturbation theory (DV80). If the loop is not smooth $z'(0) \neq z'(L) : z'(0)z'(L) = \cos\phi$, one gets

$$\frac{1}{N_c}\langle trU_C\rangle_2 = -g^2\frac{C_F}{4\pi^2}\int_0^R ds \int_0^R dt'\frac{\cos\phi}{s^2+2st'\cos\phi+t'^2+a^2} + O(a^0)$$

$$= -g^2\frac{C_F}{4\pi^2}\int_0^1 d\tau$$

$$\times \int_0^R \frac{\rho d\rho\cos\phi}{\rho^2(\tau^2+(1-\tau)^2+2\tau(1-\tau)\cos\phi)+a^2} + O(a^0)$$

$$= -g^2\frac{C_F}{2\times4\pi^2}\int_0^1 d\tau\frac{\cos\phi}{\tau^2+(1-\tau)^2+2\tau(1-\tau)\cos\phi}$$

$$\times \ln\frac{R^2(\tau^2+(1-\tau)^2+2\tau(1-\tau)\cos\phi)+a^2}{a^2} + O(a^0). \tag{P2.113}$$

The divergent contribution comes from $\ln\frac{1}{a^2}$. Hence

$$\frac{1}{N_c}\langle trU_C\rangle_2 = -g^2\frac{C_F}{4\pi^2}\ln\frac{R}{a}\int_0^1 d\tau\frac{\cos\phi}{\tau^2+(1-\tau)^2+2\tau(1-\tau)\cos\phi}$$

$$= -g^2\frac{C_F}{4\pi^2}\ln\frac{R}{a}\phi\,ctg\phi + O(a^0). \tag{P2.114}$$

Therefore the total contribution reads

$$\frac{1}{N_c}\langle trU_C\rangle = 1 - g^2\frac{C_F}{4\pi^2}\left[\frac{\pi L}{2a} + \ln\frac{L}{a}(\phi\,ctg\phi - 1)\right] + O(a^0). \quad \text{(P2.115)}$$

b) In dimensional regularization after the Wick rotation one gets

$$\frac{1}{N_c}\langle trU_C\rangle \simeq 1 - g^2\frac{C_F}{2}\frac{\Gamma(1-\varepsilon)}{4\pi^{2-\varepsilon}}\oint_C dx_E^\mu\oint_C dy_{E\mu}\frac{\mu^{2\varepsilon}}{((x-y)_E^2)^{1-\varepsilon}} \tag{P2.116}$$

Therefore in proper length parametrization (P2.107)

$$\frac{1}{N_c}\langle trU_C\rangle_1 = 1 - \frac{C_F}{2}\frac{g^2}{4\pi^{2-\varepsilon}}\int_0^L ds\int_0^L dt\,\frac{\Gamma(1-\varepsilon)\mu^{2\varepsilon}}{((s-t)^2)^{1-\varepsilon}} + O(\varepsilon^0) \tag{P2.117}$$

$$= 1 - \frac{g^2 C_F}{4\pi^{2-\varepsilon}}\int_0^L dt\int_0^t \frac{ds\,\Gamma(1-\varepsilon)\mu^{2\varepsilon}}{((s-t)^2)^{1-\varepsilon}} + O(\varepsilon^0)$$

$$= 1 - \frac{g^2 C_F}{4\pi^{2-\varepsilon}}\Gamma(1-\varepsilon)\mu^{2\varepsilon}\int_0^L dt\int_0^t \frac{ds'}{(s'^2)^{1-\varepsilon}} + O(\varepsilon^0)$$

$$= 1 - \frac{g^2 C_F}{4\pi^{2-\varepsilon}}\Gamma(1-\varepsilon)\mu^{2\varepsilon}\int_0^L dt\,\frac{t^{2\varepsilon-1}}{2\varepsilon-1} + O(\varepsilon^0)$$

$$= 1 - \frac{g^2 C_F}{4\pi^{2-\varepsilon}}\Gamma(1-\varepsilon)\mu^{2\varepsilon}\frac{L^{2\varepsilon}}{2\varepsilon(2\varepsilon-1)} + O(\varepsilon^0)$$

$$= 1 + \frac{g^2 C_F}{4\pi^2}\frac{1}{2\varepsilon} + O(\varepsilon^0). \tag{P2.118}$$

For a smooth Wilson loop changing variables as in a) one gets

$$\frac{1}{N_c}\langle trU_C\rangle_2 = -g^2\frac{C_F}{4\pi^{2-\varepsilon}}\int_0^R ds\int_{L-R}^L \frac{\mu^{2\varepsilon}\Gamma(1-\varepsilon)dt}{((s-t+L)^2)^{1-\varepsilon}} + O(\varepsilon^0)$$

$$= -g^2\frac{C_F\mu^{2\varepsilon}\Gamma(1-\varepsilon)}{4\pi^{2-\varepsilon}}\int_0^1 d\tau\int_0^R \frac{\rho d\rho}{(\rho^2)^{1-\varepsilon}} + O(\varepsilon^0)$$

$$= -g^2\frac{C_F\mu^{2\varepsilon}\Gamma(1-\varepsilon)}{4\pi^{2-\varepsilon}}\frac{(R^2)^\varepsilon}{2\varepsilon} + O(\varepsilon^0)$$

$$= -g^2\frac{C_F}{4\pi^2}\frac{1}{2\varepsilon} + O(\varepsilon^0), \tag{P2.119}$$

and the linear divergence ($\sim\frac{1}{a}$ in the cutoff regularization) cancels

$$\frac{1}{N_c}\langle trU_C\rangle \simeq 1 + g^2\frac{C_F}{4\pi^2}\frac{1}{2\varepsilon} - g^2\frac{C_F}{4\pi^2}\frac{1}{2\varepsilon} + O(\varepsilon) = 1 + g^2 O(\varepsilon). \quad \text{(P2.120)}$$

For a Wilson line with the cusp changing variables as in a) again one gets

$$\frac{1}{N_c}\langle trU_C\rangle_2 = -\frac{g^2 C_F}{4\pi^{2-\varepsilon}}\int_0^R ds \int_0^R \frac{dt'\cos\phi\,\Gamma(1-\varepsilon)\mu^{2\varepsilon}}{(s^2+2st'\cos\phi+t'^2)^{1-\varepsilon}} + O(\varepsilon^0)$$

$$= -g^2\frac{C_F}{4\pi^{2-\varepsilon}}\int_0^1 d\tau \int_0^R \frac{\rho d\rho}{\rho^{2-2\varepsilon}}$$

$$\times \frac{\cos\phi\,\Gamma(1-\varepsilon)\mu^{2\varepsilon}}{(\tau^2+(1-\tau)^2+2\tau(1-\tau)\cos\phi)^{1-\varepsilon}} + O(\varepsilon^0)$$

$$= -g^2\frac{C_F}{4\pi^2}\frac{R^{2\varepsilon}}{2\varepsilon}\int_0^1 \frac{d\tau\cos\phi}{\tau^2+(1-\tau)^2+2\tau(1-\tau)\cos\phi} + O(\varepsilon^0)$$

$$= -g^2\frac{C_F}{4\pi^2}\frac{1}{2\varepsilon}\phi\,ctg\phi + O(\varepsilon^0). \qquad (P2.121)$$

Therefore

$$\frac{1}{N_c}\langle trU_C\rangle \simeq 1 - g^2\frac{C_F}{4\pi^2}\frac{1}{2\varepsilon}(\phi\,ctg\phi - 1) + O(\varepsilon^0). \qquad (P2.122)$$

Here ϕ is the Euclidean angle between 2 length 1 vectors tangent to the branches of the cut. If the cusp had two time-like tangent vectors in the Minkowski space, than $z'(0)z'(L) = ch\phi_M$. Repeating the calculations one gets the same result with the substitution $\phi \to i\phi_M$

$$\frac{1}{N_c}\langle trU_C\rangle \simeq 1 - g^2\frac{C_F}{4\pi^2}\frac{1}{2\varepsilon}(\phi_M\,cth\phi_M - 1) + O(\varepsilon^0). \qquad (P2.123)$$

2.25 Calculate a rectangular Wilson loop lying in the $x^0 x^3$ plane with space length R and time length $T \gg R$ up to $O(g^2)$.

SOLUTION.

$$\langle\frac{1}{N_c}trU_{\Box T}\rangle = 1 + \frac{(ig)^2}{2!}C_F\frac{\Gamma(1-\varepsilon)}{4\pi^{2-\varepsilon}}\iint_C \frac{\mu^{2\varepsilon}dx^\nu dy_\nu}{(-(x-y)^2+i0)^{1-\varepsilon}} + O(g^4)$$

$$= 1 + g^2 C_F(I_1(R,T) - I_1(T,R) + I_2(T) - I_2(R)) + O(g^4). \qquad (P2.124)$$

Here

$$I_1(R, T) = \frac{1}{4\pi^2} \int_0^T dt \int_0^T ds \frac{1}{(s-t)^2 - R^2 - i0} = |\rho = s - t|$$

$$= \frac{T}{4\pi^2} \int_{-T}^T d\rho \frac{1}{\rho^2 - R^2 - i0}$$

$$= \frac{T}{4\pi^2} \left[\frac{\ln(T - R - i0) - \ln(T + R + i0)}{2R} \right.$$

$$\left. + \frac{-\ln(-T - R - i0) + \ln(-T + R + i0)}{2R} \right] = |T \gg R|$$

$$= \frac{T}{4\pi^2} \left[\frac{\ln(T - R) - \ln(T + R)}{2R} \right.$$

$$\left. + \frac{-\ln(T + R) + i\pi + \ln(T - R) + i\pi}{2R} \right]$$

$$\simeq \frac{T}{4\pi^2} \frac{i\pi}{R} + O(T^0). \tag{P2.125}$$

$$I_1(T, R) \simeq O(T^{-1}). \tag{P2.126}$$

$$I_2(T) = 2 \times \frac{-\Gamma(1 - \varepsilon)}{4\pi^{2-\varepsilon}} \int_0^T ds \int_0^s dt \frac{\mu^{2\varepsilon}}{(-(s-t)^2 + i0)^{1-\varepsilon}} + O(g^4)$$

$$= |\rho = s - t|$$

$$= -2 \frac{\Gamma(1 - \varepsilon)}{4\pi^{2-\varepsilon}} \int_0^T ds \int_0^s d\rho \frac{\mu^{2\varepsilon}}{(-\rho^2 + i0)^{1-\varepsilon}} + O(g^4)$$

$$= -2 \frac{\Gamma(1 - \varepsilon)\mu^{2\varepsilon}}{4\pi^{2-\varepsilon}(-1)^{1-\varepsilon}} \int_0^T ds \frac{s^{2\varepsilon-1}}{2\varepsilon - 1} + O(g^4)$$

$$= 2 \frac{\Gamma(1 - \varepsilon)(-\pi\mu T)^{2\varepsilon}}{4\pi^2(2\varepsilon - 1)2\varepsilon} + O(g^4). \tag{P2.127}$$

Therefore at $T \gg R$

$$\frac{1}{N_c} \langle trU_{\square T} \rangle_R \simeq e^{-iV(R)T} \simeq 1 + \frac{g^2 C_F}{4\pi^2} \frac{i\pi}{R} T + O(T^0). \tag{P2.128}$$

Hence at small g, we have the Coulomb potential

$$V(R) = -\frac{g^2 C_F}{4\pi R}. \tag{P2.129}$$

2.26 Show that one can find the potential via

$$V(R) = \lim_{T \to \infty(1 - i\varepsilon)} \frac{1}{i\Delta T} ln \frac{trU_{\square T}}{trU_{\square T + \Delta T}}. \tag{P2.130}$$

2.27 Show that the potential energy of 3 static quarks in the colorless state up to $O(g^2)$ reads

$$V(\vec{r}_1, \vec{r}_2, \vec{r}_3) = \frac{1}{2}(V(|\vec{r}_1 - \vec{r}_2|) + V(|\vec{r}_2 - \vec{r}_3|) + V(|\vec{r}_3 - \vec{r}_1|)), \quad (\text{P2.131})$$

where $V(R)$ is the $q\bar{q}$ Coulomb potential (P2.129).

HINT: calculate the baryon Wilson loop (2.23) at $T \gg r_i$ coming from the overlap of a colorless 3-quark state (2.24)

$$\psi(\vec{r}_1)^i \psi(\vec{r}_2)^j \psi(\vec{r}_3)^k \times \varepsilon_{lmn} U_{C_1}(\vec{r}_1, 0)^l_i U_{C_2}(\vec{r}_2, 0)^m_j U_{C_3}(\vec{r}_1, 0)^n_k \quad (\text{P2.132})$$

at $t = 0$ and $t = T$ and express it through the potential as was done for the $q\bar{q}$ state.

2.28 Find the potential for the $q\bar{q}$ pair in the octet state in the perturbative limit up to $O(g^2)$. Find the potential for the qq pair in the color states 3* and 6 in the perturbative limit up to $O(g^2)$.

SOLUTION. One can create the color octet state changing the color of (2.32) via t^a

$$\Psi(0)^a|0\rangle = \bar{\psi}'(R, 0)U([R, 0], [0, 0])\Gamma t^a \psi'(0, 0)|0\rangle \quad (\text{P2.133})$$

and repeat the derivation from the lecture and problem 2.25. The only difference is the color factor. One can find it from (P2.130) and (P1.9)

$$\begin{aligned} V_8(R) &= \lim_{T \to \infty(1-i\varepsilon)} \frac{1}{i\Delta T} \ln \frac{tr(t^a t^a + t^a t^b t^a t^b \frac{g^2}{4\pi^2} \frac{i\pi}{R} T)}{tr(t^a t^a + t^a t^b t^a t^b \frac{g^2}{4\pi^2} \frac{i\pi}{R} (T + \Delta T))} \\ &= \lim_{T \to \infty(1-i\varepsilon)} \frac{1}{i\Delta T} \ln \frac{tr(t^a t^a - \frac{1}{2N_c} t^a t^a \frac{g^2}{4\pi^2} \frac{i\pi}{R} T)}{tr(t^a t^a - \frac{1}{2N_c} t^a t^a \frac{g^2}{4\pi^2} \frac{i\pi}{R} (T + \Delta T))} \\ &= +\frac{1}{2N_c} \frac{g^2}{4\pi R}. \end{aligned} \quad (\text{P2.134})$$

Likewise, via completeness relation (1.19) and (P2.130) one gets

$$V_{3^*}(R) = \left(\frac{1}{2N_c} + \frac{1}{2}\right)\frac{g^2}{4\pi R}, \quad V_6(R) = \left(\frac{1}{2N_c} - \frac{1}{2}\right)\frac{g^2}{4\pi R}. \quad (\text{P2.135})$$

Note that the potential is repulsive in 3* and 8 states and attractive in 1 and 6 states.

2.29 Show that (Mak05; BGSN82)

$$\frac{\delta tr U_C(x, x)}{\delta \sigma_{\mu\nu}(x)} = ig\, tr[F^{\mu\nu}(x)U_C(x, x)], \quad (\text{P2.136})$$

$$\frac{\partial}{\partial x^\mu} \frac{\delta tr U_C(x, x)}{\delta \sigma_{\mu\nu}(x)} = ig\, tr[(D_\mu F^{\mu\nu}(x))U_C(x, x)], \quad (\text{P2.137})$$

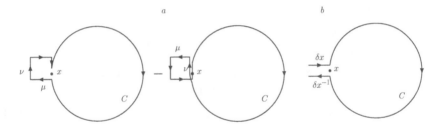

Figure 2.5 Contours for the calculation of the a) area (P2.138) and b) ordinary (P2.139) derivatives of the Wilson line with the contour C.

where the **area derivative** is defined as

$$\frac{\delta tr U_C(x,x)}{\delta \sigma_{\mu\nu}(x)} = \lim_{|\delta\sigma_{\mu\nu}|\to 0} \frac{tr U_{C\cup\delta C_{\mu\nu}}(x,x) - tr U_{C\cup\delta C_{\nu\mu}}(x,x)}{2|\delta\sigma_{\mu\nu}|}. \quad (P2.138)$$

Here the small size ε square contour $\delta C_{\mu\nu}$ in the $\mu\nu$ plane with the area $\delta\sigma_{\mu\nu}$ is attached to the initial contour C at point x, as is shown in Figure 2.5a. The ordinary derivative is defined on loops as

$$\frac{\partial tr U_C(x,x)}{\partial x^\mu} = \lim_{|\delta x_\mu|\to 0} \frac{tr U_{\delta x^{\mu-1}\cup C\cup\delta x^\mu}(x+\delta x, x+\delta x) - tr U_C(x,x)}{|\delta x_\mu|},$$
$$(P2.139)$$

where the contour for its calculation is given in Figure 2.5b.

SOLUTION. Here we follow (BGSN82). Using (P2.99), we get

$$tr U_{C\cup\delta C_{\mu\nu}}(x,x) = tr[U_{\delta C_{\mu\nu}}(x,x)U_C(x,x)]$$
$$\simeq tr[(1+ig\varepsilon^2 F^{\mu\nu}(x))U_C(x,x)]. \quad (P2.140)$$

Hence

$$\frac{\delta tr U_C(x,x)}{\delta \sigma_{\mu\nu}(x)} = \lim_{\varepsilon\to 0} \frac{2ig\varepsilon^2 tr[F^{\mu\nu}(x)U_C(x,x)]}{2\varepsilon^2}$$
$$= igtr[F^{\mu\nu}(x)U_C(x,x)]. \quad (P2.141)$$

Next,

$$U_{\delta x^\mu\cup C\cup\delta x^{\mu-1}}(x+\delta x, x+\delta x)$$
$$= (1+ig\delta x^\mu A_\mu(x))U_C(x,x)(1-ig\delta x^\mu A_\mu(x))$$
$$= U_C(x,x) - ig\delta x^\mu[U_C(x,x), A_\mu(x)]. \quad (P2.142)$$

Therefore

$$\frac{\partial tr U_C(x,x)}{\partial x^\mu} = \lim_{|\delta x_\mu|\to 0} \frac{-ig\delta x^\mu[U_C(x,x), A_\mu(x)]}{|\delta x_\mu|}$$
$$= -ig[U_C(x,x), A_\mu(x)]. \quad (P2.143)$$

Finally,

$$
\begin{aligned}
\frac{\partial}{\partial x^\mu} \frac{\delta tr U_C(x,x)}{\delta \sigma_{\mu\nu}(x)} &= ig\partial_\mu tr[F^{\mu\nu}(x)U_C(x,x)] \\
&= igtr[(\partial_\mu F^{\mu\nu}(x))U_C(x,x) - F^{\mu\nu}(x)ig[U_C(x,x), A_\mu(x)]] \\
&= igtr[(\partial_\mu F^{\mu\nu}(x) - ig[A_\mu(x), F^{\mu\nu}(x)])U_C(x,x)] \\
&= igtr[(D_\mu F_x^{\mu\nu})U_C(x,x)].
\end{aligned}
\tag{P2.144}
$$

Wilson loops are discussed in more details in (Pol80), (DV80), and (Mak10; Mak05). Their relation to the potential is explained in (BW79), (Rot12), and (TMNS01). Introduction to the temporal gauge is given in (Lei87) and (Bjo79). Vacuum effects in the temporal gauge are discussed in (JR76) and (Bjo79).

FURTHER READING

[BD65] James D. Bjorken and Sidney D. Drell. *Relativistic quantum fields*, chapter 4, (Foldy-Wouthuysen transformation). McGraw-Hill College, 1965.

[BGSN82] Richard A. Brandt, A. Gocksch, M. A. Sato, and F. Neri. LOOP SPACE. *Phys. Rev. D*, 26:3611, 1982.

[Bjo79] J. D. Bjorken. Elements of Quantum Chromodynamics. *Prog. Math. Phys.*, 4:423–561, 12 1979.

[BW79] Lowell S. Brown and William I. Weisberger. Remarks on the Static Potential in Quantum Chromodynamics. *Phys. Rev. D*, 20:3239, 1979.

[DV80] V. S. Dotsenko and S. N. Vergeles. Renormalizability of Phase Factors in the Nonabelian Gauge Theory. *Nucl. Phys. B*, 169:527–546, 1980.

[JR76] R. Jackiw and C. Rebbi. Vacuum Periodicity in a Yang-Mills Quantum Theory. *Phys. Rev. Lett.*, 37:172–175, 1976.

[Lei87] George Leibbrandt. Introduction to Noncovariant Gauges. *Rev. Mod. Phys.*, 59:1067, 1987.

[Mak05] Yu. Makeenko. *Methods of contemporary gauge theory*. Cambridge Monographs on Mathematical Physics. Cambridge University Press, 11 2005.

[Mak10] Yuri Makeenko. A Brief Introduction to Wilson Loops and Large N. *Phys. Atom. Nucl.*, 73:878–894, 2010.

[Pec99] R. D. Peccei. Discrete and global symmetries in particle physics. *Lect. Notes Phys.*, 521:1–50, 1999.

[Pol80] Alexander M. Polyakov. Gauge fields as rings of glue. *Nucl. Phys. B*, 164:171–188, 1980.

[PS95] Michael E. Peskin and Daniel V. Schroeder. *An Introduction to quantum field theory*, chapter 15.1-3, 2.4 (Causality). Addison-Wesley, Reading, USA, 1995.

[Rot12] Heinz J. Rothe. *Lattice Gauge Theories : An Introduction (Fourth Edition)*, volume 43, chapter 7 (The Wilson loop and the static quark-antiquark potential). World Scientific Publishing Company, 2012.

[TMNS01] Toru T. Takahashi, H. Matsufuru, Y. Nemoto, and H. Suganuma. The three quark potential in the SU(3) lattice QCD. *Phys. Rev. Lett.*, 86:18–21, 2001.

CHAPTER 3

Functional Integration

O NE of the most powerful methods of the quantum field theory (QFT) is the functional integration approach. This chapter introduces it in the context of quantum mechanics and QFT. We discuss both bosonic and fermionic theories, Grassmann numbers, Schwinger-Dyson equations and reduction formulas.

1. In quantum mechanics let $\hat{q}$ be a complete set of commuting operators in the basis that diagonalizes them $\hat{q}|q\rangle = q|q\rangle$. The **transition amplitude** W is a matrix element of the **evolution operator** U

$$W(q,t;q',t') = \langle q'|U(t'-t)|q\rangle, \quad U(t) = e^{-iHt}, \qquad (3.1)$$

which has a **group property**

$$U(t-t')U(t'-t'') = U(t-t''). \qquad (3.2)$$

Hence, using $1 = \int dq_n |q_n\rangle\langle q_n|$ and $\varepsilon = \frac{t'-t}{N}$, $q' = q_N$, $q = q_0$

$$W(q,t;q',t') = \langle q'|U(\varepsilon)\ldots U(\varepsilon)|q\rangle = \int dq_1...dq_{N-1} \prod_{n=1}^{N} \langle q_n|U(\varepsilon)|q_{n-1}\rangle. \qquad (3.3)$$

For $H = \frac{p^2}{2m} + V(q)$

$$U(\varepsilon) = e^{-i(\frac{p^2}{2m}+V(q))\varepsilon} = e^{-i\frac{p^2}{2m}\varepsilon}e^{-iV(q))\varepsilon} + O(\varepsilon^2) \qquad (3.4)$$

DOI: 10.1201/9781003272403-3

and

$$\langle q_n|U(\varepsilon)|q_{n-1}\rangle = \langle q_n|e^{-i\frac{p^2}{2m}\varepsilon}|q_{n-1}\rangle e^{-iV(q_{n-1}))\varepsilon}$$

$$= \langle q_n|p_n\rangle\langle p_n|e^{-i\frac{p^2}{2m}\varepsilon}|p_{n-1}\rangle\langle p_{n-1}|q_{n-1}\rangle e^{-iV(q_{n-1}))\varepsilon} \tag{3.5}$$

$$= \int \frac{dp_n}{2\pi} e^{ip_n(q_n-q_{n-1})} e^{-i\frac{p_n^2}{2m}\varepsilon} e^{-iV(q_{n-1}))\varepsilon} \tag{3.6}$$

$$= Ae^{i\frac{m(q_n-q_{n-1})^2}{2\varepsilon}} e^{-iV(q_{n-1}))\varepsilon}, \quad A = \sqrt{\frac{m}{2\pi\varepsilon i}}. \tag{3.7}$$

Therefore taking the limit $N \to \infty$ one **defines** an object called the **functional integral**

$$W(q,t;q',t') = \lim_{N\to\infty} A^N \int dq_1...dq_{N-1} e^{i\sum_{n=1}^N \left(\frac{m(q_n-q_{n-1})^2}{2\varepsilon^2} - V(q_{n-1})\right)\varepsilon}$$

$$\overset{def}{=} \int_{\substack{q(t)=q\\q(t')=q'}} \mathcal{D}q e^{iS[q]} = \int_{\substack{q(t)=q\\q(t')=q'}} \mathcal{D}q e^{i\int_{t'}^t dt L[q,\dot{q}]}. \tag{3.8}$$

2. For the **scalar field** theory one introduces discretization of space and the continuous field $\phi(x)$ is represented by a many dimensional vector $\{\phi(x_i)\}$, or $\{\phi(x)\} = \{\phi_x\}$, where x now goes through a large but finite set of lattice points. Hence, we have quantum mechanics with large number of variables. Changing these variables to a set where the Hamiltonian is diagonal in the corresponding momenta one can repeat the previous derivation. Spanning the time to infinity $t, -t' \to \infty$ and integrating over the fields at t and t', one comes to the **generating functional** $\mathcal{Z}[J]$:

$$\mathcal{Z}[0] = \mathcal{Z}_0 = \int \mathcal{D}\phi e^{i\int \mathcal{L}[\phi]d^4x} = \lim_{\Delta x\to 0} \prod_x \int d\phi_x e^{i\sum_x \mathcal{L}[\phi_x]}, \tag{3.9}$$

$$\mathcal{Z}[J] = \int \mathcal{D}\phi e^{i\int(\mathcal{L}[\phi]+J(x)\phi(x))d^4x}, \tag{3.10}$$

where we added a source field J. For a free scalar field

$$\mathcal{Z}_0 = \int \mathcal{D}\phi e^{\frac{i}{2}\int((\partial_\nu\phi)^2-m^2\phi^2)d^4x} = \int \mathcal{D}\phi e^{-\frac{i}{2}\int\phi(\partial_\nu^2+m^2)\phi d^4x}$$

$$= \frac{const}{[\det(\partial^2+m^2)]^{\frac{1}{2}}}. \tag{3.11}$$

$$\mathcal{Z}[J] = \int \mathcal{D}\phi e^{i\int(\frac{1}{2}\phi(-\partial_\nu^2-m^2+i0)\phi+J\phi)d^4x}$$

$$= \left|\phi = \phi' + i\int dz D(x-z)J(z)\right|, \tag{3.12}$$

with the **Klein-Gordon propagator**

$$(-\partial_\nu^2 - m^2 + i0)D(x - z) = i\delta(x - z). \tag{3.13}$$

Here one adds the $+i0$ prescription for convergence. Under $\int dx$ in the exponent of $\mathcal{Z}[J]$ one has

$$\frac{1}{2}\phi(-\partial_\nu^2 - m^2 + i0)\phi + J\phi = \frac{1}{2}\phi'_x(-\partial_{\nu(x)}^2 - m^2 + i0)\phi'_x + J_x\phi'_x$$

$$+ iJ_x \int dz D(x - z)J_z + i\frac{1}{2}\int dz D(x - z)J_z(-\partial_{\nu(x)}^2 - m^2 + i0)\phi'_x$$

$$+ \frac{1}{2}\phi'_x(-\partial_{\nu(x)}^2 - m^2 + i0)i \int dz D(x - z)J_z$$

$$- \frac{1}{2}\int dz D(x - z)J_z(-\partial_{\nu(x)}^2 - m^2 + i0) \int dy D(x - y)J_y$$

$$= \frac{1}{2}\phi'_x(-\partial_{\nu(x)}^2 - m^2 + i0)\phi'_x + J_x\phi'_x + iJ_x \int dz D(x - z)J_z$$

$$- \frac{1}{2}J_x\phi'_x - \frac{1}{2}J_x\phi'_x - \frac{i}{2}\int dz J_x D(x - z)J_z$$

$$= \frac{1}{2}\phi'_x(-\partial_{\nu(x)}^2 - m^2 + i0)\phi'_x + \frac{i}{2}\int dz J_x D(x - z)J_z. \tag{3.14}$$

Then,

$$\mathcal{Z}[J] = \mathcal{Z}_0 \int \mathcal{D}\phi\, e^{-\frac{1}{2}\int dx dz J_x D(x-z)J_z}. \tag{3.15}$$

3. The **functional derivative**

$$\frac{\delta}{\delta J(x)}J(y) \stackrel{def}{=} \delta^{(4)}(x - y), \qquad \frac{\delta}{\delta J(x)}\int d^4y J(y)f(y) = f(x). \tag{3.16}$$

Then

$$\frac{\delta}{\delta J(x)}e^{i\int dy J(y)f(y)} = if(x)e^{i\int dy J(y)f(y)}, \tag{3.17}$$

$$\frac{\delta}{\delta J(x)}\int d^4y(\partial_\mu J(y))f(y) = -\frac{\delta}{\delta J(x)}\int d^4y(\partial_\mu f(y))J(y) = -\partial_\mu f(x). \tag{3.18}$$

4. The **Klein-Gordon propagator as a functional derivative**

$$\frac{\delta}{\delta i J(x)}\frac{\delta}{\delta i J(y)}\frac{\mathcal{Z}[J]}{\mathcal{Z}_0}\Big|_{J=0} = \frac{\delta}{\delta i J_x}\frac{\delta}{\delta i J_y}\int \mathcal{D}\phi\, e^{-\frac{1}{2}\int dx dz J_x D(x-z)J_z}\Big|_{J=0}$$

$$= \frac{1}{2}\frac{\delta}{\delta J_x}\left(\int dx J_x D(x - y) + \int dz D(y - z)J_z\right)$$

$$\times \int \mathcal{D}\phi\, e^{-\frac{1}{2}\int dx dz J_x D(x-z)J_z}\Big|_{J=0} = D(x - y). \tag{3.19}$$

On the other hand

$$
\frac{\delta}{\delta i J_x} \frac{\delta}{\delta i J_y} \frac{Z[J]}{Z_0}\Big|_{J=0} = \frac{\delta}{\delta i J_x} \frac{\delta}{\delta i J_y} \frac{1}{Z_0} \int \mathcal{D}\phi e^{i\int(\frac{1}{2}\phi(-\partial_\nu^2 - m^2 + i0)\phi + J\phi)d^4x}
$$

$$
= \frac{1}{Z_0} \int \mathcal{D}\phi \phi_x \phi_y e^{i\frac{1}{2}\int \phi(-\partial_\nu^2 - m^2 + i0)\phi d^4x}. \qquad (3.20)
$$

Therefore the **propagator is the average of the field values at two points x and y over all possible field configurations**

$$
\langle \phi_x \phi_y \rangle = \frac{\int \mathcal{D}\phi \phi_x \phi_y e^{iS[\phi]}}{\int \mathcal{D}\phi e^{iS[\phi]}} = \frac{1}{Z_0} \int \mathcal{D}\phi \phi_x \phi_y e^{iS[\phi]} = D(x - y). \quad (3.21)
$$

In these formulas all the fields are c-**numbers** rather than operators. In the language of second quantization, the propagator is a vacuum expectation value of a $\mathcal{T}$ product of two **field operators** written in terms of creation-annihilation operators a and $a^\dagger$

$$
D(x - y) = \langle 0 | \mathcal{T}(\phi_x \phi_y) | 0 \rangle = \langle \phi_x \phi_y \rangle. \qquad (3.22)
$$

5. Hence, all **Green functions** (vacuum expectation values of $\mathcal{T}$ products of field operators) for a free theory can be calculated as **averages of the field values over all possible field configurations**

$$
\langle \phi_{x_1} ... \phi_{x_n} \rangle = \langle 0 | \mathcal{T}(\phi_{x_1} ... \phi_{x_n}) | 0 \rangle = \frac{\int \mathcal{D}\phi \phi_{x_1} ... \phi_{x_n} e^{iS[\phi]}}{\int \mathcal{D}\phi e^{iS[\phi]}}
$$

$$
= \frac{1}{Z_0} \int \mathcal{D}\phi \phi_{x_1} ... \phi_{x_n} e^{iS[\phi]} = \frac{1}{Z_0} \frac{\delta}{\delta i J_{x_1}} ... \frac{\delta}{\delta i J_{x_n}} Z[J]\big|_{J=0}.
$$

$$
\qquad (3.23)
$$

6. For the **self-interacting field theory** with

$$
S = S_0 + S_{int} = \int dx \mathcal{L}_0(\phi) + \int dx \mathcal{L}_{int}(\phi), \qquad (3.24)
$$

one writes the generating functional as

$$
Z[J] = \int \mathcal{D}\phi e^{i\int(\mathcal{L}[\phi] + J(x)\phi(x))d^4x} = \int \mathcal{D}\phi e^{i\int(\mathcal{L}_0(\phi) + \mathcal{L}_{int}(\phi) + J(x)\phi(x))d^4x}
$$

$$
= e^{i\int \mathcal{L}_{int}(\frac{\delta}{\delta i J_x})d^4x} \int \mathcal{D}\phi e^{i\int(\mathcal{L}_0(\phi) + J(x)\phi(x))d^4x}
$$

$$
= e^{i\int \mathcal{L}_{int}(\frac{\delta}{\delta i J_x})d^4x} \mathcal{Z}[J] \qquad (3.25)
$$

and **expands** $e^{i\int \mathcal{L}_{int}(\frac{\delta}{\delta iJ_x})d^4x}$ to get the perturbation series. Therefore the **Green functions in the self-interacting theory** are given by

$$\langle 0|\mathcal{T}(\phi_{x_1}...\phi_{x_n})|0\rangle = \langle \phi_{x_1}...\phi_{x_n}\rangle = \frac{\int \mathcal{D}\phi\phi_{x_1}...\phi_{x_n}e^{iS[\phi]}}{\int \mathcal{D}\phi e^{iS[\phi]}}$$

$$= \frac{1}{Z_0}\int \mathcal{D}\phi\phi_{x_1}...\phi_{x_n}e^{i\int \mathcal{L}_{int}(\frac{\delta}{\delta iJ_x})d^4x}e^{iS_0[\phi]}$$

$$= \frac{1}{Z_0}\frac{\delta}{\delta iJ_{x_1}}...\frac{\delta}{\delta iJ_{x_n}}e^{i\int \mathcal{L}_{int}(\frac{\delta}{\delta iJ_x})d^4x}\mathcal{Z}[J]|_{J=0}. \quad (3.26)$$

7. **Grassmann numbers.** Here we follow the presentation of (DGH14b). Grassmann numbers $\alpha, \beta, \gamma...$ are anti-commuting numbers

$$\{\alpha,\alpha\} = \{\alpha,\beta\} = 0, \quad \Longrightarrow \alpha^2 = \beta^2 = ... = 0. \quad (3.27)$$

Therefore any function has the general **finite expansion**

$$f(\alpha) = f_0 + \alpha f_1, \quad g(\alpha,\beta) = g_0 + g_1\alpha + g_2\beta + g_3\alpha\beta. \quad (3.28)$$

Differentiation is defined as

$$\frac{d\alpha}{d\alpha} = \frac{d\beta}{d\beta} = ... = 1, \quad \frac{d\alpha}{d\beta} = \frac{d\beta}{d\alpha} = ... = 0 \Longrightarrow \frac{d}{d\beta}\alpha\beta = -\frac{d}{d\beta}\beta\alpha = -\alpha. \quad (3.29)$$

Then, e.g. for functions (3.28)

$$\frac{d}{d\alpha}f(\alpha) = f_1, \quad \frac{d}{d\beta}g(\alpha,\beta) = g_2 - g_3\alpha. \quad (3.30)$$

and the **second derivative** has the property

$$\frac{d^2}{d\alpha d\alpha} = 0. \quad (3.31)$$

Integral is defined translationally invariant

$$\int d\alpha f(\alpha) = \int d\alpha f(\alpha + \beta) \quad \Longrightarrow \int d\alpha f_1\beta = 0 \Leftrightarrow \int d\alpha = 0. \quad (3.32)$$

Normalization is chosen to 1. Therefore

$$\int d\alpha = 0, \quad \int d\alpha\alpha = 1. \quad (3.33)$$

Then for functions (3.28)

$$\int d\alpha f(\alpha) = f_1, \quad \int d\beta g(\alpha,\beta) = g_2 - g_3\alpha, \quad \int d\alpha \int d\beta g(\alpha,\beta) = -g_3. \quad (3.34)$$

We need a **Gaussian integral over Grassmann variables** $\alpha = \{\alpha_1...\alpha_n\}$, $\bar{\alpha} = \{\bar{\alpha}_1...\bar{\alpha}_n\}$ with a c-number matrix M

$$\int d\bar{\alpha}_n...d\bar{\alpha}_1 d\alpha_n...d\alpha_1 e^{i\bar{\alpha} M \alpha} = const \det M. \tag{3.35}$$

Indeed, e.g. for a 2×2 matrix one has

$$\int d\bar{\alpha}_2 d\bar{\alpha}_1 d\alpha_2 d\alpha_1 e^{i\bar{\alpha} M \alpha}$$

$$= \int d\bar{\alpha}_2 d\bar{\alpha}_1 d\alpha_2 d\alpha_1 \left(1 + \bar{\alpha}_i i M_{ij} \alpha_j - \frac{1}{2} \bar{\alpha}_i M_{ij} \alpha_j \bar{\alpha}_{i'} M_{i'j'} \alpha_{j'}\right)$$

$$= -\frac{1}{2} \int d\bar{\alpha}_2 d\bar{\alpha}_1 d\alpha_2 d\alpha_1 (\bar{\alpha}_1 M_{11} \alpha_1 \bar{\alpha}_2 M_{22} \alpha_2 + \bar{\alpha}_1 M_{12} \alpha_2 \bar{\alpha}_2 M_{21} \alpha_1$$

$$+ \bar{\alpha}_2 M_{21} \alpha_1 \bar{\alpha}_1 M_{12} \alpha_2 + \bar{\alpha}_2 M_{22} \alpha_2 \bar{\alpha}_1 M_{11} \alpha_1)$$

$$= -\frac{1}{2} \int d\bar{\alpha}_2 d\bar{\alpha}_1 d\alpha_2 d\alpha_1 \; \alpha_1 \alpha_2 \bar{\alpha}_1 \bar{\alpha}_2$$

$$\times (-M_{11} M_{22} + M_{12} M_{21} + M_{21} M_{12} - M_{22} M_{11})$$

$$= M_{11} M_{22} - M_{12} M_{21} = \det M. \tag{3.36}$$

Complex conjugate of a Grassmann number is defined as Hermitian conjugate of operators

$$(\alpha\beta)^* = \beta^* \alpha^* = -\alpha^* \beta^*. \tag{3.37}$$

8. **Fermion fields are represented as fields of Grassmann numbers** $\{\bar{\psi}_x\}$ and $\{\psi_x\}$ and the Gaussian functional integral over these fields

$$\int \mathcal{D}\psi \mathcal{D}\bar{\psi} e^{i \int dx \bar{\psi}_x O \psi_x} = const \det O. \tag{3.38}$$

Generating functional for free fermion fields with the Lagrangian $\mathcal{L}_0 = \bar{\psi}(i\hat{\partial} - m + i0)\psi$

$$\mathcal{Z}[\eta, \bar{\eta}] = \int \mathcal{D}\psi \mathcal{D}\bar{\psi} e^{i \int dx \left(\bar{\psi}_x (i\hat{\partial}_x - m + i0)\psi_x + \bar{\eta}_x \psi_x + \bar{\psi}_x \eta_x\right)} \tag{3.39}$$

with **Grassmann** $\bar{\eta}_x$ and η_x source fields. Changing the variables

$$\psi = \psi' + i \int dy D(x - y)\eta_y, \quad \bar{\psi} = \bar{\psi}' + i \int dy \bar{\eta}_y D(y - x), \tag{3.40}$$

$$(i\hat{\partial}_x - m)D(x - y) = i\delta(x - y) \implies (-i\hat{\partial}_y - m)D(x - y) = i\delta(x - y). \tag{3.41}$$

Under $\int dx$ in the exponent of $\mathcal{Z}$ one has

$$\bar{\psi}_x(i\hat{\partial}_x - m)\psi_x + \bar{\eta}_x\psi_x + \bar{\psi}_x\eta_x = \bar{\psi}'_x(i\hat{\partial}_x - m)\psi'_x + \bar{\eta}_x\psi'_x + \bar{\psi}'_x\eta_x$$
$$+ \bar{\psi}'_x(i\hat{\partial}_x - m + i0)i \int dy D(x-y)\eta_y$$
$$+ i \int dy \bar{\eta}_y D(y-x)(i\hat{\partial}_x - m)\psi'_x$$
$$+ i \int dy \bar{\eta}_y D(y-x)(i\hat{\partial}_x - m)i \int dy D(x-y)\eta_y$$
$$+ \bar{\eta}_x i \int dy D(x-y)\eta_y + i \int dy \bar{\eta}_y D(y-x)\eta_x$$
$$= \bar{\psi}'_x(i\hat{\partial}_x - m)\psi'_x + i \int dy \bar{\eta}_y D(y-x)\eta_x. \qquad (3.42)$$

As a result,

$$\mathcal{Z}[\eta, \bar{\eta}] = \int \mathcal{D}\psi \mathcal{D}\bar{\psi} e^{i \int dx\, \bar{\psi}'_x(i\hat{\partial}_x - m + i0)\psi'_x - \int dx dy\, \bar{\eta}_y D(y-x)\eta_x}$$
$$= \mathcal{Z}[0,0] e^{-\int dx dy\, \bar{\eta}_y D(y-x)\eta_x}. \qquad (3.43)$$

Then the free Feynman fermion propagator

$$\langle \psi_x \bar{\psi}_y \rangle = \frac{1}{\mathcal{Z}[0,0]} \int \mathcal{D}\psi \mathcal{D}\bar{\psi}\, \psi_x \bar{\psi}_y\, e^{i \int \mathcal{L}_0 dx} = \frac{1}{\mathcal{Z}[0,0]} \frac{\delta}{\delta i \eta_y} \frac{\delta}{\delta i \bar{\eta}_x} \mathcal{Z}[\eta, \bar{\eta}]$$
$$= \frac{1}{\mathcal{Z}[0,0]} \frac{\delta}{\delta i \eta_y} \frac{\delta}{\delta i \bar{\eta}_x} \mathcal{Z}[\eta, \bar{\eta}]|_{\eta=\bar{\eta}=0} = D(x-y). \qquad (3.44)$$

The interacting theory is built as for the boson fields.

9. The **Schwinger-Dyson (SD) equations** are quantum equations of motion for Green functions. One derives them by infinitesimally changing integration variables in the functional integral and demanding the theory to be invariant under this reparametrization. Consider e.g. an n point scalar function

$$\langle \phi_{x_1} ... \phi_{x_n} \rangle = \frac{1}{Z_0} \int \mathcal{D}\phi\, \phi_{x_1} ... \phi_{x_n}\, e^{iS[\phi]}. \qquad (3.45)$$

and do an infinitesimal coordinate-dependent change

$$\phi_x \to \phi_x + \varepsilon_x. \qquad (3.46)$$

The correlator does not change. Hence

$$0 = \frac{1}{Z_0} \int \mathcal{D}\phi\, (\phi_{x_1} + \varepsilon_{x_1})...(\phi_{x_n} + \varepsilon_{x_n})\, e^{iS[\phi + \varepsilon]} - \frac{1}{Z_0} \int \mathcal{D}\phi\, \phi_{x_1} ... \phi_{x_n} e^{iS[\phi]}.$$
$$(3.47)$$

Expanding in ε

$$0 = \frac{1}{Z_0} \int \mathcal{D}\phi \int dr \varepsilon_r (\delta(r - x_1)\phi_{x_2}...\phi_{x_n} + \phi_{x_1}\delta(r - x_2)\phi_{x_3}...\phi_{x_n} + ...$$

$$+ \phi_{x_1}...\phi_{x_{n-1}}\delta(r - x_n) + \phi_{x_1}...\phi_{x_n}\frac{\delta iS[\phi]}{\delta\phi_r}) \, e^{iS[\phi]}, \qquad (3.48)$$

or

$$\langle\phi_{x_1}...\phi_{x_n}\frac{\delta S[\phi]}{\delta\phi_r}\rangle = i\langle\delta(r - x_1)\phi_{x_2}...\phi_{x_n} + \phi_{x_1}\delta(r - x_2)\phi_{x_3}...\phi_{x_n} + ...$$

$$+ \phi_{x_1}...\phi_{x_{n-1}}\delta(r - x_n)\rangle. \qquad (3.49)$$

Recalling functional derivative definition (3.16), it reads

$$\langle\phi_{x_1}...\phi_{x_n}\frac{\delta S[\phi]}{\delta\phi_r}\rangle = i\langle\frac{\delta}{\delta\phi_r}\phi_{x_1}...\phi_{x_n}\rangle. \qquad (3.50)$$

Therefore one can write a Swinger-Dyson equation for an arbitrary functional $F[\phi]$ of the fields

$$\langle F[\phi]\frac{\delta S[\phi]}{\delta\phi_r}\rangle = i\langle\frac{\delta}{\delta\phi_r}F[\phi]\rangle, \quad \text{or} \quad \frac{\delta S[\phi]}{\delta\phi_r} \stackrel{w.s.}{=} i\frac{\delta}{\delta\phi_r}. \qquad (3.51)$$

The second equation in (3.51) is an equivalent way of writing the first one, i.e. $\stackrel{w.s.}{=}$ means that this equality is valid in the "weak sense", i.e. after averaging when applied to a field functional $F[\phi]$. The r.h.s. of equations (3.49–3.51) are called the **contact terms**.

10. The **Lehmann-Symanzik-Zimmermann (LSZ) reduction** formula allows one to build scattering amplitudes from Green functions. We demonstrate it on the example of the Dirac fermions. For the free field

$$\psi_{in}(x) = \sum_s \int \frac{d^3p}{(2\pi)^3\sqrt{2\omega_p}}(a_{\vec{p}s,in}u_{\vec{p}s}e^{-ipx} + b^\dagger_{\vec{p}s,in}v_{\vec{p}s}e^{ipx}),$$

$$a^\dagger_{\vec{p}s,in} = \int \frac{d^3x e^{-ipx}}{\sqrt{2\omega_p}}\bar{\psi}_{in}(x)\gamma^0 u_{\vec{p}s}. \qquad (3.52)$$

We also assume the following limits

$$\lim_{x^0\to-\infty} \psi(x) \to Z^{\frac{1}{2}}\psi_{in}(x), \quad \lim_{x^0\to+\infty} \psi(x) \to Z^{\frac{1}{2}}\psi_{out}(x) \qquad (3.53)$$

for the interacting field ψ in the weak sense, i.e. in matrix elements. Here Z is the field renormalization constant and ψ_{out} is a free field. The *in* state is prepared at $t = -\infty$, where it contains a set of free particles

$$|i_{1...n}\rangle = |p_1, s_1, ...p_n, s_n, in\rangle = a^\dagger_{\vec{p}_1 s_1,in}|i_{2...n}\rangle. \qquad (3.54)$$

The *out* state is considered at $t = +\infty$, where it also contains a set of free particles

$$\langle f_{1...m}| = \langle k_1, \sigma_1...k_m, \sigma_m, out| = \langle f_{2...m}|a_{\vec{k}_1\sigma_1, out}$$
$$= \langle k_1, \sigma_1...k_m, \sigma_m, in|S = \langle \tilde{i}_{1...m}|S. \tag{3.55}$$

The S-matrix element is

$$S_{fi} = \langle f_{1...m}|i_{1...n}\rangle = \langle \tilde{i}_{1...m}|S|i_{1...n}\rangle = \langle f_{1...m}|a^\dagger_{\vec{p}_1 s_1, in}|i_{2...n}\rangle. \tag{3.56}$$

We suppose that the out state **does not have particles with the same momenta and spin as the initial ones.** Here we will consider the process without antiparticles in the in and out states. Then

$$\langle f_{1...m}|a^\dagger_{\vec{p}_1 s_1, in}|i_{2...n}\rangle = \langle f_{1...m}|a^\dagger_{\vec{p}_1 s_1, in} - a^\dagger_{\vec{p}_1 s_1, out}|i_{2...n}\rangle$$

$$= \int \frac{d^3x_1 e^{-ip_1 x_1}}{\sqrt{2\omega_{p_1}}} \langle f_{1...m}| \left(\bar{\psi}_{in}(x_1)\gamma^0 u_{\vec{p}_1 s_1} - \bar{\psi}_{out}(x_1)\gamma^0 u_{\vec{p}_1 s_1} \right) |i_{2...n}\rangle$$

$$= (\lim_{x^0 \to -\infty} - \lim_{x^0 \to +\infty}) \int \frac{d^3x_1 e^{-ip_1 x_1}}{\sqrt{2\omega_{p_1} Z}} \langle f_{1...m}|\bar{\psi}(x_1)\gamma^0 u_{\vec{p}_1 s_1}|i_{2...n}\rangle$$

$$= - \int_{-\infty}^{+\infty} dt \partial_t \int \frac{d^3x_1 e^{-ip_1 x_1}}{\sqrt{2\omega_{p_1} Z}} \langle f_{1...m}|\bar{\psi}(x_1)\gamma^0 u_{\vec{p}_1 s_1}|i_{2...n}\rangle$$

$$= i \int \frac{d^4x_1}{\sqrt{2\omega_{p_1} Z}} \langle f_{1...m}|\bar{\psi}(x_1)\gamma^0(i\overleftarrow{\partial}_t + i\vec{\partial}_t)u_{\vec{p}_1 s_1}e^{-ip_1 x_1}|i_{2...n}\rangle$$

$$= i \int \frac{d^4x_1}{\sqrt{2\omega_{p_1} Z}} \langle f_{1...m}|\bar{\psi}(x_1)(i\gamma^0\overleftarrow{\partial}_t + i\vec{\gamma}\vec{\partial} + m)u_{\vec{p}_1 s_1}e^{-ip_1 x_1}|i_{2...n}\rangle$$

$$= -i \int \frac{d^4x_1 e^{-ip_1 x_1}}{\sqrt{2\omega_{p_1} Z}} [(-i\hat{\partial}_{x_1} - m)u_{\vec{p}_1 s_1}]^{l_1} \langle f_{1...m}|\bar{\psi}(x_1)_{l_1}|i_{2...n}\rangle. \tag{3.57}$$

One can treat the *out* state in the similar manner. Extracting a particle from the out state in the matrix element in (3.57), one gets

$$\langle f_{1...m}|\bar{\psi}(x_1)_{l_1}|i_{2...n}\rangle = \langle f_{2...m}|a_{k_1\sigma_1, out}\bar{\psi}(x_1)_{l_1}|i_{2...n}\rangle$$

$$= \langle f_{2...m}|a_{k_1\sigma_1, out}\bar{\psi}(x_1)_{l_1} - \bar{\psi}(x_1)_{l_1}a_{k_1\sigma_1, in}|i_{2...n}\rangle = \int \frac{d^3y_1 e^{ik_1 y_1}}{\sqrt{2\omega_{k_1}}}$$

$$\times \langle f_{2...m}|\bar{u}_{\vec{k}_1\sigma_1}\gamma^0\psi_{out}(y_1)\bar{\psi}(x_1)_{l_1} - \bar{\psi}(x_1)_{l_1}\bar{u}_{\vec{k}_1\sigma_1}\gamma^0\psi_{in}(y_1)|i_{2...n}\rangle \tag{3.58}$$

$$= (\lim_{y^0 \to +\infty} - \lim_{y^0 \to -\infty}) \int \frac{d^3y_1 e^{ik_1 y_1}}{\sqrt{2\omega_{k_1} Z}} [\bar{u}_{\vec{k}_1\sigma_1}\gamma^0]_{r_1}$$

$$\times \langle f_{2...m}|\mathcal{T}(\psi(y_1)^{r_1}\bar{\psi}(x_1)_{l_1})|i_{2...n}\rangle \tag{3.59}$$

$$= \int_{-\infty}^{+\infty} dt \partial_{y^0} \int \frac{d^3y_1 e^{ik_1 y_1}}{\sqrt{2\omega_{k_1} Z}} [\bar{u}_{\vec{k}_1\sigma_1}\gamma^0]_{r_1}$$

$$\times \langle f_{2...m}|\mathcal{T}(\psi(y_1)^{r_1}\bar{\psi}(x_1)_{l_1})|i_{2...n}\rangle \tag{3.60}$$

$$= i \int \frac{d^4 y_1 e^{ik_1 y_1}}{\sqrt{2\omega_{k_1}} Z} [\bar{u}_{\vec{k}_1 \sigma_1} \gamma^0 (-i \overleftarrow{\partial}_{y^0} - i\vec{\partial}_{y^0})]_{r_1}$$

$$\times \langle f_{2...m} | T(\psi(y_1)^{r_1} \bar{\psi}(x_1)_{l_1}) | i_{2...n} \rangle \tag{3.61}$$

$$= i \int \frac{d^4 y_1 e^{ik_1 y_1}}{\sqrt{2\omega_{k_1}} Z} [\bar{u}_{\vec{k}_1 \sigma_1} (-i\vec{\gamma} \overleftarrow{\partial} + m - i\gamma^0 \vec{\partial}_t)]_{r_1}$$

$$\times \langle f_{2...m} | T(\psi(y_1)^{r_1} \bar{\psi}(x_1)_{l_1}) | i_{2...n} \rangle \tag{3.62}$$

$$= -i \int \frac{d^4 y_1 e^{ik_1 y_1}}{\sqrt{2\omega_{k_1}} Z} [\bar{u}_{\vec{k}_1 \sigma_1} (i \overrightarrow{\partial}_{y_1} - m)]_{r_1}$$

$$\times \langle f_{2...m} | T(\psi(y_1)^{r_1} \bar{\psi}(x_1)_{l_1}) | i_{2...n} \rangle. \tag{3.63}$$

Repeating this procedure one gets the LSZ reduction formula

$$S_{fi} = (-i)^{m+n} \prod_{i=1}^{m} \int \frac{d^4 y_i e^{ik_i y_i}}{\sqrt{2\omega_{k_i}} Z} [\bar{u}_{\vec{k}_i \sigma_i} (i\hat{\partial}_{y_i} - m)]_{r_i}$$

$$\times \prod_{j=1}^{n} \int \frac{d^4 x_j e^{-ip_j x_j}}{\sqrt{2\omega_{p_j}} Z} [(-i\hat{\partial}_{x_j} - m) u_{\vec{p}_j s_j}]^{l_j}$$

$$\times \langle 0 | T(\psi(y_1)^{r_1}...\psi(y_m)^{r_m} \bar{\psi}(x_1)_{l_1}...\bar{\psi}(x_n)_{l_n}) | 0 \rangle. \tag{3.64}$$

For the Green function in the momentum space

$$\langle 0 | T(\psi(y_1)^{r_1}...\psi(y_m)^{r_m} \bar{\psi}(x_1)_{l_1}...\bar{\psi}(x_n)_{l_n}) | 0 \rangle$$

$$= \prod_{i=1}^{m} \int \frac{d^4 k_i e^{-ik_i y_i}}{(2\pi)^4} \prod_{j=1}^{n} \int \frac{d^4 p_j e^{ip_j x_j}}{(2\pi)^4}$$

$$\times \langle 0 | T(\psi(k_1)^{r_1}...\psi(k_m)^{r_m} \bar{\psi}(p_1)_{l_1}...\bar{\psi}(p_n)_{l_n}) | 0 \rangle, \tag{3.65}$$

one has the LSZ formula in the momentum space

$$S_{fi} = (-i)^{m+n} \prod_{i=1}^{m} \frac{[\bar{u}_{\vec{k}_i \sigma_i} (\hat{k}_i - m)]_{r_i}}{\sqrt{2\omega_{k_i}} Z} \prod_{j=1}^{n} \frac{[(\hat{p}_j - m) u_{\vec{p}_j s_j}]^{l_j}}{\sqrt{2\omega_{p_j}} Z}$$

$$\times \langle 0 | T(\psi(k_1)^{r_1}...\psi(k_m)^{r_m} \bar{\psi}(p_1)_{l_1}...\bar{\psi}(p_n)_{l_n}) | 0 \rangle. \tag{3.66}$$

Therefore the S-matrix element is proportional to the residue of the Green function at the external momenta on the mass shell $\hat{p}_i = m_i$.

EXERCISES

3.1 Prove (3.7) using bracket notation

$$\langle q'|q\rangle = \delta(q - q'), \quad \langle p'|p\rangle = 2\pi\delta(p - p'), \quad \langle p|q\rangle = e^{-ipq}, \tag{P3.1}$$

$$\hat{1} = |q\rangle\langle q| = |p\rangle\langle p|, \quad |q\rangle\langle q| = \int dq |q\rangle\langle q|, \quad |p\rangle\langle p| = \int \frac{dp}{2\pi} |p\rangle\langle p|, \tag{P3.2}$$

where we often omit integrals to simplify notation.

3.2 Calculate the transition amplitude for the harmonic oscillator with the frequency w and mass m.

SOLUTION. Here we follow (Kle09) directly.

$$\langle x_b t_b | x_a t_a \rangle = \int \mathcal{D}x \, e^{iS[x]}, \quad S[x] = \int_{t_a}^{t_b} dt \frac{m}{2}(\dot{x}^2 - w^2 x^2). \quad (P3.3)$$

The classical trajectory satisfies

$$\ddot{x}_{cl} + w^2 x_{cl} = 0, \quad x_{cl}(t_a) = x_a, \quad x_{cl}(t_b) = x_b \quad \Longrightarrow$$

$$x_{cl}(t) = \frac{x_b \sin w(t - t_a) + x_a \sin w(t_b - t)}{\sin w(t_b - t_a)}. \quad (P3.4)$$

The action on the classical trajectory reads

$$S_{cl} = S[x_{cl}] = \frac{m}{2} x_{cl} \dot{x}_{cl} \Big|_{t_a}^{t_b}$$

$$= \frac{mw \left[(x_b^2 + x_a^2) \cos w(t_b - t_a) - 2 x_a x_b \right]}{2 \sin w(t_b - t_a)} \xrightarrow{w \to 0} \frac{m(x_b - x_a)^2}{2(t_b - t_a)}. \quad (P3.5)$$

The real trajectory

$$x = x_{cl} + \delta x, \quad \delta x(t_a) = \delta x(t_b) = 0 \quad \Longrightarrow \quad (P3.6)$$

$$S = S_{cl} + S_{fl}, \quad S_{fl} = \int_{t_a}^{t_b} dt \frac{m}{2}(\delta \dot{x}^2 - w^2 \delta x^2). \quad (P3.7)$$

$$\langle x_b t_b | x_a t_a \rangle = \int \mathcal{D}x \, e^{iS[x]} = e^{iS_{cl}} F_w(t_b, t_a), \quad (P3.8)$$

$$F_w(t_b, t_a) = \int \mathcal{D}\delta x \, e^{iS_{fl}[\delta x]}. \quad (P3.9)$$

One introduces $t_n = t_a + \varepsilon n$, $\varepsilon = \frac{t_b - t_a}{N+1}$, $x_n = x(t_n)$, $\bar{\nabla} x_n = \frac{x_n - x_{n-1}}{\varepsilon}$, $\nabla x_n = \frac{x_{n+1} - x_n}{\varepsilon}$ to get

$$F_w^N(t_b, t_a) = \left(\frac{m}{2\pi i \varepsilon} \right)^{\frac{N+1}{2}} \prod_{n=1}^{N} \int_{-\infty}^{+\infty} d\delta x_n \, e^{\frac{im}{2} \varepsilon \sum_{n,n'=1}^{N} \delta x_n (-\nabla \bar{\nabla} - w^2)_{nn'} \delta x_{n'}}. \quad (P3.10)$$

With $\delta x_0 = \delta x_{N+1} = 0$ one has

$$\sum_{n=0}^{N} (\nabla \delta x_n)^2 = \sum_{n=1}^{N+1} (\bar{\nabla} \delta x_n)^2 = -\sum_{n=1}^{N} \delta x_n \nabla \bar{\nabla} \delta x_n = -\sum_{n=1}^{N} \delta x_n \bar{\nabla} \nabla \delta x_n. \quad (P3.11)$$

Using the Fourier transform

$$x(t) = \int_{-\infty}^{+\infty} dw\, e^{-iwt} x(w), \qquad (\text{P3.12})$$

one gets

$$\nabla x_n = \nabla x(t_n) = \int_{-\infty}^{+\infty} dw\, \frac{e^{-iw\varepsilon} - 1}{\varepsilon} e^{-iwt_n} x(w). \qquad (\text{P3.13})$$

Therefore, on the Fourier components, the operators have the eigenvalues

$$i\nabla x(w) = \Omega x(w), \quad i\bar{\nabla} x(w) = \bar{\Omega} x(w), \quad \Omega = \bar{\Omega}^* = i\frac{e^{-iw\varepsilon} - 1}{\varepsilon}, \qquad (\text{P3.14})$$

$$-\bar{\nabla}\nabla x(w) = -\nabla\bar{\nabla} x(w) = \Omega\bar{\Omega} x(w), \quad \Omega\bar{\Omega} = \frac{2(1 - \cos w\varepsilon)}{\varepsilon^2} > 0. \qquad (\text{P3.15})$$

For the boundary conditions $\delta x(t_a) = \delta x(t_b) = 0$, the Fourier transform reduces to the Fourier series

$$\delta x(t) = \sqrt{\frac{2}{t_b - t_a}} \sum_{m=1}^{\infty} x(\nu_m) \sin \nu_m (t - t_a), \quad \nu_m = \frac{\pi m}{t_b - t_a}. \qquad (\text{P3.16})$$

In the lattice points t_n, this series has only N independent terms and reduces to the discrete sine Fourier transform

$$\delta x(t_n) = \sqrt{\frac{2}{N + 1}} \sum_{m=1}^{N} \delta x(\nu_m) \sin \nu_m (t_n - t_a), \qquad (\text{P3.17})$$

with the orthogonality and completeness relations

$$\frac{2}{N + 1} \sum_{n=1}^{N} \sin \nu_m (t_n - t_a) \sin \nu_{m'} (t_n - t_a) = \delta_{mm'}, \qquad (\text{P3.18})$$

$$\frac{2}{N + 1} \sum_{m=1}^{N} \sin \nu_m (t_n - t_a) \sin \nu_m (t_{n'} - t_a) = \delta_{nn'}. \qquad (\text{P3.19})$$

Then

$$(-\nabla\bar{\nabla} - w^2)_{nn'} \delta x_{n'} = \delta_{nn'} \sqrt{\frac{2}{N + 1}} \sum_{m=1}^{N} (\Omega_m \bar{\Omega}_m - w^2)$$

$$\times\, \delta x(\nu_m) \sin \nu_m (t_n - t_a). \qquad (\text{P3.20})$$

Therefore using (P3.16) and (P3.18) one has for the fluctuation part of the action

$$S_{fl}^N = \frac{m}{2}\varepsilon \sum_{n,n'=1}^{N} \delta x_n (-\nabla\bar{\nabla} - w^2)_{nn'} \delta x_{n'}$$

$$= \frac{m}{2}\varepsilon \sum_{m=1}^{N} \delta x(\nu_m)^2 (\Omega_m \bar{\Omega}_m - w^2). \qquad (P3.21)$$

Changing variables in functional integral (P3.10) one gets the Jacobian

$$\left| \det \frac{\partial \delta x(t_n)}{\partial \delta x(\nu_m)} \right| = \left| \det \sqrt{\frac{2}{N+1}} \sin \nu_m(t_n - t_a) \right| = 1. \qquad (P3.22)$$

It follows from (P3.18), which one can see as $\frac{\partial \delta x(t_n)}{\partial \delta x(\nu_m)} \frac{\partial \delta x(t_{n'})}{\partial \delta x(\nu_m)} = \delta_{nn'}$. Finally,

$$F_w^N(t_b, t_a) = (\frac{m}{2\pi i \varepsilon})^{\frac{N+1}{2}} \prod_{m=1}^{N} \int_{-\infty}^{+\infty} d\delta x(\nu_m) e^{\frac{m}{2} i \varepsilon \sum_{m=1}^{N} \delta x(\nu_m)^2 (\Omega_m \bar{\Omega}_m - w^2)}$$

$$= (\frac{m}{2\pi i \varepsilon})^{\frac{N+1}{2}} \prod_{m=1}^{N} \left(\frac{2\pi}{-i\varepsilon m(\Omega_m \bar{\Omega}_m - w^2)} \right)^{\frac{1}{2}}$$

$$= (\frac{m}{2\pi i \varepsilon})^{\frac{1}{2}} \prod_{m=1}^{N} \frac{1}{\sqrt{\varepsilon^2 (\Omega_m \bar{\Omega}_m - w^2)}}. \qquad (P3.23)$$

For $w = 0$ one uses the relation

$$\prod_{m=1}^{N} 2(1 - \cos \frac{\pi m}{N+1}) = N + 1 \qquad (P3.24)$$

to get

$$F_0(t_b, t_a) = \sqrt{\frac{m}{2\pi i(t_b - t_a)}}. \qquad (P3.25)$$

One can also introduce the **fluctuation determinant**

$$\det(-\varepsilon^2 \nabla\bar{\nabla}) = \prod_{m=1}^{N} \varepsilon^2 \Omega_m \bar{\Omega}_m = \det \begin{pmatrix} 2 & -1 & & \\ -1 & 2 & -1 & \\ & -1 & 2 & -1 \\ & & & \cdots \end{pmatrix}, \qquad (P3.26)$$

which has the property

$$\det_N = 2\det_{N-1} - \det_{N-2}, \quad \det_2 = 3, \ \det_3 = 4 \implies \det_N = N + 1.$$
$$(P3.27)$$

For $w \neq 0$ using

$$\frac{\varepsilon^2(\Omega_m \bar{\Omega}_m - w^2)}{\varepsilon^2 \Omega_m \bar{\Omega}_m} = 1 - \frac{\varepsilon^2 w^2}{4 \sin^2 \frac{\varepsilon^2 \nu_m^2}{2}} \xrightarrow[\varepsilon \to 0]{} 1 - \frac{w^2}{\nu_m^2}$$

$$= 1 - \frac{(t_b - t_a)^2 w^2}{\pi^2 m^2} \tag{P3.28}$$

and the identity

$$\frac{\sin x}{x} = \prod_{m=1}^{\infty} (1 - \frac{x^2}{m^2 \pi^2}), \tag{P3.29}$$

one gets

$$\prod_{m=1}^{\infty} \frac{\varepsilon^2(\Omega_m \bar{\Omega}_m - w^2)}{\varepsilon^2 \Omega_m \bar{\Omega}_m} \xrightarrow[\varepsilon \to 0]{} \frac{\sin w(t_b - t_a)}{w(t_b - t_a)}. \tag{P3.30}$$

Combining it with (P3.27), one gets

$$F_w(t_b, t_a) = \lim_{\varepsilon \to 0} F_w^N(t_b, t_a) = \sqrt{\frac{m}{2\pi i \varepsilon}} \prod_{m=1}^{N} \frac{1}{\sqrt{\varepsilon^2(\Omega_m \bar{\Omega}_m - w^2)}}$$

$$= \sqrt{\frac{m}{2\pi i \varepsilon(N+1)}} \sqrt{\frac{w(t_b - t_a)}{\sin w(t_b - t_a)}}$$

$$= \sqrt{\frac{m}{2\pi i}} \sqrt{\frac{w}{\sin w(t_b - t_a)}}, \tag{P3.31}$$

which has the correct $w \to 0$ limit (P3.25). Finally, the transition amplitude reads

$$\langle x_b t_b | x_a t_a \rangle = \int \mathcal{D}x \, e^{i \int_{t_a}^{t_b} dt \frac{m}{2}(\dot{x}^2 - w^2 x^2)}$$

$$= \left(\frac{m}{2\pi i}\right)^{\frac{d}{2}} \left(\frac{w}{\sin w(t_b - t_a)}\right)^{\frac{d}{2}} e^{i \frac{mw\left[(x_b^2 + x_a^2)\cos w(t_b - t_a) - 2x_a x_b\right]}{2 \sin w(t_b - t_a)}},$$

$$\tag{P3.32}$$

where we generalized the result for the d-dimensional space. In the momentum space

$$\langle p_b t_b | p_a t_a \rangle = \int d^d x_b d^d x_a e^{-i[p_b x_b - p_a x_a]} \langle x_b t_b | x_a t_a \rangle$$

$$= \left(\frac{2\pi}{imw \sin w(t_b - t_a)}\right)^{\frac{d}{2}} e^{i \frac{(p_b^2 + p_a^2)\cos w(t_b - t_a) - 2p_a p_b}{2mw \sin w(t_b - t_a)}}. \tag{P3.33}$$

3.3 Calculate the transition amplitude for the harmonic oscillator with the time-dependent frequency $\Omega(t)$ and mass m.

SOLUTION. Here we again follow (Kle09). Repeating the steps described in the solution to the previous problem, one gets

$$\langle x_b t_b | x_a t_a \rangle = \int \mathcal{D}x \, e^{iS[x]} = e^{iS_{cl}} F_\Omega(t_b, t_a), \tag{P3.34}$$

$$S[x] = \int_{t_a}^{t_b} dt \frac{m}{2} (\dot{x}^2 - \Omega(t)^2 x^2), \tag{P3.35}$$

$$F_\Omega(t_b, t_a) = \int \mathcal{D}\delta x \, e^{iS[\delta x]}, \tag{P3.36}$$

with the discretized approximation for the functional integral

$$F_\Omega^N(t_b, t_a) = \sqrt{\frac{m}{2\pi i \varepsilon}} \frac{1}{\sqrt{\det(-\varepsilon^2 \nabla \bar{\nabla})}} \left[\frac{\det(-\varepsilon^2 (\nabla \bar{\nabla} + \Omega(t)^2))}{\det(-\varepsilon^2 \nabla \bar{\nabla})} \right]^{-\frac{1}{2}}$$

$$= \sqrt{\frac{m}{2\pi i (t_b - t_a)}} \left[\frac{\det(-\varepsilon^2 (\nabla \bar{\nabla} + \Omega(t)^2))}{\det(-\varepsilon^2 \nabla \bar{\nabla})} \right]^{-\frac{1}{2}}, \tag{P3.37}$$

$$\Omega(t)^2 = \begin{pmatrix} \Omega_N^2 & & & \\ & \Omega_{N-1}^2 & & \\ & & \ddots & \\ & & & \Omega_1^2 \end{pmatrix}, \quad \Omega_i^2 = \Omega(t_i)^2. \tag{P3.38}$$

Then the determinant

$$D_N = \det{}_N(-\varepsilon^2 \nabla \bar{\nabla} - \Omega^2)$$

$$= \det \begin{pmatrix} 2 - \varepsilon^2 \Omega_N^2 & -1 & & \\ -1 & 2 - \varepsilon^2 \Omega_{N-1}^2 & -1 & \\ & -1 & \ddots & -1 \\ & & -1 & 2 - \varepsilon^2 \Omega_1^2 \end{pmatrix} \tag{P3.39}$$

$$= (2 - \varepsilon^2 \Omega_N^2) D_{N-1} - D_{N-2}. \tag{P3.40}$$

Changing $N \to N+1$ the latter equation can be written as the **Gelfand-Yaglom formula**

$$\nabla \bar{\nabla} D_N + \Omega_{N+1}^2 D_N = 0, \tag{P3.41}$$

with the initial conditions

$$D_1 = 2 - \varepsilon^2 \Omega_1^2, \quad D_2 = (2 - \varepsilon^2 \Omega_1^2)(2 - \varepsilon^2 \Omega_2^2) - 1. \tag{P3.42}$$

In the continuous limit $\nabla D_1 \to \frac{1}{\varepsilon}$. Therefore, it is reasonable to define D_{ren} :

$$\varepsilon D_i \to D_{ren}(t_i). \tag{P3.43}$$

It satisfies

$$(\partial_t^2 + \Omega^2(t))D_{ren}(t) = 0, \quad D_{ren}(t_a) = 0, \quad \dot{D}_{ren}(t_a) = 1. \quad \text{(P3.44)}$$

In general, one has to solve this initial value problem for the specific $\Omega(t)$ and use the solution to calculate the tunctional integral via (P3.27)

$$F_\Omega(t_b, t_a) = \sqrt{\frac{m}{2\pi i(t_b - t_a)}} \left[\frac{D_{ren}(t_b)}{\varepsilon(N+1)}\right]^{-\frac{1}{2}} = \sqrt{\frac{m}{2\pi i}} \frac{1}{\sqrt{D_{ren}(t_b)}}. \quad \text{(P3.45)}$$

However, one can write the general solution for this determinant via the classical trajectory. To this end one writes the solution for (P3.44) through 2 arbitrary independent solutions ξ and η of this equation using their Wronskian

$$D_{ren}(t) = \frac{\xi(t_a)\eta(t) - \xi(t)\eta(t_a)}{W}, \quad \text{(P3.46)}$$

$$W = \xi(t_a)\dot{\eta}(t_a) - \dot{\xi}(t_a)\eta(t_a) = const. \quad \text{(P3.47)}$$

One needs

$$D_{ren}(t_b) = \frac{\xi(t_a)\eta(t_b) - \xi(t_b)\eta(t_a)}{W}, \quad \text{(P3.48)}$$

which one can also find as $\tilde{D}_{ren}(t_a)$ for

$$\tilde{D}_{ren}(t) = \frac{\xi(t)\eta(t_b) - \xi(t_b)\eta(t)}{W}. \quad \text{(P3.49)}$$

Introducing the notation

$$D_a(t) = D_{ren}(t), \quad D_b(t) = \tilde{D}_{ren}(t), \quad \text{(P3.50)}$$

one has

$$(\partial_t^2 + \Omega^2(t))D_a(t) = 0, \quad D_a(t_a) = 0, \quad \dot{D}_a(t_a) = 1, \quad \text{(P3.51)}$$

$$(\partial_t^2 + \Omega^2(t))D_b(t) = 0, \quad D_b(t_b) = 0, \quad \dot{D}_b(t_b) = -1, \quad \text{(P3.52)}$$

$$D_{ren}(t_b) = D_a(t_b) = D_b(t_a). \quad \text{(P3.53)}$$

The solution to

$$(\partial_t^2 + \Omega^2(t))x(t) = 0, \quad x(t_b) = x_b, \quad x(t_a) = x_a, \quad \text{(P3.54)}$$

which is the equation for the classical trajectory minimizing the action, reads

$$x(t) = \frac{D_b(t)}{D_b(t_a)}x_a + \frac{D_a(t)}{D_a(t_b)}x_b \quad \Longrightarrow \quad \text{(P3.55)}$$

$$\frac{\partial \dot{x}(t_a)}{\partial x_b} = \frac{1}{D_a(t_b)}, \quad \frac{\partial \dot{x}(t_b)}{\partial x_a} = \frac{-1}{D_b(t_a)}, \quad \text{(P3.56)}$$

$$D_{ren}(t_b) = \frac{\partial x_b}{\partial \dot{x}_a} = -\frac{\partial x_a}{\partial \dot{x}_b}. \quad \text{(P3.57)}$$

The action on the classical trajectory reads

$$S_{cl} = \frac{m}{2}x\dot{x}\Big|_{t_a}^{t_b} = \frac{m}{2D_{ren}(t_b)}\left[\dot{D}_a(t_b)x_b^2 - \dot{D}_b(t_a)x_a^2 - 2x_ax_b\right]. \quad \text{(P3.58)}$$

Hence

$$\frac{\partial^2 S_{cl}}{\partial x_a \partial x_b} = \frac{-m}{D_{ren}(t_b)} \quad \Longrightarrow \quad D_{ren}(t_b) = -m\left(\frac{\partial^2 S_{cl}}{\partial x_a \partial x_b}\right)^{-1}. \quad \text{(P3.59)}$$

Finally, we get the transition amplitude

$$\langle x_b t_b | x_a t_a \rangle = \int \mathcal{D}x\, e^{iS[x]} = \sqrt{\frac{m}{2\pi i}}\frac{e^{iS_{cl}}}{\sqrt{D_{ren}(t_b)}}, \quad \text{(P3.60)}$$

$$D_{ren}(t_b) = -m\left(\frac{\partial^2 S_{cl}}{\partial x_a \partial x_b}\right)^{-1} = \frac{\partial x_b}{\partial \dot{x}_a} = -\frac{\partial x_a}{\partial \dot{x}_b}. \quad \text{(P3.61)}$$

One can check that for $\Omega = const$ this result reproduces transition amplitude (P3.32) for the oscillator with the constant frequency.

3.4 Calculate $\frac{\det(-\varepsilon^2(\nabla\bar{\nabla}+w^2))}{\det(-\varepsilon^2\nabla\bar{\nabla})}$ with $w = const$ using the Gelfand-Yaglom formula.

SOLUTION. $\varepsilon D_N = \varepsilon \det_N(-\varepsilon^2\nabla\bar{\nabla} - w^2) \to D_{ren}(t)$, which satisfies the equation

$$(\partial_t^2 + w^2)D_{ren}(t) = 0, \quad D_{ren}(t_a) = 0, \quad \dot{D}_{ren}(t_a) = 1. \quad \text{(P3.62)}$$

Its solution reads

$$D_{ren}(t_b) = \frac{\sin w(t_b - t_a)}{w} \underset{w\to 0}{\to} (t_b - t_a) \quad \Longrightarrow \quad \text{(P3.63)}$$

$$\frac{\det(-\varepsilon^2(\nabla\bar{\nabla} + w^2))}{\det(-\varepsilon^2\nabla\bar{\nabla})} \underset{\varepsilon\to 0}{\to} \frac{\sin w(t_b - t_a)}{w(t_b - t_a)}, \quad \text{(P3.64)}$$

which coincides with (P3.30).

3.5 Find $D_{ren}(t_b)$ from the solution to the following problems

$$(\partial_t^2 + \Omega^2(t))x(t) = 0, \quad x(t_a) = x_a, \quad \dot{x}(t_a) = \dot{x}_a, \quad \text{(P3.65)}$$
$$(\partial_t^2 + \Omega^2(t))x(t) = 0, \quad x(t_b) = x_b, \quad \dot{x}(t_b) = \dot{x}_b. \quad \text{(P3.66)}$$

SOLUTION. The solution to the initial value problem can be expressed via $D_{a,b}$ as

$$x(t) = \frac{D_b(t) - D_a(t)\dot{D}_b(t_a)}{D_b(t_a)}x_a + D_a(t)\dot{x}_a \quad \Longrightarrow \quad \text{(P3.67)}$$

$$D_a(t) = \frac{\partial x(t)}{\partial \dot{x}_a} \quad \Longrightarrow \quad D_{ren}(t_b) = \frac{\partial x_b}{\partial \dot{x}_a}. \quad \text{(P3.68)}$$

Analogously, the solution to (P3.66) reads

$$x(t) = \frac{D_a(t) + D_b(t)\dot{D}_a(t_b)}{D_a(t_b)} x_b - D_b(t)\dot{x}_b \qquad \Longrightarrow \qquad \text{(P3.69)}$$

$$D_{ren}(t_b) = -\frac{\partial x_a}{\partial \dot{x}_b}. \qquad \text{(P3.70)}$$

3.6 Show that the functional integral

$$\int_{z(0)=y,\, z(\tau)=x} \mathcal{D}z_\mu(t) e^{-\frac{i}{2}\int_0^\tau dt\, \dot{z}_\mu^2(t)}$$

$$\stackrel{def}{=} \lim_{N\to\infty} \frac{i}{(2\pi i\varepsilon)^{\frac{d}{2}}} \prod_{i=1}^{N-1} \int \frac{i d^d z_i}{(2\pi i\varepsilon)^{\frac{d}{2}}} e^{-\frac{i}{2}\sum_{i=1}^N \frac{(z_i - z_{i-1})^2}{2\varepsilon}}$$

$$= \frac{i}{(2\pi i\tau)^{\frac{d}{2}}} e^{-\frac{i(x-y)^2}{2\tau}}, \qquad \text{(P3.71)}$$

where

$$\varepsilon = \frac{\tau}{N}, \quad z_0 = y, \quad z_N = x, \quad z_j = z(j\varepsilon). \qquad \text{(P3.72)}$$

SOLUTION. Using

$$\int \frac{i d^d z}{(2\pi i)^{\frac{d}{2}}} e^{-\frac{i}{2a}(x-z)^2 - \frac{i}{2b}(y-z)^2} = \frac{i(2\pi i)^{\frac{d-1}{2}}}{(2\pi i)^{\frac{d}{2}}} \left(\frac{2\pi}{i}\right)^{\frac{1}{2}} \left(\frac{ab}{a+b}\right)^{\frac{d}{2}} e^{-\frac{i(x-y)^2}{2(a+b)}}$$

$$= \left(\frac{ab}{a+b}\right)^{\frac{d}{2}} e^{-\frac{i(x-y)^2}{2(a+b)}}, \qquad \text{(P3.73)}$$

one gets

$$\int \mathcal{D}z_\mu(t) e^{-\frac{i}{2}\int_0^\tau dt\, \dot{z}_\mu^2(t)} = \frac{i}{(2\pi i\varepsilon)^{\frac{d}{2}}} \prod_{i=1}^{N-1} \int \frac{i d^d z_i}{(2\pi i\varepsilon)^{\frac{d}{2}}} e^{-\frac{i}{2}\sum_{i=1}^N \frac{(z_i - z_{i-1})^2}{2\varepsilon}}$$

$$= \frac{i}{(2\pi i\varepsilon)^{\frac{d}{2}} \varepsilon^{(N-1)\frac{d}{2}}} \left(\frac{\varepsilon^2}{2\varepsilon}\right)^{\frac{d}{2}} \left(\frac{2\varepsilon^2}{3\varepsilon}\right)^{\frac{d}{2}} \cdots \left(\frac{(N-1)\varepsilon^2}{N\varepsilon}\right)^{\frac{d}{2}}$$

$$= \frac{i}{(2\pi i\varepsilon N)^{\frac{d}{2}}} = \frac{i}{(2\pi i\tau)^{\frac{d}{2}}}. \qquad \text{(P3.74)}$$

3.7 Derive the free quark propagator as a functional integral over all paths.

SOLUTION.

$$D(x-y) = \langle x| \frac{i}{\hat{p} - m + i0} |y\rangle = \langle x| \frac{i(\hat{p}+m)}{p^2 - m^2 + i0} |y\rangle$$

$$= \frac{1}{2} \int_0^{+\infty} d\tau \langle x|(\hat{p}+m) e^{i\frac{\tau}{2}(p^2 - m^2 + i0)} |y\rangle. \qquad \text{(P3.75)}$$

Introducing $\varepsilon = \frac{\tau}{N}$,

$$\langle x|(\hat{p}+m)e^{i\frac{\tau}{2}p^2}|y\rangle = \int d^d z_1 \dots d^d z_{N-1}\langle x|(\hat{p}+m)e^{i\frac{\varepsilon}{2}p^2}|z_{N-1}\rangle$$

$$\times \prod_{n=1}^{N-1}\langle z_n|e^{i\frac{\varepsilon}{2}p^2}|z_{n-1}\rangle, \qquad (P3.76)$$

$$\langle z_n|e^{i\frac{\varepsilon}{2}p^2}|z_{n-1}\rangle = \langle z_n|p_n\rangle\langle p_n|e^{i\frac{\varepsilon}{2}p^2}|p'_n\rangle\langle p'_n|z_{n-1}\rangle$$

$$= \int \frac{d^d p_n}{(2\pi)^d}e^{-ip_n(z_n-z_{n-1})}e^{i\frac{\varepsilon}{2}p_n^2}$$

$$= \frac{i}{(2\pi\varepsilon i)^{\frac{d}{2}}}e^{-\frac{i}{2\varepsilon}(z_n-z_{n-1})^2}. \qquad (P3.77)$$

It is important to stress here that in the d-dimensional space the scalar product is defined as $px = p^0 x^0 - \vec{p}\vec{x}$. Hence

$$\langle p|x\rangle = e^{ipx} = e^{ip^0 x^0 - i\vec{p}\vec{x}}, \qquad (P3.78)$$

which is consistent with (P3.1), where we had only space coordinates.

$$\langle x|(\hat{p}+m)e^{i\frac{\varepsilon}{2}p^2}|z_{N-1}\rangle = \langle x|p_n\rangle\langle p_n|(\hat{p}+m)e^{i\frac{\varepsilon}{2}p^2}|p'_n\rangle\langle p'_n|z_{N-1}\rangle$$

$$= \int \frac{d^d p_n}{(2\pi)^d}e^{-ip_n(x-z_{N-1})}(\hat{p}_n+m)e^{i\frac{\varepsilon}{2}p_n^2}$$

$$= \int \frac{d^d p_n}{(2\pi)^d}(i\gamma^\mu\frac{\partial}{\partial x^\mu}+m)e^{-ip_n(x-z_{N-1})}e^{i\frac{\varepsilon}{2}p_n^2}$$

$$= (i\gamma^\mu\frac{\partial}{\partial x^\mu}+m)\frac{i}{(2\pi\varepsilon i)^{\frac{d}{2}}}e^{-\frac{i}{2\varepsilon}(x-z_{N-1})^2}$$

$$\qquad (P3.79)$$

$$= (\gamma^\mu\frac{(x-z_{N-1})_\mu}{\varepsilon}+m)\frac{i}{(2\pi\varepsilon i)^{\frac{d}{2}}}e^{-\frac{i}{2\varepsilon}(x-z_{N-1})^2}.$$

$$\qquad (P3.80)$$

Therefore

$$D(x-y) = \frac{1}{2}\int_0^{+\infty} d\tau(i\gamma^\mu\frac{\partial}{\partial x^\mu}+m)e^{i\frac{\tau}{2}(-m^2+i0)}$$

$$\times \int d^d z_1 \dots d^d z_{N-1}\prod_{n=1}^{N}\frac{i}{(2\pi\varepsilon i)^{\frac{d}{2}}}e^{-\frac{i}{2\varepsilon}(z_n-z_{n-1})^2}. \qquad (P3.81)$$

In the limit $N \to \infty$, one can write the propagator as the functional integral

$$D(x-y) = \frac{1}{2}\int_0^{+\infty} d\tau\, e^{i\frac{\tau}{2}(-m^2+i0)}$$

$$\times \int_{z(0)=y,\,z(\tau)=x} \mathcal{D}z_\mu(t)(\gamma^\mu \dot{z}(\tau)_\mu + m)e^{-\frac{i}{2}\int_0^\tau \dot{z}^2(t)dt}. \quad \text{(P3.82)}$$

One can calculate this integral straightforwardly via (P3.71)

$$D(x-y) = \frac{1}{2}\int_0^{+\infty} d\tau (i\gamma^\mu \frac{\partial}{\partial x^\mu} + m)\frac{i}{(2\pi i \tau)^{\frac{d}{2}}}e^{-\frac{i(x-y)^2}{2\tau}}e^{i\frac{\tau}{2}(-m^2+i0)}. \quad \text{(P3.83)}$$

In the massless limit, we met this integral in (P2.103)

$$D(x-y) = -\gamma^\mu \frac{\partial}{\partial x^\mu}\frac{1}{2}\int_0^{+\infty}\frac{d\tau}{(2\pi i \tau)^{\frac{d}{2}}}e^{-\frac{i(x-y)^2}{2\tau}}$$

$$= -\gamma^\mu \frac{\partial}{\partial x^\mu}\frac{i}{4\pi^2(x-y)^2} = \frac{i(\hat{x}-\hat{y})}{2\pi^2(x-y)^4}. \quad \text{(P3.84)}$$

3.8 Show that

$$(i\hat{D})^2 = (iD)^2 + \frac{1}{2}g\sigma F. \quad \text{(P3.85)}$$

SOLUTION.

$$(i\hat{D})^2 = -D_\mu D_\nu \gamma^\mu \gamma^\nu = -D_\mu D_\nu \left(\frac{\{\gamma^\mu \gamma^\nu\} + [\gamma^\mu \gamma^\nu]}{2}\right)$$

$$= -D_\mu D_\nu [g^{\mu\nu} - i\sigma^{\mu\nu}] = (iD)^2 + i\sigma^{\mu\nu}D_\mu D_\nu$$

$$= (iD)^2 + i\sigma^{\mu\nu}\left(\frac{\{D_\mu D_\nu\} + [D_\mu D_\nu]}{2}\right)$$

$$= (iD)^2 + \frac{i}{2}\sigma^{\mu\nu}[D_\mu D_\nu] = (iD)^2 + \frac{1}{2}g\sigma^{\mu\nu}F^a_{\mu\nu}t^a$$

$$= (iD)^2 + \frac{1}{2}g\sigma F. \quad \text{(P3.86)}$$

Here we used

$$\sigma_{\mu\nu} = \frac{i}{2}[\gamma_\mu \gamma_\nu]. \quad \text{(P3.87)}$$

3.9 Derive the quark propagator in the external classical field A as a functional integral over all paths.

SOLUTION.

$$D(x-y) = \langle x|\frac{i}{i\hat{D}-m+i0}|y\rangle = \langle x|\frac{i}{i\hat{\partial}+g\hat{A}-m+i0}|y\rangle$$

$$= (i\gamma^\mu \frac{\partial}{\partial x^\mu} + g\hat{A}(x) + m)\langle x|\frac{i}{(iD)^2+\frac{1}{2}g\sigma F-m^2+i0}|y\rangle$$

$$= (i\gamma^\mu \frac{\partial}{\partial x^\mu} + g\hat{A}(x) + m)\frac{1}{2}\int_0^{+\infty} d\tau \langle x|e^{i\frac{\tau}{2}((iD)^2+\frac{1}{2}g\sigma F-m^2)}|y\rangle.$$

$$\text{(P3.88)}$$

Introducing $\varepsilon = \frac{\tau}{N}$, $z_0 = y$, $z_N = x$, $z_j = z(j\varepsilon)$, $A_j = A(z_j)$, and $z_{ik} = z_i - z_k$,

$$\langle x|e^{i\frac{\tau}{2}((iD)^2+\frac{1}{2}g\sigma F)}|y\rangle = \int d^d z_1 \ldots d^d z_{N-1} \prod_{n=1}^{N} \langle z_n|e^{i\frac{\varepsilon}{2}((iD)^2+\frac{1}{2}g\sigma F)}|z_{n-1}\rangle.$$

$$\text{(P3.89)}$$

Here

$$\langle z_n|e^{i\frac{\varepsilon}{2}((iD)^2+\frac{1}{2}g\sigma F)}|z_{n-1}\rangle = \langle z_n|e^{i\frac{\varepsilon}{2}(p^2+gpA+gAp+g^2A^2)}|z_{n-1}\rangle$$
$$\times e^{i\frac{\varepsilon}{4}g\sigma F(z_{n-1})} + O(\varepsilon^2), \qquad \text{(P3.90)}$$

and

$$\langle z_n|e^{i\frac{\varepsilon}{2}(p^2+gpA+gAp+g^2A^2)}|z_{n-1}\rangle$$

$$= \langle z_n|p_n\rangle\langle p_n|e^{i\frac{\varepsilon}{2}(p^2+gpA+gAp+g^2A^2)}|p_n'\rangle\langle p_n'|z_{n-1}\rangle$$

$$= \langle z_n|p_n\rangle\langle p_n|1+i\frac{\varepsilon}{2}(p^2+gpA+gAp+g^2A^2)|p_n'\rangle\langle p_n'|z_{n-1}\rangle + O(\varepsilon^2)$$

$$= \int \frac{dp_n}{(2\pi)^d} e^{-ip_n z_{nn-1}} \left[1 + i\frac{\varepsilon}{2}\left(p_n^2 + p_n g(A_{n-1}+A_n)\right.\right.$$
$$\left.\left. + g^2(\frac{A_{n-1}+A_n}{2})^2\right)\right] + O(\varepsilon^2)$$

$$= \int \frac{dp_n}{(2\pi)^d} \exp\left[-ip_n z_{nn-1} + i\frac{\varepsilon}{2}\left(p_n + g\frac{A_{n-1}+A_n}{2}\right)^2\right] + O(\varepsilon^2)$$

$$= \left|\tilde{p} = p_n + g\frac{A_{n-1}+A_n}{2} - \frac{z_{nn-1}}{\varepsilon}\right|$$

$$= \int \frac{d\tilde{p}}{(2\pi)^d} \exp\left[i\frac{\varepsilon}{2}\tilde{p}^2 - (\frac{z_{nn-1}}{\varepsilon})^2 + 2g\frac{A_{n-1}+A_n}{2}\frac{z_{nn-1}}{\varepsilon}\right] + O(\varepsilon^2)$$

$$= \frac{i}{(2\pi\varepsilon i)^{\frac{d}{2}}} \exp\left[-i\frac{\varepsilon}{2}(\frac{z_{nn-1}}{\varepsilon})^2 + i\varepsilon g\frac{A_{n-1}+A_n}{2}\frac{z_{nn-1}}{\varepsilon}\right] + O(\varepsilon^2).$$

$$\text{(P3.91)}$$

Therefore

$$
D(x - y) = (i\gamma^\mu \frac{\partial}{\partial x^\mu} + g\hat{A}(x) + m)\frac{1}{2}\int_0^{+\infty} d\tau e^{i\frac{\tau}{2}(-m^2 + i0)}
$$

$$
\times \frac{i^N}{(2\pi\varepsilon i)^{\frac{d}{2}N}}\int d^d z_1 \ldots d^d z_{N-1}
$$

$$
\times \exp\left[-i\frac{\varepsilon}{2}(\frac{x - z_{N-1}}{\varepsilon})^2 + i\varepsilon\, g\frac{A_{N-1} + A(x)}{2}\frac{x - z_{N-1}}{\varepsilon}\right]
$$

$$
\times \exp\left[i\frac{\varepsilon}{4}g\sigma F(z_{N-1})\right] \times \ldots
$$

$$
\times \exp\left[-i\frac{\varepsilon}{2}(\frac{z_{10}}{\varepsilon})^2 + i\varepsilon\, g\frac{A_0 + A_1}{2}\frac{z_{10}}{\varepsilon}\right]
$$

$$
\times \exp\left[i\frac{\varepsilon}{4}g\sigma F(y)\right] + O(\varepsilon). \tag{P3.92}
$$

In the limit $N \to \infty$

$$
D(x - y) = (i\gamma^\mu \frac{\partial}{\partial x^\mu} + g\hat{A}(x) + m)\frac{1}{2}\int_0^{+\infty} d\tau e^{i\frac{\tau}{2}(-m^2 + i0)}
$$

$$
\times \int_{z(0) = y, z(\tau) = x} \mathcal{D}z(t)\, e^{-\frac{i}{2}\int_0^\tau dt \dot{z}^2(t)} \mathcal{P} e^{ig\int_0^\tau dt[A(z(t))_\mu \dot{z}(t)^\mu + \frac{1}{4}\sigma F(z(t))]}.
$$

$$
\tag{P3.93}
$$

3.10 Show that for a complex scalar field with $\mathcal{L} = |\partial\phi|^2 - m\,|\phi|^2$

$$
\int \mathcal{D}\phi \mathcal{D}\phi^* e^{i\int \mathcal{L}[\phi]dx} = \frac{const}{\det(\partial^2 + m^2)}. \tag{P3.94}
$$

3.11 Show that the change of variables in the Grassmann integral reads

$$
\int d\alpha_1 \ldots d\alpha_n = \int d\beta_1 \ldots d\beta_n \det \frac{\partial\beta}{\partial\alpha}. \tag{P3.95}
$$

Note that the Jacobian for the Grassmann variables is in the inverse power compared to the Jacobian for the commuting variables.

SOLUTION. We will demonstrate the variable change in the integral over Grassmann variables on the example of 2 variables. For the arbitrary

$$
f(\alpha_1, \alpha_2) = f_0 + f_1\alpha_1 + f_2\alpha_2 + f_{12}\alpha_1\alpha_2, \tag{P3.96}
$$

and the substitution

$$
\begin{cases} \alpha_1 = M_{11}\beta_1 + M_{12}\beta_2 \\ \alpha_2 = M_{21}\beta_1 + M_{22}\beta_2 \end{cases}, \tag{P3.97}
$$

$$f_{12} = \int d\alpha_1 d\alpha_2 f(\alpha_1, \alpha_2) = \int d\beta_1 d\beta_2 f(M_{1i}\beta_i, M_{2j}\beta_j) J$$

$$= \int d\beta_1 d\beta_2 (f_0 + f_1 M_{1i}\beta_i + f_2 M_{2j}\beta_j + f_{12} M_{1i}\beta_i M_{2j}\beta_j) J$$

$$= f_{12}(M_{11}M_{22} - M_{12}M_{21}) J = f_{12} \det M \, J. \qquad \text{(P3.98)}$$

Therefore

$$J = (\det M)^{-1} = \det M^{-1} = \det \frac{\partial \beta}{\partial \alpha}, \qquad \text{(P3.99)}$$

$$\int d\alpha_1 d\alpha_2 f(\alpha_1, \alpha_2) = \int d\beta_1 d\beta_2 f(\beta_1, \beta_2) \det \frac{\partial \beta}{\partial \alpha}. \qquad \text{(P3.100)}$$

The generalization of this formula for the n-fold integral is straightforward. Therefore the variable change law in the functional integral for bosonic fields

$$\int \mathcal{D}\phi \mathcal{D}\phi^* = \int \mathcal{D}\phi' \mathcal{D}\phi'^* \det(\frac{\delta\phi}{\delta\phi'}) \det(\frac{\delta\phi^*}{\delta\phi'^*}), \qquad \text{(P3.101)}$$

while for fermions

$$\int \mathcal{D}\psi \mathcal{D}\bar{\psi} = \int \mathcal{D}\psi' \mathcal{D}\bar{\psi}' \det(\frac{\delta\psi'}{\delta\psi}) \det(\frac{\delta\bar{\psi}'}{\delta\bar{\psi}}). \qquad \text{(P3.102)}$$

3.12 Show that

$$\frac{\delta S[\phi]}{\delta\phi_r} = \frac{\partial \mathcal{L}}{\partial \phi_r} - \partial_\mu \frac{\partial \mathcal{L}}{\partial(\partial_\mu \phi_r)}. \qquad \text{(P3.103)}$$

3.13 Show that in classical limit $\hbar \to 0$ Dyson-Schwinger equations go to the classical Lagrange equations of motion derived from the extremality of the action.

SOLUTION. The inverse Plank constant is the coefficient of the action in the generating functional. Therefore in SD equation (3.50) the l. h. s. is proportional to $\hbar$ and vanishes in the classical limit. Since this equation is valid for any Green function with any number of fields, one comes to classic equations of motion (P3.103)

$$\frac{\delta S[\phi]}{\delta\phi_r} = 0. \qquad \text{(P3.104)}$$

3.14 Show that the topological charge (P2.68) is gauge invariant under local gauge transformations

$$A_\mu^a \to A_\mu^{\theta a} = A_\mu^a + \frac{1}{g}(\partial_\mu \delta^{ac} - ig T_{ac}^b A_\mu^b)\theta^c = A_\mu^a + \frac{1}{g} D^{ac} \theta^c, \qquad \text{(P3.105)}$$

with the infinitesimal θ vanishing at infinity.

SOLUTION.

$$\frac{\delta F_{\mu\nu}^a (x)}{\delta A_\rho^d (y)} = \frac{\delta[(\partial_\mu A_\nu^a) - (\partial_\nu A_\mu^a) + g f^{abc} A_\mu^b A_\nu^c]}{\delta A_\rho^d (y)}$$

$$= [\delta^{ad} g_\nu^\rho \partial_\mu - \delta^{ad} g_\mu^\rho \partial_\nu + g f^{adc} g_\mu^\rho A_\nu^c + g f^{abd} g_\nu^\rho A_\mu^b] \delta(x - y)$$

$$= [g_\nu^\rho (\delta^{ad} \partial_\mu + g f^{abd} A_\mu^b) - g_\mu^\rho (\delta^{ad} \partial_\nu + g f^{acd} A_\nu^c)] \delta(x - y)$$

$$= [g_\nu^\rho D_\mu^{ad} - g_\mu^\rho D_\nu^{ad}] \delta(x - y). \tag{P3.106}$$

Therefore via (P2.6)

$$\frac{\delta \nu[A]}{\delta A_\rho^d (y)} = \frac{g^2}{64\pi^2} 2 \int \varepsilon^{\mu\nu\alpha\beta} F_{\alpha\beta}^a [g_\nu^\rho D_\mu^{ad} - g_\mu^\rho D_\nu^{ad}] \delta(x - y) d^4 x$$

$$= -\frac{g^2}{64\pi^2} 4 \varepsilon^{\mu\rho\alpha\beta} D_\mu^{da} F_{\alpha\beta}^a = -\frac{g^2}{16\pi^2} \varepsilon^{\mu\rho\alpha\beta} D_\mu^{da} F_{\alpha\beta}^a \equiv 0. \tag{P3.107}$$

As a result the topological charge a) is invariant under any local transformations including gauge ones b) does not contribute to the equations of motion, i.e. Euler-Lagrange equations.

3.15 Consider the action for a scalar field invariant under transformation $\phi_x \to \phi_x + \varepsilon \phi_x$ with constant ε. Derive the corresponding Schwinger-Dyson equations assuming that the integration measure in the functional integral is invariant under $\phi_x \to \phi_x + \varepsilon_x \phi_x$.

SOLUTION. Let us consider e.g. the 3-point correlator

$$\langle \phi_x \phi_y \phi_z \rangle = \frac{1}{Z_0} \int \mathcal{D}\phi \, \phi_x \phi_y \phi_z \, e^{iS[\phi]}. \tag{P3.108}$$

and do an infinitesimal coordinate dependent change

$$\phi_x = \phi_x' + \varepsilon_x \phi_x'. \tag{P3.109}$$

The correlator does not change. Hence

$$0 = \frac{1}{Z_0} \int \mathcal{D}\phi' \, (\phi_x' + \varepsilon_x \phi_x')(\phi_y' + \varepsilon_y \phi_y')(\phi_z' + \varepsilon_z \phi_z') \, e^{i \int dx' \mathcal{L}(\phi_{x'}' + \varepsilon_{x'} \phi_{x'}')}$$

$$- \frac{1}{Z_0} \int \mathcal{D}\phi \, \phi_x \phi_y \phi_z \, e^{iS[\phi]}, \tag{P3.110}$$

Since action is invariant under this transformation with constant ε, the Lagrangian changes to a full derivative

$$\mathcal{L}(\phi_{x'}' + \varepsilon \phi_{x'}') \simeq \mathcal{L}(\phi_{x'}') + \varepsilon \partial_\mu J^\mu. \tag{P3.111}$$

For a position dependent ε_x one gets an extra term with $\partial_\mu \varepsilon_{x'}$

$$\int dx' \mathcal{L}(\phi_{x'}' + \varepsilon_{x'} \phi_{x'}') \simeq \int dx' \left(\mathcal{L}(\phi_{x'}') + \varepsilon_{x'} \partial_\mu J^\mu + \frac{\partial \mathcal{L}}{\partial \partial_\mu \phi_{x'}'} \partial_\mu (\varepsilon_{x'}) \phi_{x'}' \right). \tag{P3.112}$$

Then

$$\frac{\delta}{\delta\varepsilon_x}\int dx'\mathcal{L}(\phi'_{x'}+\varepsilon_{x'}\phi'_{x'})=-\partial_{\mu(x)}j^\mu(x)=-\partial_\mu\left(\frac{\partial\mathcal{L}}{\partial\partial_\mu\phi'_x}\phi'_x-J^\mu(x)\right).$$
(P3.113)

Here j^μ is a classical Noether current associated with the symmetry. One gets

$$0=\frac{1}{Z_0}\int\mathcal{D}\phi'\ \phi'_x\phi'_y\phi'_z\ e^{i\int dx'\mathcal{L}(\phi'_{x'}+\varepsilon_{x'}\phi'_{x'})}$$
$$\times\left\{\varepsilon_x+\varepsilon_y+\varepsilon_z+i\int dr\left(\varepsilon_r\partial_\mu J^\mu+\phi'_r\frac{\partial\mathcal{L}}{\partial\partial_\mu\phi'_r}\partial_\mu\varepsilon_r\right)\right\},$$ (P3.114)

$$0=\frac{1}{Z_0}\int\mathcal{D}\phi'\ \phi'_x\phi'_y\phi'_z\ e^{i\int dx'\mathcal{L}(\phi'_{x'}+\varepsilon_{x'}\phi'_{x'})}$$
$$\times\left\{\varepsilon_x+\varepsilon_y+\varepsilon_z-i\int dr\left(\varepsilon_r\partial_\mu j^\mu(r)\right)\right\},$$ (P3.115)

$$0=\frac{1}{Z_0}\int\mathcal{D}\phi'\ \phi'_x\phi'_y\phi'_z\int dr\varepsilon_r\ e^{i\int dx'\mathcal{L}(\phi'_{x'}+\varepsilon_{x'}\phi'_{x'})}$$
$$\times\{\delta(x-r)+\delta(y-r)+\delta(z-r)-i\partial_\mu j^\mu(r)\}.$$ (P3.116)

Since ε_r is arbitrary one has

$$\langle\phi_x\phi_y\phi_z\partial_\mu j^\mu(r)\rangle=-i\langle\phi_x\phi_y\phi_z(\delta(r-x)+\delta(r-y)+\delta(r-z))\rangle.$$ (P3.117)

A similar SD equation is valid for a Green function with a different number of fields.

3.16 Derive the Swinger-Dyson equations for QED associated with the fermion field transformation

$$\psi_x\to e^{i\alpha_x}\psi_x,\quad\bar\psi_x\to\bar\psi_xe^{-i\alpha_x}.$$ (P3.118)

Show that these equations are the QED Ward-Takahashi identities.
SOLUTION. For

$$S[\psi]=\int dx\bar\psi_x(i\hat\partial_x+e\hat A-m)\psi_x,$$ (P3.119)

$$\int\mathcal{D}\psi'\mathcal{D}\bar\psi'e^{iS[\psi']}\psi'_z\bar\psi'_y=\left|\begin{array}{l}\psi'=e^{i\alpha}\psi\\\bar\psi'=\bar\psi e^{-i\alpha}\end{array}\right.$$
$$=\int\mathcal{D}\psi\mathcal{D}\bar\psi e^{iS[\psi]-i\int dx\bar\psi_x\gamma^\mu\psi_x\partial_\mu\alpha_x}(1+i\alpha_z-i\alpha_y)\psi_z\bar\psi_y$$
$$=\int\mathcal{D}\psi\mathcal{D}\bar\psi e^{iS[\psi]}(1+i\alpha_z-i\alpha_y+i\int dx\alpha_x\frac{\partial\bar\psi_x\gamma^\mu\psi_x}{\partial x^\mu})\psi_z\bar\psi_y.$$ (P3.120)

Therefore

$$\int \mathcal{D}\psi \mathcal{D}\bar{\psi} e^{iS[\psi]} \left(i\delta(z-x) - i\delta(y-x) + i\frac{\partial \bar{\psi}_x \gamma^\mu \psi_x}{\partial x^\mu} \right) \psi_z \bar{\psi}_y = 0,$$

(P3.121)

$$i\frac{\partial}{\partial x^\mu} \langle \psi_z \bar{\psi}_x \gamma^\mu \psi_x \bar{\psi}_y \rangle = i\delta(y-x)\langle \psi_z \bar{\psi}_y \rangle - i\delta(z-x)\langle \psi_x \bar{\psi}_y \rangle.$$

(P3.122)

In the momentum space

$$i\int dydz e^{ip(z-x)+iq(x-y)} \frac{\partial}{\partial x^\mu} \langle \psi_z \bar{\psi}_x \gamma^\mu \psi_x \bar{\psi}_y \rangle$$

$$= (q-p)_\mu \int dydz e^{ip(z-x)+iq(x-y)} \langle \psi_z \bar{\psi}_x \gamma^\mu \psi_x \bar{\psi}_y \rangle$$

$$= \int dz e^{ip(z-x)} i \langle \psi_z \bar{\psi}_x \rangle - \int dy e^{iq(x-y)} i \langle \psi_x \bar{\psi}_y \rangle$$

$$= iD(p) - iD(q).$$

(P3.123)

One can rewrite this expression as

$$iD^{-1}(p)[D(p) - D(q)]D^{-1}(q) = iD^{-1}(q) - iD^{-1}(p)$$

$$= (q-p)_\mu D^{-1}(p) \int dydz e^{ip(z-x)+iq(x-y)} \langle \psi_z \bar{\psi}_x \gamma^\mu \psi_x \bar{\psi}_y \rangle D^{-1}(q).$$

(P3.124)

One can check it in the Born level of perturbation theory

$$(\hat{q} - m) - (\hat{p} - m) = (q-p)_\mu \gamma^\mu.$$

(P3.125)

Note that without $\psi_z \bar{\psi}_y$ we would have found

$$i\frac{\partial}{\partial x^\mu} \langle \bar{\psi}_x \gamma^\mu \psi_x \rangle = 0.$$

(P3.126)

3.17 Express the fermion creation operator through the field.

SOLUTION.

$$\frac{1}{\sqrt{2\omega_k}} \int d^3x e^{-ikx} \bar{\psi}(x) \gamma^0 u_{\vec{k}\sigma} \Big|_{\omega_k = k^0 = \sqrt{\vec{k}^2 + m^2}}$$

$$= \int \frac{d^3x e^{-ikx}}{\sqrt{2\omega_k}} \left[\sum_s \int \frac{d^3p}{(2\pi)^3 \sqrt{2\omega_p}} (b_{\vec{p}s} \bar{v}_{\vec{p}s} e^{-ipx} + a_{\vec{p}s}^\dagger \bar{u}_{\vec{p}s} e^{ipx}) \right] \gamma^0 u_{\vec{k}\sigma}$$

$$= \frac{1}{2\omega_k} \sum_s (b_{-\vec{k}s} e^{-2ik^0x^0} v_{-\vec{k}s}^\dagger u_{\vec{k}\sigma} + a_{\vec{k}s}^\dagger u_{\vec{k}s}^\dagger u_{\vec{k}\sigma}) = a_{\vec{k}\sigma}^\dagger,$$

(P3.127)

since $v_{-\vec{k}s}^\dagger u_{\vec{k}\sigma} = 0$.

3.18 Show that the substitution of the anti-fermion for the fermion in LSZ formula (3.64) can be done via

$$u_{\vec{k}}e^{-ikx} \to -v_{\vec{k}}e^{ikx}, \quad \bar{u}_{\vec{k}}e^{ikx} \to -\bar{v}_{\vec{k}}e^{-ikx}. \tag{P3.128}$$

3.19 Find the Born scattering amplitude for the $e^-\mu^- \to e^-\mu^-$ from $\langle 0|\mathcal{T}\psi(y_1)\phi(y_2)\bar{\psi}(x_1)\bar{\phi}(x_2)|0\rangle$, where ψ is the electron field and ϕ is the muon field.

3.20 Derive the LSZ reduction formula for the scalar fields ϕ.

ANSWER.

$$S_{fi} = i^{m+n} \prod_{i=1}^{m} \int \frac{d^4 y_i e^{ik_i y_i}}{\sqrt{2\omega_{k_i} Z}} (\partial_{y_i}^2 + m^2) \prod_{j=1}^{n} \int \frac{d^4 x_j e^{-ip_j x_j}}{\sqrt{2\omega_{p_j} Z}} (\partial_{x_j}^2 + m^2)$$

$$\times \langle 0|\mathcal{T}(\phi(y_1)...\phi(y_m)\phi(x_1)...\phi(x_n))|0\rangle. \tag{P3.129}$$

For the Green function in the momentum space

$$\langle 0|\mathcal{T}(\phi(y_1)...\phi(y_m)\phi(x_1)...\phi(x_n))|0\rangle = \prod_{i=1}^{m} \int \frac{d^4 k_i e^{-ik_i y_i}}{(2\pi)^4}$$

$$\times \prod_{j=1}^{n} \int \frac{d^4 p_j e^{ip_j x_j}}{(2\pi)^4} \langle 0|\mathcal{T}(\phi(k_1)...\phi(k_m)\phi(p_1)...\phi(p_n))|0\rangle, \tag{P3.130}$$

one has the LSZ formula in the momentum space

$$S_{fi} = (-i)^{m+n} \prod_{i=1}^{m} \frac{k_i^2 - m^2}{\sqrt{2\omega_{k_i} Z}} \prod_{j=1}^{n} \frac{p_j^2 - m^2}{\sqrt{2\omega_{p_j} Z}}$$

$$\times \langle 0|\mathcal{T}(\phi(k_1)...\phi(k_m)\phi(p_1)...\phi(p_n))|0\rangle. \tag{P3.131}$$

Here k_i are the outgoing and p_j are the incoming momenta.

One of the clearest introductions to the functional methods in quantum mechanics the reader can find in (Kle09). In the lecture we mostly followed (DGH14a) and (PS95). Systematic description of non-commuting numbers and their applications is given in (DeW12), while (IZ80) gives a comprehensive review of LSZ reduction and (Mak05) discusses more advanced topics of functional methods in QFT.

FURTHER READING

[DeW12] Bryce S. DeWitt. *Supermanifolds*. Cambridge Monographs on Mathematical Physics. Cambridge University Press, Cambridge, UK, 5, 2012.

[DGH14a] J. F. Donoghue, E. Golowich, and Barry R. Holstein. *Dynamics of the standard model*, volume 2, chapter Appendix A: Functional Integration. CUP, 2014.

[DGH14b] J. F. Donoghue, E. Golowich, and Barry R. Holstein. *Dynamics of the standard model*, volume 2, chapter Appendix A: Functional Integration. CUP, 2014.

[IZ80] C. Itzykson and J. B. Zuber. *Quantum field theory*, chapter Vol.1, 5.1. International Series In Pure and Applied Physics. McGraw-Hill, New York, 1980.

[Kle09] H. Kleinert. *path integrals in quantum mechanics, statistics, polymer physics, and financial markets*. EBL-Schweitzer. World Scientific, 2009.

[Mak05] Yu. Makeenko. *Methods of contemporary gauge theory*, chapter 1.2.1–1.2.7. Cambridge Monographs on Mathematical Physics. Cambridge University Press, 11, 2005.

[PS95] Michael E. Peskin and Daniel V. Schroeder. *An introduction to quantum field theory*, chapter 9.1, 9.2, 9.5, 9.6, 7.2. Addison-Wesley, Reading, USA, 1995.

Gauge Fixing and Calculation Rules

THIS chapter shows how to use functional integral with the gluon fields. We derive the gluon propagator, introduce ghosts in different gauges via the Faddeev-Popov method. Problems discuss the asymptotic behavior of the basic QCD processes, Becchi-Rouet-Stora-Tyutin symmetry, Slavnov-Taylor identities, and spinor-helicity techniques.

1. **Naive generating functional for the YM theory**.

$$\begin{aligned}
\mathcal{L}_{YM} &= -\frac{1}{2}tr(F_{\mu\nu}F^{\mu\nu}) = \mathcal{L}_{YM}^0 + \mathcal{L}_{YM}^{int} \\
&= -\frac{1}{4}(\partial_\mu A_\nu^a - \partial_\nu A_\mu^a)(\partial^\mu A^{a\nu} - \partial^\nu A^{a\mu}) + \mathcal{L}_{YM}^{int} \\
&\to \frac{1}{2}A^{\mu a}(\partial_\alpha \partial^\alpha g_{\mu\nu} - \partial_\mu \partial_\nu)A^{\nu a} + \mathcal{L}_{YM}^{int}.
\end{aligned} \tag{4.1}$$

As in a scalar theory (3.15), one may write

$$\mathcal{Z}[J_\mu] = \int \mathcal{D}A_x^{a\mu} e^{iS_{YM}[A] + i\int dx J_\mu^a A^{\mu a}} \tag{4.2}$$

and try calculating the functional integral shifting $A^{\mu a}$ to get

$$\mathcal{Z}[J_\mu] \overset{?}{=} \mathcal{Z}[0]e^{-\frac{1}{2}\int dx dy J_x^\mu D_{\mu\nu}(x-y)J_y^\nu}, \tag{4.3}$$

with $D_{\mu\nu}$:

$$\delta^{ab}(\partial_\alpha \partial^\alpha g_{\mu\nu} - \partial_\mu \partial_\nu)D_\rho^\nu(x-y)^{bc} = i\delta^{ac}g_{\mu\rho}\delta(x-y). \tag{4.4}$$

However, the operator $(\partial_\alpha \partial^\alpha g_{\mu\nu} - \partial_\mu \partial_\nu)$ has zero modes, e.g. $\partial^\nu \alpha_x$. Hence, it cannot be inverted and the equation above does not have a

DOI: 10.1201/9781003272403-4

unique solution. The zero modes indicate that there are the field configurations gauge equivalent to 0. On them the action vanishes and the functional integral diverges.

Therefore one has to diagonalize the functional integration over all field configurations $A_x^{\mu a}$ into the product of integrals over the gauge nonequivalent fields and over the components responsible for all possible gauge transformations. The corresponding differential operator in the action in the former functional integral will have no zero modes and it will be taken as in the scalar theory. The latter integral will be factored out as an infinite constant since the action is gauge invariant.

2. **Gauge fixing.** Suppose we have a gauge fixing condition in each point x

$$G^i(A_\mu^\theta(x)) = 0, \quad \theta = \{\theta^1 \dots \theta^8\}, \quad i = 1, \dots, 8,$$

$$A_\mu^\theta = U^\theta(A_\mu - \frac{\partial_\mu}{ig})U^{\theta\dagger}, \quad A_\mu = A_\mu^{\theta=0}. \tag{4.5}$$

Let us integrate it over the group in each point x

$$\int \mathcal{D}U^\theta(x)\delta(G^i(A_\mu^\theta(x))) = \frac{1}{\Delta_G(A)}. \tag{4.6}$$

The result is gauge invariant. Indeed, we could have started from a different $\theta' \neq 0$

$$\frac{1}{\Delta_G(A^{\theta'})} = \int \mathcal{D}U^\theta(x)\delta(G^i(A_\mu^{\theta'\theta}(x)))$$

$$= \int \mathcal{D}\underbrace{U^\theta(x)U^{\theta'}(x)}_{U^{\theta''}(x)}\delta(G^i(\underbrace{A_\mu^{\theta'\theta}(x)}_{A_\mu^{\theta''}(x)})) = \frac{1}{\Delta_G(A)}. \tag{4.7}$$

Here

$$A_\mu^{\theta'\theta} = U^\theta(A_\mu^{\theta'} - \frac{\partial_\mu}{ig})U^{\theta\dagger} = U^\theta([U^{\theta'}(A_\mu - \frac{\partial_\mu}{ig})U^{\theta'\dagger}] - \frac{\partial_\mu}{ig})U^{\theta\dagger}$$

$$= U^\theta U^{\theta'}(A_\mu - \frac{\partial_\mu}{ig})U^{\theta'\dagger}U^{\theta\dagger} = U^{\theta''}(A_\mu - \frac{\partial_\mu}{ig})U^{\theta''\dagger}. \tag{4.8}$$

One can calculate Δ_G changing the variables $U^\theta \to G^i$

$$\frac{1}{\Delta_G(A)} = \int \mathcal{D}U^\theta\delta(G^i(A_\mu^\theta(x))) = \int \mathcal{D}G^i\delta(G^i(x)) \det \frac{\delta U^\theta(y)}{\delta G^i(x)}$$

$$= \frac{1}{\det \frac{\delta G^i(A_\mu^\theta(x))}{\delta U^\theta(y)}\big|_{G^i=0}}. \tag{4.9}$$

While doing these manipulations we assumed that gauge fixing condition (4.5) has only one solution, i.e. the variable change $U^\theta \to G^i$ is well defined and nonsingular. In other words for each set of 8 functions G^i there is only one gauge transformation $U^{\{\theta^1 \dots \theta^8\}}$ and vice versa.

Using gauge invariance of $\Delta_G(A)$ we can calculate it for $A_\mu(x)$ such that the gauge condition is satisfied at $\theta = 0$

$$G^i(A_\mu^{\theta(x)=0}(x)) = 0 \quad \Longrightarrow \quad \Delta_G(A) = \det \frac{\delta(G^i(A_\mu^\theta(x)))}{\delta(\theta^k(y))}|_{\theta=0}, \quad (4.10)$$

and use the infinitesimal transformation law for A^θ (2.11)

$$A_\mu^{\theta a} = A_\mu^a + \frac{1}{g}(\partial_\mu \delta^{ac} - ig T_{ac}^b A_\mu^b)\theta^c = A_\mu^a + \frac{1}{g} D_\mu^{ac} \theta^c. \quad (4.11)$$

Then

$$\Delta_G(A) = \det \frac{\delta(G^i(A_\mu^a(x) + \frac{1}{g} D_\mu^{ac} \theta^c(x)))}{\delta(\theta^k(y))}|_{\theta=0}. \quad (4.12)$$

For the **Lorentz gauge**

$$G^a(A_\mu) = \partial^\mu A_\mu^a = 0 \quad \Longrightarrow \quad \det \frac{\delta(\frac{1}{g}\partial^\mu D_\mu^{ac}\theta^c(x))}{\delta(\theta^k(y))} = \det(\frac{1}{g}\partial^\mu D_\mu^{ac}), \quad (4.13)$$

for the **axial gauge**

$$G^a(A_\mu) = n^\mu A_\mu^a = 0 \quad \Longrightarrow \quad \det \frac{\delta(\frac{1}{g}n^\mu D_\mu^{ac}\theta^c(x))}{\delta(\theta^k(y))} = \det(\frac{1}{g}n^\mu D_\mu^{ac}). \quad (4.14)$$

A functional determinant can be expressed as a **functional integral over the Grassmann scalar fields** called Faddeev-Popov **ghosts** $c_x^a, \bar{c}_x^c$

$$\det(\frac{1}{g}\partial^\mu D_\mu^{ac}) = const \int \mathcal{D}c\mathcal{D}\bar{c}e^{-i\int dx \bar{c}_x^a(\partial^\mu \partial_\mu \delta^{ac} - ig T_{ac}^b \partial^\mu A_\mu^b)c_x^c}, \quad (4.15)$$

$$\det(\frac{1}{g}n^\mu D_\mu^{ac}) = const \int \mathcal{D}c\mathcal{D}\bar{c}e^{-i\int dx \bar{c}_x^a(n^\mu \partial_\mu \delta^{ac} - ig T_{ac}^b n^\mu A_\mu^b)c_x^c}. \quad (4.16)$$

3. One can **separate the functional integral into the integral over the gauge transformations and the non-equivalent fields via** $\Delta_G(A)$. Indeed,

$$\mathcal{Z}_{YM}[0] = \int \mathcal{D}A_x^{a\mu} e^{iS_{YM}[A]} = \int \mathcal{D}A_x^{a\mu} e^{iS_{YM}[A]}$$

$$\times \left[1 = \Delta_G(A) \int \mathcal{D}U^\theta(x)\delta(G^i(A_\mu^\theta(x)))\right]$$

$$= \int \underbrace{\mathcal{D}A_x^{a\mu} \Delta_G(A(x)) e^{iS_{YM}[A]}}_{gauge\ invariant} \int \mathcal{D}U^\theta(x)\delta(G^i(A_\mu^\theta(x))), \quad (4.17)$$

where the measure is gauge invariant since gauge transformation (4.5) is shift plus a unitary transformation. Hence, one can change the integration variable $A \to A^\theta$

$$
\begin{aligned}
\mathcal{Z}_{YM}[0] &= \int \mathcal{D}A_x^{\theta a \mu} \Delta_G(A^\theta(x)) e^{iS_{YM}[A^\theta]} \int \mathcal{D}U^\theta(x) \delta(G^i(A_\mu^\theta(x))) \\
&= |A^\theta \to A| \\
&= \int \mathcal{D}A_x^{a\mu} \Delta_G(A(x)) e^{iS_{YM}[A]} \delta(G^i(A_\mu(x))) \quad \times \underbrace{\int \mathcal{D}U^\theta(x)}_{(const \,\times\, \text{group volume})^\infty} .
\end{aligned}
$$

$$(4.18)$$

Here for convenience $A^\theta \to A$ and we factored out the infinite constant $\sim$ group volume in every point x. If we choose a nonzero gauge condition

$$G^i(A_\mu^\theta(x)) - b^i(x) = 0, \tag{4.19}$$

we get

$$
\begin{aligned}
\mathcal{Z}_{YM}[0] &= const \int \mathcal{D}A \Delta_G(A_x) e^{iS_{YM}[A]} \delta(G^i(A_\mu(x)) - b^i(x)) \\
&= const' \int \mathcal{D}b e^{-\frac{i}{2\xi} \int dx (b_x^i)^2} \int \mathcal{D}A \Delta_G(A_x) e^{iS_{YM}[A]} \\
&\quad \times \delta(G^i(A_\mu(x)) - b^i(x)) \\
&= const \int \mathcal{D}A \Delta_G(A_x) e^{iS_{YM}[A] + iS_{fix}},
\end{aligned}
$$

$$(4.20)$$

$$S_{fix} = -\frac{1}{2\xi} \int dx (G^i(A_x))^2. \tag{4.21}$$

Changing b does not change the result. Therefore the integration w. r. t. b in the second line of this equation gives only an extra overall constant. Addition of S_{fix} modifies the differential operator in (4.4) so that it does not have a zero mode. As a result, one can find the propagator depending on the gauge.

4. **Feynman rules.** Now we have a well-defined generating functional

$$\mathcal{Z}_{YM}[0] = const \int \mathcal{D}A \mathcal{D}c \mathcal{D}\bar{c} \, e^{iS_{QCD} + iS_{fix} + iS_{ghost}}, \tag{4.22}$$

$$S_{QCD} = S_{YM} + \int dx \bar{\psi}(i\hat{D} - m)\psi, \quad S_{YM} = -\frac{1}{4} \int dx F^2. \tag{4.23}$$

The quark propagator reads (see Figure 4.1)

$$D_j^i(x - y) = \langle \psi_x^i \bar{\psi}_{jy} \rangle = \int \frac{dk}{(2\pi)^4} e^{-ik(x-y)} \frac{(\hat{k} - m)\delta_j^i}{k^2 - m^2 + i0}. \tag{4.24}$$

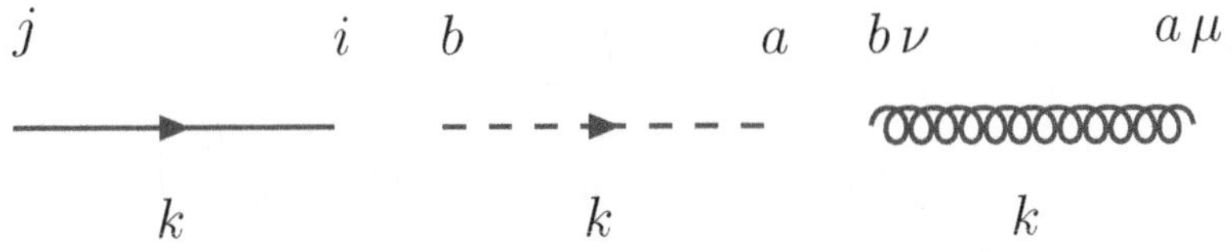

$$j \qquad\qquad i \quad b \qquad\qquad a \quad b\,\nu \qquad\qquad a\,\mu$$

$$\underbrace{\longrightarrow}_{k} \qquad\qquad \underbrace{---\blacktriangleright---}_{k} \qquad\qquad \underbrace{\text{0000000000000}}_{k}$$

Figure 4.1 Propagators: quark, ghost, gluon.

In the **Lorentz gauge**

$$S_{fix} = -\frac{1}{2\xi}\int dx (\partial^\mu A_\mu^a)^2, \ S_{ghost} = \int dx \bar{c}_x^a(-\partial^\mu \partial_\mu \delta^{ac} - gf^{abc}\partial^\mu A_\mu^b)c_x^c,$$

$$(4.25)$$

and the ghost and gluon propagators read (see Figure 4.1)

$$\langle c_x^a \bar{c}_y^b \rangle = \int \frac{dk}{(2\pi)^4} e^{-ik(x-y)} \frac{i\delta^{ab}}{k^2+i0},$$

$$(4.26)$$

$$D_{\mu\nu}^{ab}(x-y) = \langle A_{x\mu}^a A_{y\nu}^b \rangle$$

$$= \int \frac{dk}{(2\pi)^4} e^{-ik(x-y)} \frac{-i\delta^{ab}}{k^2+i0}\left(g_{\mu\nu} - (1-\xi)\frac{k_\mu k_\nu}{k^2}\right). \quad (4.27)$$

The ghost-ghost-gluon vertex reads (see Figure 4.2) $(p^\mu = i\partial^\mu)$

$$-gp_\mu f^{abc} = igp_\mu T_{ac}^b.$$

$$(4.28)$$

In the **axial gauge**

$$S_{fix} = -\frac{1}{2\xi}\int dx(n^\mu A_\mu^a)^2, \ S_{ghost} = \int dx \bar{c}_x^a(-n^\mu\partial_\mu\delta^{ac} - gf^{abc}n^\mu A_\mu^b)c_x^c,$$

$$(4.29)$$

and the ghost and gluon propagators read (see Figure 4.1)

$$\langle c_x^a \bar{c}_y^b \rangle = \int \frac{dk}{(2\pi)^4} e^{-ik(x-y)} \frac{\delta^{ab}}{nk},$$

$$(4.30)$$

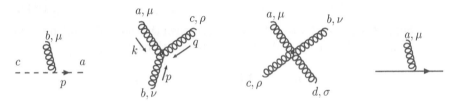

Figure 4.2 Vertices: ghost-gluon, 3-gluon, 4-gluon, quark-gluon.

$$D_{\mu\nu}^{ab}(x-y) = \langle A_{x\mu}^a A_{y\nu}^b \rangle = \int \frac{dk}{(2\pi)^4} e^{-ik(x-y)} \frac{-i\delta^{ab}}{k^2 + i0} \rangle$$

$$\times \left(g_{\mu\nu} - \frac{k_\mu n_\nu + n_\mu k_\nu}{kn} + k_\mu k_\nu \frac{n^2 + \xi k^2}{(kn)^2} \right). \qquad (4.31)$$

The ghost-ghost-gluon vertex reads (see Figure 4.2)

$$-ig f^{abc} n_\mu = -g n_\mu T_{ac}^b. \qquad (4.32)$$

The following vertices are the same for both gauges (see Figure 4.2).
The 3-gluon vertex $g f^{abc} V^{\mu\nu\rho}(k, p, q)$:

$$V^{\mu\nu\rho}(k, p, q) = g^{\mu\nu}(k - p)^\rho + g^{\nu\rho}(p - q)^\mu + g^{\rho\mu}(q - k)^\nu. \qquad (4.33)$$

The 4-gluon vertex

$$-ig^2[f^{abe} f^{cde}(g^{\mu\rho}g^{\nu\sigma} - g^{\mu\sigma}g^{\nu\rho}) + f^{ace} f^{bde}(g^{\mu\nu}g^{\rho\sigma} - g^{\mu\sigma}g^{\nu\rho})$$

$$+ f^{ade} f^{bce}(g^{\mu\nu}g^{\rho\sigma} - g^{\mu\rho}g^{\nu\sigma})]. \qquad (4.34)$$

The quark-gluon vertex

$$ig\gamma^\mu t^a. \qquad (4.35)$$

Both quark and ghost loops must be multiplied by (-1) since these particles are fermions.

EXERCISES

4.1 Apply the Faddev-Popov procedure to the **finite-dimensional integral**

$$Z[O] = \left[\prod_{i=1}^N \int_{-\infty}^{+\infty} dx_i \right] e^{-\sum_{k,l} x_k O_{kl} x_l} \qquad (P4.1)$$

with zero modes (DGH14).

SOLUTION.

$$Z[O] = \left| diag[O] = O^D = ROR^{-1}, \quad \vec{y} = R\vec{x}, \quad \left| \frac{\partial \vec{x}}{\partial \vec{y}} \right| = \det R = 1 \right|$$

$$= \left[\prod_{i=1}^N \int_{-\infty}^{+\infty} dy_i \right] e^{-\sum_k y_k^2 O_{kk}^D} = \left[\prod_{i=1}^N \left(\frac{\pi}{O_{ii}^D} \right)^{\frac{1}{2}} \right] = \frac{\pi^{\frac{N}{2}}}{(\det O)^{\frac{1}{2}}}. \qquad (P4.2)$$

Suppose $O_{kk} = 0$ for $k = N-n+1, ...N$. then the integral is independent of y_k, $k = N - n + 1, ...N$ and these integrations diverge

$$Z[O] = \left[\prod_{i=N-n+1}^N \int_{-\infty}^{+\infty} dy_i \right] Z^f[O],$$

$$Z^f[O] = \left[\prod_{i=1}^{N-n} \int_{-\infty}^{+\infty} dy_i \right] e^{-\sum_{k=1}^{N-n} y_k^2 O_{kk}^D}, \qquad (P4.3)$$

while the restricted functional $Z^f[O]$ is finite. One can rewrite the $Z^f[O]$ as an integral over the whole set of N variables

$$z_i = \begin{cases} y_i(x), & i = 1, \ldots, N - n \\ anything(x), & i = N - n + 1, \ldots, N, \end{cases} \tag{P4.4}$$

as

$$Z^f[O] = \left[\prod_{i=1}^{N} \int_{-\infty}^{+\infty} dz_i \right] \delta(z_{N-n+1}) \ldots \delta(z_N) e^{-\sum_{k=1}^{N} z_k^2 O_{kk}^D}$$

$$= \left[\prod_{i=1}^{N} \int_{-\infty}^{+\infty} dx_i \right] \left| \frac{\partial \vec{z}}{\partial \vec{x}} \right| \delta(z_{N-n+1}(x)) \ldots \delta(z_N(x)) e^{-\sum_{k,l} x_k O_{kl} x_l},$$

$$\tag{P4.5}$$

since

$$\sum_{k=1}^{N} z_k^2 O_{kk}^D = \sum_{k,l=1}^{N} z_k O_{kl}^D z_l = \sum_{i,j=1}^{N} (R^{-1}z)_i^T (R^{-1} O^D R)_{ij} (R^{-1}z)_j$$

$$= \sum_{k,l} x_k O_{kl} x_l. \tag{P4.6}$$

$$z_i = \begin{cases} y_i(x), & i = 1, \ldots, N - n \\ anything(x), & i = N - n + 1, \ldots, N, \end{cases} \tag{P4.7}$$

as

$$Z^f[O] = \left[\prod_{i=1}^{N} \int_{-\infty}^{+\infty} dz_i \right] \delta(z_{N-n+1}) \ldots \delta(z_N) e^{-\sum_{k=1}^{N} z_k^2 O_{kk}^D}$$

$$= \left[\prod_{i=1}^{N} \int_{-\infty}^{+\infty} dx_i \right] \left| \frac{\partial \vec{z}}{\partial \vec{x}} \right| \delta(z_{N-n+1}(x)) \ldots \delta(z_N(x)) e^{-\sum_{k,l} x_k O_{kl} x_l},$$

$$\tag{P4.8}$$

since

$$\sum_{k=1}^{N} z_k^2 O_{kk}^D = \sum_{k,l=1}^{N} z_k O_{kl}^D z_l = \sum_{i,j=1}^{N} (R^{-1}z)_i^T (R^{-1} O^D R)_{ij} (R^{-1}z)_j$$

$$= \sum_{k,l} x_k O_{kl} x_l. \tag{P4.9}$$

4.2 Find dimension of the space of gauge non-equivalent $A_x^{\mu a}$ configurations at a fixed point x.

SOLUTION. $A_x^{\mu a}$ has 4×8 d.o.f. -8 d.o.f. of an $SU(3)$ gauge transformation.

4.3 Show that a gauge transformation $A_\mu^\theta = U^\theta(A_\mu - \frac{\partial_\mu}{ig})U^{\theta\dagger}$ does not change the integration measure in $\int \mathcal{D}A$.

SOLUTION. Indeed,

$$\det \frac{\partial(A^\theta)_j^i}{\partial A_l^k} = \det \frac{\partial(U^\theta)_q^i A_s^q(U^{\theta\dagger})_j^s}{\partial A_l^k} = \det\left\{ (U^\theta)_q^i \delta^{kq} \delta_{sl}(U^{\theta\dagger})_j^s \right\}$$

$$= \det\left\{ (U^\theta)_k^i (U^{\theta\dagger})_j^l \right\} = 1. \tag{P4.10}$$

4.4 Show that the limit $\xi \to 0$ in (4.21) defines a pure (homogeneous) gauge $G^i(A_x) = 0$.

SOLUTION. Indeed, one may use the δ-function representation

$$\delta[G^i(A_x)] = \lim_{\xi\to 0}\left(\Pi_x \frac{1}{\sqrt{2\pi\xi}} \right) e^{-\frac{i}{2\xi}\int dx (G^i(A_x))^2}. \tag{P4.11}$$

4.5 Derive the propagators in the a) Lorentz and b) axial gauges.

SOLUTION. a) Taking into account gauge-fixing term (4.25) and doing the Fourier transform,

$$(\partial^2 g_{\mu\nu} - \partial_\mu\partial_\nu(1 - \frac{1}{\xi}))D_\rho^\nu(x - y) = ig_{\mu\rho}\delta(x - y), \tag{P4.12}$$

$$(-k^2 g_{\mu\nu} + k_\mu k_\nu(1 - \frac{1}{\xi}))D_\rho^\nu(k) = ig_{\mu\rho}. \tag{P4.13}$$

Next, we parameterize the propagator as

$$D_\rho^\nu(k) = A g_\rho^\nu + B k^\nu k_\rho \qquad \Longrightarrow \tag{P4.14}$$

$$\begin{cases} -k^2 A = i \\ \frac{1}{\xi}B = (1 - \frac{1}{\xi})A \end{cases} \Longrightarrow \quad D_\rho^\nu(k) = \frac{-i[g_\rho^\nu - (1 - \xi)\frac{k^\nu k_\rho}{k^2}]}{k^2}. \tag{P4.15}$$

b) Taking into account gauge-fixing term (4.29) and doing the Fourier transform,

$$(\partial^2 g_{\mu\nu} - \partial_\mu\partial_\nu - \frac{1}{\xi}n_\mu n_\nu)D_\rho^\nu(x - y) = ig_{\mu\rho}\delta(x - y), \tag{P4.16}$$

$$(-k^2 g_{\mu\nu} + k_\mu k_\nu - \frac{1}{\xi}n_\mu n_\nu)D_\rho^\nu(k) = ig_{\mu\rho}. \tag{P4.17}$$

We parameterize the propagator as

$$D_\rho^\nu(k) = A g_\rho^\nu + B k^\nu k_\rho + C(k^\nu n_\rho + k_\rho n^\nu) + D n^\nu n_\rho \qquad \Longrightarrow \tag{P4.18}$$

$$\begin{cases} -k^2 A = i, \quad D = 0 \\ C(k^2 + \frac{1}{\xi}n^2) + \frac{1}{\xi}B(kn) = 0 \\ A + C(kn) = 0 \end{cases} \Longrightarrow \tag{P4.19}$$

$$D_\rho^\nu(k) = \frac{-i}{k^2}\left(g_\rho^\nu - \frac{k^\nu n_\rho + n^\nu k_\rho}{(kn)} + k^\nu k_\rho \frac{n^2 + \xi k^2}{(kn)^2} \right). \tag{P4.20}$$

4.6 Show that in the pure axial gauge $nA = 0 \Longleftrightarrow \xi \to 0$ ghosts decouple.

SOLUTION. Ghost-gluon vertex (4.32) is proportional to n_ν. Therefore the amplitude with such vertices has $(nA) = 0$. One may also say that ghost action (4.29) does not contain the gluon field in this gauge. It corresponds to the $\xi \to 0$ limit in propagator (4.31) since

$$n_\nu \left(g_\rho^\nu - \frac{k^\nu n_\rho + n^\nu k_\rho}{(kn)} + k^\nu k_\rho \frac{n^2 + \xi k^2}{(kn)^2} \right) = k_\rho \frac{\xi k^2}{(kn)} \underset{\xi \to 0}{\to} 0. \quad \text{(P4.21)}$$

4.7 **Faddeev-Popov action in the external field.**

a. Show that transformation $gA^\mu \to A^\mu$ (non-perturbative field normalization) leads to

$$\mathcal{L}_{QCD} = -\frac{1}{4g^2} F^2 + \bar{\psi}(i\hat{D} - m)\psi, \quad \text{(P4.22)}$$

$$D_\mu = \partial_\mu - iA_\mu, \quad F_{\mu\nu}^a = \partial_\mu A_\nu^a - \partial_\nu A_\mu^a + f^{abc} A_\mu^b A_\nu^c, \quad \text{(P4.23)}$$

$$(A_\mu^a)^\theta = A_\mu^a + \partial_\mu \theta^a + f^{abc} A_\mu^b \theta^c, \quad \psi^\theta = (1 + i\theta)\psi, \quad \text{(P4.24)}$$

i.e. that with such normalization the charge stands only in front of F^2 and is absent in the covariant derivative, strength tensor and gauge transformations. Where does the charge come from e.g. in the qq scattering diagram?

b. Derive the QCD Lagrangian in the presence of classical field b

$$\mathcal{L}_{QCD}^b = -\frac{1}{4g^2}(F_{\mu\nu}^a + D_\mu^{ab} a_\nu^b - D_\nu^{ab} a_\mu^b + f^{abc} a_\mu^b a_\nu^c)^2 + \bar{\psi}(i\hat{D} + \hat{a} - m)\psi. \quad \text{(P4.25)}$$

Here $A^\mu = a^\mu + b^\mu$, where a^μ is a quantum field and b^μ is the external classical field,

$$D^\mu = \partial^\mu - ib^\mu, \quad F_{\mu\nu}^a = \partial_\mu b_\nu^a - \partial_\nu b_\mu^a + f^{abc} b_\mu^b b_\nu^c. \quad \text{(P4.26)}$$

c. Show that $\mathcal{L}_{QCD}^b$ at fixed external field b is invariant under gauge transformations

$$(a_\mu^a)^\theta = a_\mu^a + D_\mu^{ab} \theta^b + f^{abc} a_\mu^b \theta^c, \quad \psi^\theta = (1 + i\theta)\psi. \quad \text{(P4.27)}$$

d. Show that Faddeev-Popov determinant for the gauge fixing condition $0 = G(a) = D^{\mu ab} a_\mu^b - \omega^a$ reads

$$\Delta_G(a) = \det \frac{\delta(G^a(a^\theta(x)))}{\delta(\theta^b(y))}\Big|_{\theta=0} = \det D^{\mu ac}(D_\mu^{cb} + f^{cdb} a_\mu^d). \quad \text{(P4.28)}$$

e. Derive Faddeev-Popov Lagrangian with the gauge fixing condition $0 = G(a) = D^{\mu ab} a_\mu^b - \omega^a$

$$\mathcal{L}_{FP}^b = \mathcal{L}_{QCD}^b - \frac{1}{2\xi g^2}(D^{ab\mu} a_\mu^b)^2 + \bar{c}^a(-(D^2)^{ac} - (D^{ab\mu} a_\mu^d) f^{bdc})c^c. \quad \text{(P4.29)}$$

f. Show that $\mathcal{L}_{FP}^b$ is invariant under residual gauge transformation including the field b

$$b_\mu^a \rightarrow b_\mu^a + (D_\mu \beta)^a, \tag{P4.30}$$

$$a_\mu^a \rightarrow a_\mu^a - f^{abc}\beta^b a_\mu^c = a_\mu^a + i\beta^b T_{ac}^b a_\mu^c = a_\mu^a + i\left(\beta a_\mu\right)^a, \tag{P4.31}$$

$$\psi \rightarrow \psi + i\beta^a t^a \psi, \tag{P4.32}$$

$$c^a \rightarrow c^a - f^{abc}\beta^b c^c = c^a + i\beta^b T_{ac}^b c^c = c^a + i(\beta c)^a. \tag{P4.33}$$

SOLUTION.

(f) This transformation is a gauge transformation for b and ψ with

$$U = e^{i\beta}, \tag{P4.34}$$

i.e.

$$D_\mu \rightarrow U D_\mu U^\dagger, \quad F^b \rightarrow U^{ba} F^a, \quad c^b \rightarrow U^{ba} c^a, \tag{P4.35}$$

$$a_\mu^b \rightarrow U^{ba} a_\mu^a \quad \Longrightarrow \quad a_\mu = a_\mu^a T^a \rightarrow U a_\mu U^\dagger, \tag{P4.36}$$

$$(D^{ab\mu} a_\mu^b)^2 \sim a_\nu D^\nu D^\mu a_\mu = a_\nu U^\dagger U D^\nu U^\dagger U D^\mu U^\dagger U a_\mu, \tag{P4.37}$$

$$\bar{c}((iD)^2 - D^\mu a_\mu)c = \bar{c} U^\dagger (U(iD)^2 U^\dagger - U D^\mu U^\dagger U a_\mu U^\dagger) U c, \tag{P4.38}$$

$$\bar{\psi}(i\hat{D} + \hat{a} - m)\psi = \bar{\psi} U^\dagger (U i\hat{D} U^\dagger + U\hat{a}U^\dagger - m)U\psi. \tag{P4.39}$$

$(F_{\mu\nu}^a + D_\mu^{ab} a_\nu^b - D_\nu^{ab} a_\mu^b - iT_{ac}^b a_\mu^b a_\nu^c)^2$ contains the following terms as well

$$a^c a_{cd} a_{de} a^e = a U^\dagger U a U^\dagger U a U^\dagger U a, \tag{P4.40}$$

$$a^c D_{cd} a_{de} a^e = a U^\dagger U D U^\dagger U a U^\dagger U a, \tag{P4.41}$$

$$F^c D_{cd} a^d = F U^\dagger U D U^\dagger U a, \tag{P4.42}$$

$$F^c a_{cd} a^d = F U^\dagger U a U^\dagger U a. \tag{P4.43}$$

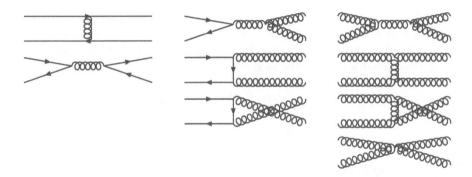

Figure 4.3 Elementary processes in QCD: $q\bar{q} \to q\bar{q}$, $q\bar{q} \to gg$, $gg \to gg$. Amplitudes for other elementary processes can be obtained from these ones via crossing symmetry.

Or straightforwardly for fermions

$$\delta\left(\bar{\psi}(i\hat{D} + \hat{a} - m)\psi\right)$$
$$= \bar{\psi}\left\{(i\hat{D} + \hat{a} - m)i\beta^a t^a - i\beta^a t^a(i\hat{D} + \hat{a} - m)\right\}\psi$$
$$+ \bar{\psi}(t^a(\hat{D}^{ab}\beta^b) - t^a f^{abc}\beta^b \hat{a}^c)\psi$$
$$= \bar{\psi}\left\{(i\hat{D}i\beta^a t^a - i\beta^a t^a i\hat{D} + t^a(\hat{D}^{ab}\beta^b)\right.$$
$$\left. + \hat{a}i\beta^a t^a - i\beta^a t^a\hat{a} - t^a f^{abc}\beta^b\hat{a}^c\right\}\psi$$
$$= \bar{\psi}\left\{([i\hat{D}, i\beta] - (i\hat{D}i\beta) + i[\hat{a}, \beta] - t^a f^{abc}\beta^b\hat{a}^c\right\}\psi$$
$$= \bar{\psi}\left\{(i\hat{\partial} + \hat{b})i\beta - i\beta(i\hat{\partial} + \hat{b}) - ((i\hat{\partial} + \hat{b})i\beta)\right\}$$
$$+ ii f^{cba} t^a \hat{a}^c \beta^b - t^a f^{abc}\beta^b\hat{a}^c\right\}\psi = 0. \tag{P4.44}$$

4.8 Estimate the asymptotic behavior of the QCD elementary processes shown in Figure 4.3 in the **Regge region** $s \gg |t| \gg m_q^2$.

SOLUTION. We denote the incoming particles' momenta p_1 and p_2 and the outgoing particles' momenta p'_1 and p'_2:

$$p_1 = \sqrt{\frac{s}{2}}n_1, \quad p_2 = \sqrt{\frac{s}{2}}n_2, \quad n_1 n_2 = 1, \quad p_i^2 = p_i'^2 = n_i^2 = 0, \tag{P4.45}$$
$$s = (p_1 + p_2)^2 = 2p_1 p_2 = (p'_1 + p'_2)^2, \quad t = (p_1 - p'_1)^2 = (p_2 - p'_2)^2. \tag{P4.46}$$

The metric tensor in the gluon propagator in the Lorenz gauge with $\xi = 1$ can be written as

$$g_{\mu\nu} = n_{1\mu}n_{2\nu} + n_{1\nu}n_{2\mu} + g_{\mu\nu\perp}. \tag{P4.47}$$

For the $qq \to qq$ process the dominant diagram in the Regge region is the first diagram in Figure 4.3 after the crossing transformation, i.e. changing the antiquark line to the quark one since it has the small denominator $\sim \frac{1}{t}$. In the Regge limit, it reads

$$iM = -i^3 g^2 t^a \otimes t^a \frac{\bar{u}'_1 \gamma^\mu u_1 g_{\mu\nu} \bar{u}'_2 \gamma^\mu u_2}{t} \simeq -i^3 g^2 t^a \otimes t^a \frac{\bar{u}_1 \gamma^\mu u_1 g_{\mu\nu} \bar{u}_2 \gamma^\mu u_2}{t}$$

$$= -i^3 g^2 t^a \otimes t^a \frac{2p_1^\mu g_{\mu\nu} 2p_2^\nu}{t} \simeq -i^3 g^2 t^a \otimes t^a \frac{4(p_1 n_2)(n_1 p_2)}{t}$$

$$= i g^2 t^a \otimes t^a \frac{2s}{t}, \tag{P4.48}$$

$$\frac{1}{4N^2} \sum_{\substack{spin\ q\bar{q}gg \\ col}} |M|^2 = \frac{g^4}{4N^2} tr[t^a t^b]^2 \left(\frac{2s}{t}\right)^2 = \frac{g^4 C_F}{8N} \left(\frac{2s}{t}\right)^2. \tag{P4.49}$$

For the $gg \to gg$ process the dominant diagram in the Regge region is the second diagram in the third column of Figure 4.3 since it also has the small denominator $\sim \frac{1}{t}$. In the Regge limit, it reads

$$iM = -ig^2 f^{aba'} f^{c'bc} \varepsilon_1^{*\mu} \varepsilon_1'^\rho V_{\mu\nu\rho}(p_1, p_2 - p'_2, -p'_1) \frac{g_{\nu\mu'}}{t}$$

$$\times V_{\mu'\nu'\rho'}(p'_2 - p_2, p_2, -p'_2) \varepsilon_2^{*\nu'} \varepsilon_2^{\rho'\prime}$$

$$\simeq ig^2 f^{aba'} f^{cbc'} \varepsilon_1^{*\mu} \varepsilon_1'^\rho V_{\mu\nu\rho}(p_1, 0, -p_1) \frac{n_{2\nu} n_{1\mu'}}{t}$$

$$\times V_{\mu'\nu'\rho'}(0, p_2, -p_2) \varepsilon_2^{*\nu'} \varepsilon_2^{\rho'\prime}$$

$$\simeq ig^2 f^{aba'} f^{cbc'} \frac{-2(p_1 n_2)(\varepsilon_1^* \varepsilon_1') 2(p_2 n_1)(\varepsilon_2^* \varepsilon_2')}{t}$$

$$\simeq -ig^2 f^{aba'} f^{cbc'} \frac{-2(p_1 n_2) 2(p_2 n_1)}{t}$$

$$= -ig^2 f^{aba'} f^{cbc'} \frac{2s}{t} = ig^2 T_{aa'}^b T_{cc'}^b \frac{2s}{t}, \tag{P4.50}$$

$$\frac{1}{4(N^2-1)^2} \sum_{\substack{spin\ gggg \\ col}} |M|^2 = \frac{g^4}{4(N^2-1)^2} T_{aa'}^b T_{a'a}^{b'} T_{cc'}^b T_{c'c}^{b'} \left(\frac{2s}{t}\right)^2$$

$$= \frac{N^2}{4(N^2-1)} g^4 \left(\frac{2s}{t}\right)^2. \tag{P4.51}$$

For the $q\bar{q} \to gg$ process the dominant diagram in the Regge region is the second diagram in the second column of Figure 4.3 since it again has the small denominator $\sim \frac{1}{t}$. In the Regge limit $p'_i \varepsilon_i \simeq p_i \varepsilon_i = 0$ and it reads

$$iM = i^3 g^2 t^b t^a \frac{\bar{v}_2 \hat{\epsilon}_2^*(\hat{p}_1 - \hat{p}_1') \hat{\epsilon}_1^* u_1}{t}, \qquad (\text{P4.52})$$

$$\frac{1}{4N^2} \sum_{\substack{spin\ q\bar{q}gg \\ col}} |M|^2 = g^4 \frac{C_F^2}{4N} \frac{tr[\hat{p}_2 \hat{\epsilon}_2^*(\hat{p}_1 - \hat{p}_1') \hat{\epsilon}_1^* \hat{p}_1 \hat{\epsilon}_1 (\hat{p}_1 - \hat{p}_1') \hat{\epsilon}_2]}{t^2}$$

$$\simeq g^4 \frac{C_F^2}{4N} \frac{tr[\hat{p}_2 (\hat{p}_1 - \hat{p}_1') \hat{p}_1 (\hat{p}_1 - \hat{p}_1')]}{t^2}$$

$$\simeq g^4 \frac{C_F^2}{N} \frac{-(p_1 p_2) t}{t^2} = g^4 \frac{C_F^2}{N} \frac{s}{2|t|}. \qquad (\text{P4.53})$$

4.9 Estimate the asymptotic behavior of the $2 \to 2$ processes in the **Regge limit** $s \gg |t| \gg m^2$.

SOLUTION. Consider the amplitude for the $2 \to 2$ process in the t-channel, i.e. with large positive t. One can write its partial wave expansion as

$$M(s,t) = \sum_{l=0}^{\infty} (2l+1) a_l(t) P_l(\cos\theta), \qquad (\text{P4.54})$$

$$t = 4E^2, \quad s = -2E^2(1 - \cos\theta) \quad \Longrightarrow \quad \cos\theta = 1 + \frac{2s}{t}. \qquad (\text{P4.55})$$

Here the scattering angle in the center of mass frame is θ. If the Born process goes via a particle with spin j in the t-channel, the partial wave expansion has the leading contribution for $l = j$. In the s-channel, i.e. with large positive s then

$$M(s,t) = \sum_{l=0}^{\infty} (2l+1) a_l(t) P_l(1 + \frac{2s}{t}) \sim \left(\frac{2s}{t}\right)^j, \quad s \gg |t|. \qquad (\text{P4.56})$$

Hence, the s-channel cross-section has the Regge asymptotics

$$\frac{d\sigma}{dt} \sim \frac{1}{s^2} \left(\frac{s}{|t|}\right)^{2j}. \qquad (\text{P4.57})$$

Here an extra $\frac{1}{s^2}$ comes from the s-channel flux invariant $I = s$ and j is the spin of the mediator particle in the t-channel. Therefore the cross-sections for $gg \to gg$, $qq \to qq$, $gq \to gq$ behave as

$$\sigma \sim \frac{1}{|t_{min}|}, \qquad (\text{P4.58})$$

since they have the gluon ($j = 1$) in the t-channel, while $q\bar{q} \to gg$ behaves as

$$\sigma \sim \frac{1}{s} \ln \frac{s}{|t_{min}|}, \qquad (\text{P4.59})$$

since this process has the quark $(j = \frac{1}{2})$ in the t-channel, as one can see from Figure 4.3.

4.10 Find solution to the massless Dirac equation normalized to

$$u_p^\dagger u_p = v_p^\dagger v_p = 2p^0 \tag{P4.60}$$

in the spinor or Weyl γ-matrix rep

$$\gamma^0 = \begin{pmatrix} 0 & 1 \\ 1 & 0 \end{pmatrix}, \quad \gamma^i = \begin{pmatrix} 0 & -\sigma^i \\ \sigma^i & 0 \end{pmatrix}, \quad \gamma^5 = \begin{pmatrix} 1 & 0 \\ 0 & -1 \end{pmatrix},$$

$$\frac{1+\gamma^5}{2} = \begin{pmatrix} 1 & 0 \\ 0 & 0 \end{pmatrix} = P_R, \quad \frac{1-\gamma^5}{2} = \begin{pmatrix} 1 & 0 \\ 0 & 0 \end{pmatrix} = P_L. \tag{P4.61}$$

SOLUTION. For the positive energy solution, $u_R = u_+ = P_R u$, $u_L = u_- = P_L u$,

$$\hat{p}u = \begin{pmatrix} 0 & p^0 + \vec{p}\vec{\sigma} \\ p^0 - \vec{p}\vec{\sigma} & 0 \end{pmatrix} \begin{pmatrix} w_R \\ w_L \end{pmatrix}, \tag{P4.62}$$

$$(p^0 + \vec{p}\vec{\sigma})w_L = 0, \quad (p^0 - \vec{p}\vec{\sigma})w_R = 0. \tag{P4.63}$$

$$p^0 + \vec{p}\vec{\sigma} = \begin{pmatrix} p^+ & p^1 - ip^2 \\ p^1 + ip^2 & p^- \end{pmatrix}$$

$$= \begin{pmatrix} p^+ & \sqrt{p^+p^-}\,e^{-i\phi_p} \\ \sqrt{p^+p^-}\,e^{i\phi_p} & p^- \end{pmatrix} \quad \Longrightarrow \tag{P4.64}$$

$$w_L = w_- = const \begin{pmatrix} \sqrt{p^-}\,e^{-i\phi_p} \\ -\sqrt{p^+} \end{pmatrix}. \tag{P4.65}$$

Here

$$p^\pm = p^0 \pm p^3, \quad p^2 = 0 \quad \Longrightarrow \quad p^+p^- = p_1^2 + p_2^2,$$

$$e^{\pm i\phi_p} = \frac{p^1 \pm ip^2}{\sqrt{p^+p^-}}. \tag{P4.66}$$

To satisfy normalization condition (P4.60), one usually takes $const = 1$. Likewise one gets

$$w_R = w_+ = \begin{pmatrix} \sqrt{p^+} \\ \sqrt{p^-}\,e^{i\phi_p} \end{pmatrix}. \tag{P4.67}$$

Finally,

$$u_R = u_+ = \begin{pmatrix} w_+ \\ 0 \end{pmatrix} = \begin{pmatrix} \sqrt{p^+} \\ \sqrt{p^-}e^{i\phi_p} \\ 0 \\ 0 \end{pmatrix}, \tag{P4.68}$$

$$u_L = u_- = \begin{pmatrix} 0 \\ w_- \end{pmatrix} = \begin{pmatrix} 0 \\ 0 \\ \sqrt{p^-}e^{-i\phi_p} \\ -\sqrt{p^+} \end{pmatrix}. \tag{P4.69}$$

For the negative energy solution $p^0 \to -p^0$ in (P4.62) and one can take

$$v_R = u_L, \quad v_L = u_R, \quad \text{or} \quad v_\mp = u_\pm \tag{P4.70}$$

as solutions.

4.11 Show how $u_{R,L}$ change under the charge conjugation C

SOLUTION.

$$Cu_R = (\bar{u}_R i\gamma^0\gamma^2)^T = i\gamma^{2T}u_R^* = \begin{pmatrix} 0 & -i\sigma^2 \\ i\sigma^2 & 0 \end{pmatrix} u_R^* = u_L, \tag{P4.71}$$

$$Cu_L = (\bar{u}_L i\gamma^0\gamma^2)^T = i\gamma^{2T}u_L^* = \begin{pmatrix} 0 & -i\sigma^2 \\ i\sigma^2 & 0 \end{pmatrix} u_L^* = u_R. \tag{P4.72}$$

4.12 Show that

$$\bar{u}_R(p_i)\gamma^\mu u_R(p_j) = \bar{u}_L(p_j)\gamma^\mu u_L(p_i). \tag{P4.73}$$

SOLUTION. Via (P4.71)

$$\begin{aligned}
\bar{u}_R(p_i)\gamma^\mu u_R(p_j) &= (\bar{u}_L(p_i)i\gamma^0\gamma^2)^{T\dagger}\gamma^0\gamma^\mu(\bar{u}_L(p_j)i\gamma^0\gamma^2)^T \\
&= \bar{u}_L(p_j)i\gamma^0\gamma^2\gamma^{\mu T}\gamma^0(\bar{u}_L(p_i)i\gamma^0\gamma^2)^\dagger \\
&= \bar{u}_L(p_j)\gamma^0\gamma^2\gamma^{\mu T}\gamma^0\gamma^{2\dagger}u_L(p_i) \\
&= -\bar{u}_L(p_j)\gamma^0\gamma^2\gamma^{\mu T}\gamma^0\gamma^2 u_L(p_i) \\
&= \bar{u}_L(p_j)\gamma^\mu u_L(p_i). \tag{P4.74}
\end{aligned}$$

4.13 **Spinor-helicity formalism.** We follow (Dix96). Using the helicity spinors $u_\pm = v_\mp$ (P4.68-P4.69) from problem 4.10 and introducing the notations

$$|i^\pm\rangle = |p_i^\pm\rangle = u_\pm(p_i) = v_\mp(p_i), \tag{P4.75}$$

$$\langle i^\pm| = \langle p_i^\pm| = u_\pm^\dagger(p_i)\gamma^0 = \bar{u}_\pm(p_i) = v_\mp^\dagger(p_i)\gamma^0 = \bar{v}_\mp(p_i), \tag{P4.76}$$

$$\langle ij\rangle = \langle i^-|j^+\rangle = \bar{u}_-(p_i)u_+(p_j), \tag{P4.77}$$

$$[ij] = \langle i^+|j^-\rangle = \bar{u}_+(p_i)u_-(p_j), \tag{P4.78}$$

demonstrate

a.

$$\langle i^+|j^+\rangle = \langle i^-|j^-\rangle = [ii] = \langle ii \rangle = 0; \qquad \text{(P4.79)}$$

$$\langle ij\rangle[ji] = s_{ij}; \quad \langle i^\pm|\gamma^\mu|i^\pm\rangle = 2p_i^\mu; \quad |i^\pm\rangle\langle i^\pm| = \frac{1\pm\gamma^5}{2}\hat{p}_i; \quad \text{(P4.80)}$$

$$[ij] = -[ji], \quad \langle ij\rangle = -\langle ji\rangle; \qquad \text{(P4.81)}$$

for $p_{i,j}^0 > 0$

$$\langle ij\rangle = \sqrt{p_i^- p_j^+}\,e^{i\phi_{p_i}} - \sqrt{p_i^+ p_j^-}\,e^{i\phi_{p_j}} = \sqrt{|s_{ij}|}e^{i\phi_{ij}}, \qquad \text{(P4.82)}$$

$$[ij] = -\sqrt{p_i^- p_j^+}\,e^{-i\phi_{p_i}} + \sqrt{p_i^+ p_j^-}\,e^{-i\phi_{p_j}} = -\sqrt{|s_{ij}|}e^{-i\phi_{ij}}, \quad \text{(P4.83)}$$

$$s_{ij} = (p_i + p_j)^2, \qquad \text{(P4.84)}$$

$$\cos\phi_{ij} = \frac{p_j^+ p_i^1 - p_j^1 p_i^+}{\sqrt{p_i^+ p_j^+}\,|s_{ij}|}, \quad \sin\phi_{ij} = \frac{p_j^+ p_i^2 - p_j^2 p_i^+}{\sqrt{p_i^+ p_j^+}\,|s_{ij}|}; \qquad \text{(P4.85)}$$

b. the current charge conjugation identity

$$\langle i^-|\gamma_\rho|j^-\rangle = \langle j^+|\gamma_\rho|i^+\rangle; \qquad \text{(P4.86)}$$

the Fierz identities

$$2(|i^+\rangle\langle j^+| + |j^-\rangle\langle i^-|) = \langle i^-|\gamma_\rho|j^-\rangle\gamma^\rho = \langle j^+|\gamma_\rho|i^+\rangle\gamma^\rho; \qquad \text{(P4.87)}$$

$$\langle i^+|\gamma^\mu|j^+\rangle\langle k^+|\gamma_\mu|l^+\rangle = 2[ik]\langle lj\rangle; \qquad \text{(P4.88)}$$

the **Schouten identity**

$$\langle ij\rangle\langle kl\rangle = \langle ik\rangle\langle jl\rangle + \langle il\rangle\langle kj\rangle; \qquad \text{(P4.89)}$$

the momentum conservation identity

$$\sum_{\substack{i=1 \\ i\neq j,k}}^{n} [ji]\langle ik\rangle = 0, \quad \text{for} \quad \sum_{i=1}^{n} p_i = 0; \qquad \text{(P4.90)}$$

c. the properties of the polarization vector for

$$\varepsilon_\mu^\pm(k,q) = \pm\frac{\langle q^\mp|\gamma_\mu|k^\mp\rangle}{\sqrt{2}\langle q^\mp|k^\pm\rangle}, \qquad \text{(P4.91)}$$

where the gluon has momentum k and the gauge vector $q : q^2 = 0$. One needs the following properties

$$\varepsilon^\pm(k,q)k = \varepsilon^\pm(k,q)q = 0, \quad (\varepsilon_\mu^+)^* = \varepsilon_\mu^-, \qquad \text{(P4.92)}$$

$$\varepsilon^+(\varepsilon^+)^* = \varepsilon^+\varepsilon^- = -1, \quad \varepsilon^+(\varepsilon^-)^* = \varepsilon^+\varepsilon^+ = 0, \qquad \text{(P4.93)}$$

$$\varepsilon_\mu^+(k, q_1) - \varepsilon_\mu^+(k, q_2) = -\frac{\sqrt{2}\langle q_1 | q_2 \rangle}{\langle q_1 | k \rangle \langle q_2 | k \rangle} k_\mu, \tag{P4.94}$$

$$\sum_{\lambda=\pm} \varepsilon_\mu^\lambda(k, q) \varepsilon_\nu^\lambda(k, q)^* = -g_{\mu\nu} + \frac{k_\mu q_\nu + k_\nu q_\mu}{(kq)}, \tag{P4.95}$$

which is the numerator of the propagator in the light cone gauge $(Aq) = 0$,

$$\hat{\varepsilon}^+(k, q) = \sqrt{2} \frac{|q^+\rangle\langle k^+| + |k^-\rangle\langle q^-|}{\langle qk \rangle}, \tag{P4.96}$$

$$\hat{\varepsilon}^-(k, q) = -\sqrt{2} \frac{|q^-\rangle\langle k^-| + |k^+\rangle\langle q^+|}{[qk]}, \tag{P4.97}$$

$$\varepsilon^+(k_1, q)\varepsilon^+(k_2, q) = \varepsilon^-(k_1, q)\varepsilon^-(k_2, q) = 0, \tag{P4.98}$$

$$\varepsilon^+(k_1, k_2)\varepsilon^-(k_2, q) = \varepsilon^-(k_1, q)\varepsilon^-(k_2, k_1) = 0, \tag{P4.99}$$

$$\hat{\varepsilon}^+(k_1, k_2)|2^+\rangle = \hat{\varepsilon}^-(k_1, k_2)|2^-\rangle = 0, \tag{P4.100}$$

$$\langle 2^+|\hat{\varepsilon}^-(k_1, k_2) = \langle 2^-|\hat{\varepsilon}^+(k_1, k_2) = 0. \tag{P4.101}$$

SOLUTION.

a.

$$\langle i^+|j^+\rangle = \bar{u}_+(p_i)u_+(p_j) = u_i^\dagger P_R^\dagger \gamma^0 P_R u_j$$

$$= u_i^\dagger \gamma^0 P_L P_R u_j = 0, \tag{P4.102}$$

$$[ii] = \bar{u}_+(p_i)u_-(p_i) = 0; \tag{P4.103}$$

$$\langle ij\rangle[ji] = \bar{u}_-(p_i)u_+(p_j)\bar{u}_+(p_j)u_-(p_i)$$

$$= u(p_i)P_L\gamma^0 P_R u(p_j)u^\dagger(p_j)P_R\gamma^0 P_L u(p_i)$$

$$= tr[\hat{p}_i \frac{1+\gamma^5}{2}\hat{p}_j] = 2(p_ip_j) = s_{ij}; \tag{P4.104}$$

$$|i^\pm\rangle\langle i^\pm| = u_\pm(p_i)\bar{u}_\pm(p_i) = P_\pm u(p_i)u^\dagger(p_i)P_\pm\gamma^0$$

$$= P_\pm u(p_i)u^\dagger(p_i)\gamma^0 P_\mp = P_\pm\hat{p}_i; \tag{P4.105}$$

$$\phi_{ij} = \phi_{ji} - \pi; \tag{P4.106}$$

$$\langle ij \rangle = \bar{u}_-(p_i)u_+(p_j) = \sqrt{p_i^- p_j^+}\, e^{i\phi_{p_i}} - \sqrt{p_i^+ p_j^-}\, e^{i\phi_{p_j}}, \quad (P4.107)$$

$$\langle ij \rangle \langle ij \rangle^* = p_i^- p_j^+ + p_i^+ p_j^-$$
$$- \sqrt{p_i^- p_j^+}\sqrt{p_i^+ p_j^-}\,(e^{i\phi_{p_i}-i\phi_{p_j}} + e^{i\phi_{p_j}-i\phi_{p_i}})$$
$$= p_i^- p_j^+ + p_i^+ p_j^- - 2p_i^1 p_j^1 - 2p_i^2 p_j^2$$
$$= 2p_i p_j = (p_i + p_j)^2 = s_{ij}^2, \quad (P4.108)$$

$$\langle ij \rangle = \sqrt{p_i^- p_j^+}\, \frac{p_i^1}{\sqrt{p_i^+ p_i^-}} - \sqrt{p_i^+ p_j^-}\, \frac{p_j^1}{\sqrt{p_j^+ p_j^-}}$$
$$+ i\left(\sqrt{p_i^- p_j^+}\, \frac{p_i^2}{\sqrt{p_i^+ p_i^-}} - \sqrt{p_i^+ p_j^-}\, \frac{p_j^2}{\sqrt{p_j^+ p_j^-}} \right)$$
$$= \sqrt{|s_{ij}|}\, e^{i\phi_{ij}}. \quad (P4.109)$$

b. One can decompose $|i^+\rangle\langle j^+| + |j^-\rangle\langle i^-|$ into the basis of gamma matrices anticommuting with γ^5

$$|i^+\rangle\langle j^+| + |j^-\rangle\langle i^-| = c_1 1 + c_5 \gamma^5 + c_\rho \gamma^\rho + c'_\rho \gamma^\rho \gamma^5 + c_{\mu\nu}\sigma^{\mu\nu}, \quad (P4.110)$$

$$(|i^+\rangle\langle j^+| + |j^-\rangle\langle i^-|)\gamma^5 = -\gamma^5(|i^+\rangle\langle j^+| + |j^-\rangle\langle i^-|) \implies \quad (P4.111)$$

$$|i^+\rangle\langle j^+| + |j^-\rangle\langle i^-| = d_\rho \gamma^\rho P_R + e_\rho \gamma^\rho P_L, \quad (P4.112)$$

$$tr[(|i^+\rangle\langle j^+| + |j^-\rangle\langle i^-|)\gamma^\alpha P_L] = \langle i^-|\gamma^\alpha|j^-\rangle$$
$$= d_\rho tr[\gamma^\rho P_R \gamma^\alpha P_L] = 2d^\alpha, \quad (P4.113)$$

$$tr[(|i^+\rangle\langle j^+| + |j^-\rangle\langle i^-|)\gamma^\alpha P_R] = \langle j^+|\gamma^\alpha|i^+\rangle$$
$$= e_\rho tr[\gamma^\rho P_L \gamma^\alpha P_R] = 2e^\alpha, \quad (P4.114)$$

$$2(|i^+\rangle\langle j^+| + |j^-\rangle\langle i^-|) = \langle i^-|\gamma_\rho|j^-\rangle \gamma^\rho P_R$$
$$+ \langle j^+|\gamma_\rho|i^+\rangle \gamma^\rho P_L. \quad (P4.115)$$

Next, via (P4.73)

$$\langle i^-|\gamma_\rho|j^-\rangle = \langle j^+|\gamma_\rho|i^+\rangle, \quad (P4.116)$$

Therefore

$$2(|i^+\rangle\langle j^+| + |j^-\rangle\langle i^-|) = \langle i^-|\gamma_\rho|j^-\rangle \gamma^\rho = \langle j^+|\gamma_\rho|i^+\rangle \gamma^\rho. \quad (P4.117)$$

Then

$$2\langle k^+|(|i^+\rangle\langle j^+| + |j^-\rangle\langle i^-|)|l^+\rangle = 2\langle k^+|j^-\rangle\langle i^-|l^+\rangle \quad (P4.118)$$
$$= 2[kj]\langle il \rangle = \langle j^+|\gamma_\rho|i^+\rangle\langle k^+|\gamma^\rho|l^+\rangle. \quad (P4.119)$$

The Shouten identity

$$\langle ij\rangle\langle kl\rangle - \langle il\rangle\langle kj\rangle = \langle ik\rangle\langle jl\rangle \quad | \times [kj][li] \quad \Longrightarrow$$

$$(\langle ik\rangle\langle jl\rangle - \langle ij\rangle\langle kl\rangle)[kj][li] = tr[P_-\hat{p}_j\{\hat{p}_l\hat{p}_i\}\hat{p}_k] = 4(p_ip_l)(p_jp_k),$$

$$-\langle il\rangle\langle kj\rangle[kj][li] = tr[P_-\hat{p}_j\hat{p}_k]tr[P_-\hat{p}_i\hat{p}_l]$$

$$= 4(p_ip_l)(p_jp_k). \tag{P4.120}$$

$$\sum_{\substack{i=1\\i\neq j,k}}^{n} [ji]\langle ik\rangle = \bar{u}(p_j)P_- \sum_{\substack{i=1\\i\neq j,k}}^{n} \hat{p}_i P_+ u(p_k)$$

$$= \bar{u}(p_j)P_- \sum_{i=1}^{n} \hat{p}_i P_+ u(p_k) = 0. \tag{P4.121}$$

c. Thanks to the Dirac equation $\hat{k}u_k = 0 = \bar{u}_q\hat{q}$

$$\varepsilon^\pm(k,q)k = \pm\frac{\langle q^\mp|\hat{k}|k^\mp\rangle}{\sqrt{2}\langle q^\mp|k^\pm\rangle} = 0 = \varepsilon^\pm(k,q)q. \tag{P4.122}$$

$$\varepsilon^+(\varepsilon^+)^* = \varepsilon^+\varepsilon^- = -\frac{\langle q^-|\gamma^\mu|k^-\rangle\langle q^+|\gamma_\mu|k^+\rangle}{2\langle qk\rangle[qk]}$$

$$= -\frac{\langle k^+|\gamma^\mu|q^+\rangle\langle q^+|\gamma_\mu|k^+\rangle}{2\langle qk\rangle[qk]} = -\frac{2[kq]\langle kq\rangle}{2\langle qk\rangle[qk]} = -1, \tag{P4.123}$$

$$\varepsilon^+(\varepsilon^-)^* = \varepsilon^+\varepsilon^+ = \frac{\langle q^-|\gamma^\mu|k^-\rangle\langle q^-|\gamma_\mu|k^-\rangle}{2\langle qk\rangle^2}$$

$$= \frac{\langle k^+|\gamma^\mu|q^+\rangle\langle k^+|\gamma_\mu|q^+\rangle}{2\langle qk\rangle^2} = \frac{2[kk]\langle qq\rangle}{2\langle qk\rangle^2} = 0. \tag{P4.124}$$

$$\varepsilon_\mu^+(k,q_1) - \varepsilon_\mu^+(k,q_2) = \frac{\langle q_1^-|\gamma_\mu|k^-\rangle}{\sqrt{2}\langle q_1^-|k^+\rangle} - \frac{\langle q_2^-|\gamma_\mu|k^-\rangle}{\sqrt{2}\langle q_2^-|k^+\rangle}$$

$$= \frac{\langle q_1^-|\gamma_\mu|k^-\rangle\langle q_2^-|k^+\rangle - \langle q_2^-|\gamma_\mu|k^-\rangle\langle q_1^-|k^+\rangle}{\sqrt{2}\langle q_1^-|k^+\rangle\langle q_2^-|k^+\rangle}$$

$$= \frac{\langle q_2^-|\hat{k}\gamma_\mu|q_1^+\rangle + \langle q_2^-|\gamma_\mu\hat{k}|q_1^+\rangle}{\sqrt{2}\langle q_1^-|k^+\rangle\langle q_2^-|k^+\rangle}$$

$$= -\frac{\sqrt{2}\langle q_1|q_2\rangle}{\langle q_1|k\rangle\langle q_2|k\rangle}k_\mu. \tag{P4.125}$$

$$\sum_{\lambda=\pm} \varepsilon_\mu^\lambda(k,q)\varepsilon_\nu^\lambda(k,q)^* = -\frac{\langle q^-|\gamma_\mu|k^-\rangle\langle q^+|\gamma_\nu|k^+\rangle}{2\langle qk\rangle[qk]} + (\mu \leftrightarrow \nu)$$

$$= -\frac{\langle q^-|\gamma_\mu\hat{k}\gamma_\nu + \gamma_\nu\hat{k}\gamma_\mu|q^-\rangle}{2\langle qk\rangle[qk]}$$

$$= -\frac{\langle q^-|\gamma_\mu\hat{k}\gamma_\nu + \gamma_\nu\hat{k}\gamma_\mu|q^-\rangle}{2\langle qk\rangle[qk]}$$

$$= -\frac{\langle q^-|\gamma_\mu k_\nu + \gamma_\nu k_\mu - g_{\mu\nu}\hat{k}|q^-\rangle}{\langle qk\rangle[qk]}$$

$$= g_{\mu\nu} - \frac{k_\nu\langle q^-|\gamma_\mu|q^-\rangle + k_\mu\langle q^-|\gamma_\nu|q^-\rangle}{2(kq)}$$

$$= g_{\mu\nu} - \frac{k_\nu q_\mu + k_\mu q_\nu}{(kq)}. \qquad \text{(P4.126)}$$

$$\varepsilon^+(k_1,q)\varepsilon^+(k_2,q) = \frac{\langle q^-|\gamma^\mu|k_1^-\rangle\langle q^-|\gamma_\mu|k_2^-\rangle}{2\langle qk_1\rangle\langle qk_2\rangle}$$

$$= \frac{2[k_1k_2]\langle qq\rangle}{2\langle qk_1\rangle\langle qk_2\rangle} = 0, \qquad \text{(P4.127)}$$

$$\varepsilon^+(k_1,k_2)\varepsilon^-(k_2,q) = -\frac{\langle k_2^-|\gamma^\mu|k_1^-\rangle\langle q^+|\gamma_\mu|k_2^+\rangle}{2\langle k_2k_1\rangle\langle qk_2\rangle}$$

$$= -\frac{\langle k_1^+|\gamma^\mu|k_2^+\rangle\langle q^+|\gamma_\mu|k_2^+\rangle}{2\langle k_2k_1\rangle\langle qk_2\rangle}$$

$$= -\frac{2[k_1q]\langle k_2k_2\rangle}{2\langle k_2k_1\rangle\langle qk_2\rangle} = 0, \qquad \text{(P4.128)}$$

$$\hat{\varepsilon}^+(k_1,k_2)|2^+\rangle = \sqrt{2}\frac{|k_2^+\rangle\langle k_1^+| + |k_1^-\rangle\langle k_2^-|}{\langle k_2k_1\rangle}|2^+\rangle = 0, \quad \text{(P4.129)}$$

$$\langle 2^-|\hat{\varepsilon}^+(k_1,k_2) = \langle 2^-|\sqrt{2}\frac{|k_2^+\rangle\langle k_1^+| + |k_1^-\rangle\langle k_2^-|}{\langle k_2k_1\rangle} = 0. \quad \text{(P4.130)}$$

4.14 Derive the **color-ordered Feynman rules** for the generators $\tau^a = t^a\sqrt{2}$ for the tree n-gluon amplitude.

SOLUTION. The generators τ obey the following relations

$$[\tau^a\tau^b] = i\sqrt{2}f^{abc}, \quad tr[\tau^a\tau^b] = \delta^{ab}, \qquad \text{(P4.131)}$$

$$f^{abc} = -\frac{i}{\sqrt{2}}tr([\tau^a\tau^b]\tau^c), \qquad \text{(P4.132)}$$

$$(\tau^a)_i^j(\tau^a)_k^l = \delta_i^l\delta_k^j - \frac{1}{N_c}\delta_i^j\delta_k^l, \qquad \text{(P4.133)}$$

which follow from (1.10, 1.13, 1.19) in Chapter 1.

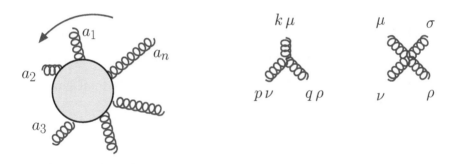

Figure 4.4 Color ordering of a tree amplitude, 3- and 4-gluon vertices. All momenta are outgoing.

The 3-gluon vertex (4.33) in Figure 4.2 then reads

$$-\frac{ig}{\sqrt{2}}tr([\tau^a\tau^b]\tau^c)V^{\mu\nu\rho}(k,p,q), \qquad (P4.134)$$

and the 4-guon vertex (4.34)

$$\begin{aligned}
&- ig^2[f^{abe}f^{cde}(g^{\mu\rho}g^{\nu\sigma} - g^{\mu\sigma}g^{\nu\rho}) + f^{ace}f^{bde}(g^{\mu\nu}g^{\rho\sigma} - g^{\mu\sigma}g^{\nu\rho})\\
&+ f^{ade}f^{bce}(g^{\mu\nu}g^{\rho\sigma} - g^{\mu\rho}g^{\nu\sigma})]\\
=&\frac{ig^2}{2}[tr([\tau^a\tau^b][\tau^c\tau^d])(g^{\mu\rho}g^{\nu\sigma} - g^{\mu\sigma}g^{\nu\rho})\\
&+ tr([\tau^a\tau^c][\tau^b\tau^d])(g^{\mu\nu}g^{\rho\sigma} - g^{\mu\sigma}g^{\nu\rho})\\
&+ tr([\tau^a\tau^d][\tau^b\tau^c])(g^{\mu\nu}g^{\rho\sigma} - g^{\mu\rho}g^{\nu\sigma})] \qquad (P4.135)
\end{aligned}$$

$$\begin{aligned}
=&\frac{ig^2}{2}[(tr(\tau^a\tau^b\tau^c\tau^d) + tr(\tau^a\tau^d\tau^c\tau^b))(2g^{\mu\rho}g^{\nu\sigma} - g^{\mu\sigma}g^{\nu\rho} - g^{\mu\nu}g^{\rho\sigma})\\
&+ (tr(\tau^a\tau^c\tau^d\tau^b) + tr(\tau^a\tau^b\tau^d\tau^c))(-g^{\mu\rho}g^{\nu\sigma} + 2g^{\mu\sigma}g^{\nu\rho} - g^{\mu\nu}g^{\rho\sigma})\\
&+ (tr(\tau^a\tau^d\tau^b\tau^c) + tr(\tau^a\tau^c\tau^b\tau^d))(-g^{\mu\rho}g^{\nu\sigma} - g^{\mu\sigma}g^{\nu\rho} + 2g^{\mu\nu}g^{\rho\sigma})].\\
&\qquad (P4.136)
\end{aligned}$$

For a chosen up to cyclic permutations color order (see Figure 4.4) one has the following kinematic structures for the 3-gluon vertex

$$\frac{i}{\sqrt{2}}V^{\mu\nu\rho}(k,p,q) = \frac{i}{\sqrt{2}}[g^{\mu\nu}(k-p)^\rho + g^{\nu\rho}(p-q)^\mu + g^{\rho\mu}(q-k)^\nu], \qquad (P4.137)$$

and for the 4-guon vertex

$$\frac{i}{2}(2g^{\mu\rho}g^{\nu\sigma} - g^{\mu\sigma}g^{\nu\rho} - g^{\mu\nu}g^{\rho\sigma}). \qquad (P4.138)$$

Compared to Figure 4.2 in the lecture we changed all momenta to outgoing in (P4.137). Any internal gluon line will have the convolution like

$$tr[A\tau^a B]tr[C\tau^a D] = tr[ADCB] \qquad (P4.139)$$

and after taking all such convolutions one comes to the total amplitude as a sum of partial color ordered amplitudes

$$M_n(\{p_i\}, \{a_i\}, \{\lambda_i\}) = g^{n-2} \sum_{\sigma \in S_n/Z_n} tr[\tau^{a_{\sigma(1)}} \tau^{a_{\sigma(2)}} ... \tau^{a_{\sigma(n)}}]$$
$$\times m(\sigma(1^{\lambda_1}), \sigma(2^{\lambda_2}), ..., \sigma(n^{\lambda_n})), \qquad (P4.140)$$

where the sum goes over al permutations σ in the set S_n/Z_n of all permutations excluding the cyclic ones, and p_i, λ_i and a_i are the momenta, helicities, and color indices of the external gluons, which are all considered outgoing.

4.15 Show that the color-stripped partial amplitudes

a. obey the **reflection identity**

$$m(1, 2, ..., n) = (-1)^n m(n, ..., 2, 1), \qquad (P4.141)$$

b. obey the **decoupling identity**

$$m(1, 2, 3, ..., n) + m(2, 1, 3, ..., n) + m(2, 3, 1, ..., n)$$
$$+ ... + m(2, 3, ..., 1, n) = 0, \qquad (P4.142)$$

c. vanish for all and all but one identical helicities

$$m(1^{\pm}, 2^+, ..., n^+) = m(1^{\pm}, 2^-, 3^-, ..., n^-) = 0. \qquad (P4.143)$$

SOLUTION.

a. Reversal of the order of momenta changes the vertex sign for the 3-gluon vertex and does not change it for the 4-gluon vertex. Therefore for the even number of external gluons the partial amplitude does not change, while for the odd number of gluons it changes sign.

b. Color decomposition of tree amplitudes (P4.140) is valid for the $U(N_c)$ group as well as for the $SU(N_c)$. Therefore one can introduce the N_c^2-th generator $\tau^a \sim \delta_i^j$, which commutes with the remaining generators of the $SU(N_c)$. Therefore for this generator all structure constants are zero, i.e. the corresponding particle does not interact with the gluons. It means that an amplitude with such a particle as particle 1 and $n-1$ gluons vanish. The color decomposition for such an amplitude has exactly the decoupling identity as a coefficient of the color trace since one can place the identity generator corresponding to particle 1 in any position in the trace.

c. Each vertex in the amplitude contributes no more than one external momentum to the numerator. The amplitude with n gluons has no more than $n - 2$ vertices. Therefore it has n polarization vectors contracted with no more than $n - 2$ external momenta. Hence, it has at least one convolution $(\varepsilon_i \varepsilon_j)$. For the amplitude with all helicities of the same kind one can choose the same gauge vector for all particles. Then, thanks to (P4.127) all convolutions $(\varepsilon_i^\pm \varepsilon_j^\pm) = 0$. For the amplitude with all but the first particle's helicity of the same kind one can take the gauge momentum $q_2 = \ldots = q_n = p_1$, $q_1 = p_2$. Then $(\varepsilon_1^+ \varepsilon_j^-) = 0$ thanks to (P4.128).

4.16 Calculate the tree $gg \to gg$ cross-section.

SOLUTION. Here we follow (MP91). We start from $m(1^-, 2^-, 3^+, 4^+)$. We will take the momenta $p_{1,2,3,4}$ of all 4 gluons outgoing at the beginning and take the gauge vectors q_i :

$$q_1 = q_2 = p_4, \quad q_3 = q_4 = p_1. \tag{P4.144}$$

Then there is only one nonzero convolution of ε : $(\varepsilon_2^- \varepsilon_3^+) \neq 0$, and only one nonzero diagram $\sim \frac{1}{s_{12}}$ out of 4 in the third column of Figure 4.3

$$m(1^-, 2^-, 3^+, 4^+) = -i(\frac{i}{\sqrt{2}})^2 \frac{-i}{s_{12}}$$

$$\times \varepsilon_{1\mu}^- \varepsilon_{2\nu}^- \varepsilon_{3\alpha}^+ \varepsilon_{4\beta}^+ V^{\mu\nu\rho}(p_1, p_2, -p_1 - p_2) V^{\alpha\beta}{}_\rho(p_3, p_4, -p_3 - p_4)$$

$$= \frac{1}{2s_{12}} \left[(\varepsilon_1^- \varepsilon_2^-)(p_1 - p_2)^\rho + \varepsilon_2^{-\rho}(2p_2 + p_1)\varepsilon_1^- + \varepsilon_1^{-\rho}(-2p_1 - p_2)\varepsilon_2^- \right]$$

$$\times \left[(\varepsilon_4^+ \varepsilon_3^+)(p_3 - p_4)_\rho + \varepsilon_{4\rho}^+(2p_4 + p_3)\varepsilon_3^+ + \varepsilon_{3\rho}^+(-2p_3 - p_4)\varepsilon_4^+ \right]$$

$$= \frac{-2}{s_{12}} (\varepsilon_2^- \varepsilon_3^+)(p_2 \varepsilon_1^-)(p_3 \varepsilon_4^+) \tag{P4.145}$$

$$= \frac{2}{[12]\langle 21 \rangle} \frac{\langle p_4^+ | \gamma_\mu | p_2^+ \rangle}{\sqrt{2}[p_4 p_2]} \frac{\langle p_1^- | \gamma^\mu | p_3^- \rangle}{\sqrt{2}\langle p_1 p_3 \rangle} \frac{\langle p_4^+ | \hat{p}_2 | p_1^+ \rangle}{\sqrt{2}[p_4 p_1]} \frac{\langle p_1^- | \hat{p}_3 | p_4^- \rangle}{\sqrt{2}\langle p_1 p_4 \rangle}$$

$$= \frac{-2}{[12]\langle 21 \rangle} \frac{2[43]\langle 12 \rangle}{4[42]\langle 13 \rangle} \frac{[42]\langle 21 \rangle \langle 13 \rangle [34]}{[41]\langle 14 \rangle}$$

$$= \frac{[43]^2 \langle 12 \rangle}{[12][41]\langle 14 \rangle} \times \frac{\langle 23 \rangle \langle 34 \rangle}{\langle 23 \rangle \langle 34 \rangle} = \frac{\langle 12 \rangle^4}{\langle 12 \rangle \langle 23 \rangle \langle 34 \rangle \langle 41 \rangle}. \tag{P4.146}$$

In the last line we used momentum conservation (P4.90)

$$\langle 34 \rangle [43] = [12]\langle 21 \rangle = s_{12} = s_{34}, \tag{P4.147}$$

$$[43]\langle 32 \rangle + [41]\langle 12 \rangle = 0. \tag{P4.148}$$

Thanks to Shouten (P4.89) and decoupling identities (P4.142)

$$m(1^-, 2^+, 3^-, 4^+) = -m(2^+, 1^-, 3^-, 4^+) - m(2^+, 3^-, 1^-, 4^+)$$

$$= -\frac{\langle 13 \rangle^3}{\langle 34 \rangle \langle 42 \rangle \langle 21 \rangle} + \frac{\langle 13 \rangle^3}{\langle 14 \rangle \langle 42 \rangle \langle 23 \rangle}$$

$$= \frac{\langle 13 \rangle^3 (-\langle 14 \rangle \langle 23 \rangle + \langle 34 \rangle \langle 21 \rangle)}{\langle 34 \rangle \langle 21 \rangle \langle 14 \rangle \langle 42 \rangle \langle 23 \rangle}$$

$$= |\langle 34 \rangle \langle 21 \rangle = \langle 32 \rangle \langle 41 \rangle + \langle 31 \rangle \langle 24 \rangle|$$

$$= \frac{\langle 13 \rangle^4}{\langle 12 \rangle \langle 23 \rangle \langle 34 \rangle \langle 41 \rangle}. \qquad \text{(P4.149)}$$

Parity conservation of strong interactions means that under P transformation the gluons change helicities and the amplitude gets and overall sign $(-1)^n$. Therefore, e.g.

$$m(1^+, 2^-, 3^+, 4^-) = \frac{[13]^4}{[12][23][34][41]}. \qquad \text{(P4.150)}$$

Finally, the amplitude for the specific helicities reads

$$M_4 = g^2 tr[\tau^{a_1} \tau^{a_2} \tau^{a_3} \tau^{a_4}] m(1, 2, 3, 4)$$
$$+ \text{ all noncyclic premutations of 1234} \qquad \text{(P4.151)}$$
$$= g^2 \{[a_1, a_2, a_3, a_4] m(1, 2, 3, 4) + [a_1, a_3, a_2, a_4] m(1, 3, 2, 4)$$
$$+ [a_1, a_2, a_4, a_3] m(1, 2, 4, 3)\}, \qquad \text{(P4.152)}$$

$$[a_1, a_2, a_3, a_4] \overset{def}{=} tr[\tau^{a_1} \tau^{a_2} \tau^{a_3} \tau^{a_4}] + tr[\tau^{a_4} \tau^{a_3} \tau^{a_2} \tau^{a_1}]. \qquad \text{(P4.153)}$$

Here we used reflection identity (P4.141). Then

$$\sum_{\{2,3,4\}} |M_4|^2 = g^4 \sum_{\{2,3,4\}'} m(1, 2, 3, 4) \sum_{\text{col}} [a_1, a_2, a_3, a_4]$$
$$\times ([a_1, a_2, a_3, a_4]^* m(1, 2, 3, 4)^*$$
$$+ [a_1, a_3, a_2, a_4]^* m(1, 3, 2, 4)^*$$
$$+ [a_1, a_2, a_4, a_3]^* m(1, 2, 4, 3)^*), \qquad \text{(P4.154)}$$

Where $\{\}$ stands for all permutations and $\{\}'$ stands for permutations different under reflection $((1234) \to (4321))$.

$$\sum_{\text{col}} [a_1, a_2, a_3, a_4][a_1, a_2, a_3, a_4]^* = \sum_{\text{col}} [a_1, a_2, a_3, a_4]^2,$$

$$\sum_{\text{col}} [a_1, a_2, a_3, a_4][a_1, a_3, a_2, a_4]^* = \sum_{\text{col}} [a_1, a_2, a_3, a_4]$$
$$\times ([a_1, a_2, a_3, a_4]^* + i\sqrt{2} f^{a_3 a_2 c} (tr[\tau^{a_1} \tau^c \tau^{a_4}] - tr[\tau^{a_4} \tau^c \tau^{a_1}])) \qquad \text{(P4.155)}$$

$$= \sum_{\text{col}} [a_1, a_2, a_3, a_4]^2 + i f^{a_2 a_3 d} (tr[\tau^{a_1} \tau^d \tau^{a_4}] - tr[\tau^{a_4} \tau^d \tau^{a_1}])$$

$$\times i f^{a_3 a_2 c} (tr[\tau^{a_1} \tau^c \tau^{a_4}] - tr[\tau^{a_4} \tau^c \tau^{a_1}])$$

$$= \sum_{\text{col}} [a_1, a_2, a_3, a_4]^2 + N_c (tr[\tau^{a_1} \tau^c \tau^{a_4}] - tr[\tau^{a_4} \tau^c \tau^{a_1}])^2$$

$$= \sum_{\text{col}} [a_1, a_2, a_3, a_4]^2 - 2N_c^2 (N_c^2 - 1), \tag{P4.156}$$

$$\sum_{\text{col}} [a_1, a_2, a_3, a_4][a_1, a_2, a_4, a_3]^* = \sum_{\text{col}} [a_1, a_2, a_3, a_4]^2 - 2N_c^2 (N_c^2 - 1).$$
$$\tag{P4.157}$$

Then, using decoupling identity (P4.142), $\sum_{\text{col}} [a_1, a_2, a_3, a_4]^2$ cancels

$$\sum_{\{2,3,4\}} |M_4|^2 = 2N_c^2 (N_c^2 - 1) g^4 \sum_{\{2,3,4\}'} |m(1,2,3,4)|^2$$

$$= 2N_c^2 (N_c^2 - 1) g^4 \times 2 \Big[\frac{s_{12}^4 + s_{13}^4 + s_{14}^4}{s_{12} s_{23} s_{34} s_{41}}$$

$$+ \frac{s_{12}^4 + s_{13}^4 + s_{14}^4}{s_{12} s_{24} s_{34} s_{31}} + \frac{s_{12}^4 + s_{13}^4 + s_{14}^4}{s_{13} s_{23} s_{24} s_{41}} \Big]$$

$$= 4N_c^2 (N_c^2 - 1) g^4 [s^4 + t^4 + u^4] \Big[\frac{1}{s^2 t^2} + \frac{1}{s^2 u^2} + \frac{1}{t^2 u^2} \Big]$$

$$= 16 N_c^2 (N_c^2 - 1) g^4 \Big[3 - \frac{us}{t^2} - \frac{ut}{s^2} - \frac{st}{u^2} \Big]. \tag{P4.158}$$

Here

$$s = (-p_1 - p_2)^2 = (p_3 + p_4)^2 = s_{12} = s_{34},$$

$$t = (-p_1 + p_4)^2 = (p_3 - p_2)^2 = s_{14} = s_{23},$$

$$u = (-p_1 + p_3)^2 = (p_4 - p_2)^2 = s_{13} = s_{24},$$

$$s + t + u = 0. \tag{P4.159}$$

The cross-section reads

$$\frac{d\sigma}{d(-t)} = \frac{1}{2} \frac{1}{64\pi} \frac{4}{s^2} \frac{1}{4(N_c^2 - 1)^2} 16 N_c^2 (N_c^2 - 1) g^4 \Big[3 - \frac{us}{t^2} - \frac{ut}{s^2} - \frac{st}{u^2} \Big]$$

$$= \frac{1}{32\pi s^2} \frac{9}{2} (4\pi\alpha_s)^2 \Big[3 - \frac{us}{t^2} - \frac{ut}{s^2} - \frac{st}{u^2} \Big]. \tag{P4.160}$$

Here $\frac{1}{2}$ takes into account the identity of the final gluons, $\frac{s^2}{4} = I^2$ is the flux invariant, $\frac{1}{4(N_c^2 - 1)^2}$ is the averaging over the initial gluons' helicities and color. For fixed $\alpha_s = 0.118$ with the cut $-t \simeq p_\perp^2 > 0.01s$, one gets

$$s\sigma = s \int_{0.01s}^{s - 0.01s} d\sigma \simeq 19. \tag{P4.161}$$

TABLE 4.1   The stripped of charge squared matrix elements summed over the final and averaged over the initial colors and polarizations, see e. g. (Man98).

$qQ \to qQ$	$\frac{4}{9}\frac{s^2+u^2}{t^2}$
$qq \to qq$	$\frac{1}{2}\left[\frac{4}{9}\frac{s^2+u^2}{t^2} + \frac{4}{9}\frac{s^2+t^2}{u^2} - \frac{8}{27}\frac{s^2}{ut}\right]$
$q\bar{q} \to Q\bar{Q}$	$\frac{4}{9}\frac{t^2+u^2}{s^2}$
$q\bar{q} \to q\bar{q}$	$\frac{4}{9}\frac{s^2+u^2}{t^2} + \frac{4}{9}\frac{u^2+t^2}{s^2} - \frac{8}{27}\frac{u^2}{st}$
$q\bar{q} \to gg$	$\frac{1}{2}\left[\frac{32}{27}\frac{t^2+u^2}{tu} - \frac{8}{3}\frac{t^2+u^2}{s^2}\right]$
$gg \to q\bar{q}$	$\frac{1}{6}\frac{t^2+u^2}{tu} - \frac{3}{8}\frac{t^2+u^2}{s^2}$
$gq \to gq$	$-\frac{4}{9}\frac{s^2+u^2}{su} + \frac{s^2+u^2}{t^2}$
$gg \to gg$	$\frac{1}{2}\frac{9}{2}\left[3 - \frac{us}{t^2} - \frac{ut}{s^2} - \frac{st}{u^2}\right]$

4.17   Calculate the other parton level cross-sections $s\sigma$ shown in Figure 4.3 for massless partons with the cut $p_\perp^2 > 0.01s$ and the fixed $\alpha_s = 0.118$.

ANSWER. See Figure 11.9. The stripped of charge squared matrix elements summed over the final and averaged over the initial colors and polarizations are given in Table 4.1.

4.18   Show the **Britto-Cachazo-Feng-Witten recursion relation** for tree gluon amplitudes

$$M = \sum_{ij}\sum_{h} M_L^h(z_{ij})\frac{1}{P_{ij}^2}M_R^{-h}(z_{ij}), \quad z_{ij} = \frac{2P_{ij}^2}{\langle k^-|\hat{P}_{ij}|n^-\rangle}, \quad \text{(P4.162)}$$

$$P_{ij} = p_i + ... + p_j, \quad j > i. \quad \text{(P4.163)}$$

Here $M_{L,R}$ are the amplitudes on the left and on the right of the propagator $\frac{1}{P_{ij}^2}$ as is shown in Figure 4.5. The first sum goes over the helicities h and the second sum goes over the all separations of the amplitude so that the gluon with the momentum p_k and helicity $(-)$ is in the right sum and the gluon with the momentum p_n and helicity $(+)$ is in the left sum.

SOLUTION. We follow (BCFW05). One considers a tree on-shell amplitude $M(1, 2, ..., n)$ of n cyclically ordered outgoing gluons and shifts the k-th and n-th momenta with a complex z and a null vector q so that

$$q^2 = 0, \quad qp_k = qp_n = 0, \qquad \text{(P4.164)}$$

$$p_k \to p_k(z) = p_k + zq, \quad p_k^2(z) = p_k^2, \qquad \text{(P4.165)}$$

$$p_n \to p_n(z) = p_n - zq, \quad p_n^2(z) = p_n^2, \qquad \text{(P4.166)}$$

$$p_n + p_k = p_n(z) + p_k(z), \qquad \text{(P4.167)}$$

$$p_s(z) = p_s, s \neq k, n. \qquad \text{(P4.168)}$$

Then, one defines the amplitude as a complex function of the shifted momenta

$$M(z) = M(p_1, ..., p_k(z), p_{k+1}, ..., p_{n-1}, p_n(z)). \qquad \text{(P4.169)}$$

In this amplitude for all z all external momenta are on-shell and momentum is conserved $\sum p_j = \sum p_j(z) = 0$. It has only simple poles in z coming from internal propagators since any internal line carries a sum of external momenta ajacent to it, e. g. the internal line in Figure 4.5 has momentum P_{ij} :

$$P_{ij} = p_i + ... + p_j = -(p_{j+1} + ... + p_{i-1}) \to P_{ij}(z), \quad j > i. \qquad \text{(P4.170)}$$

If both p_k and p_n are present or absent in the sum $p_i + ... + p_j$ then $P_{ij}(z)$ does not depend in z. Suppose n is present in this sum and k is not. Then the propagator of this internal line will have a simple pole at

$$P_{ij}(z)^2 = P_{ij}^2 - zP_{ij}q = 0 \quad \implies \quad z_{ij} = \frac{P_{ij}^2}{P_{ij}q}. \qquad \text{(P4.171)}$$

Assume that $M(z) \to 0$ as $z \to \infty$. Then

$$M(z) = \sum_{ij} \frac{c_{ij}}{z - z_{ij}} \quad \implies \quad M(0) = -\sum_{ij} \frac{c_{ij}}{z_{ij}}, \qquad \text{(P4.172)}$$

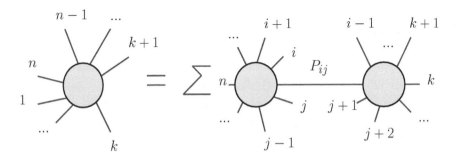

Figure 4.5 Separation of the cyclically ordered gluon tree amplitude into the left and right amplitudes connected via the propagator with the momentum P_{ij}. All momenta are outgoing. Each part contains at least 2 external particles.

where c_{ij} are the residues of the poles and $M(0)$ is the physical amplitude. To get a pole one takes the internal propagator separating the diagram into 2 parts left and right as in Figure 4.5. The contribution with this propagator reads

$$\sum_h \frac{M_L^h(z)M_R^{-h}(z)}{P_{ij}(z)^2} \implies c_{ij} = \sum_h \frac{M_L^h(z_{ij})M_R^{-h}(z_{ij})}{-qP_{ij}}, \quad \text{(P4.173)}$$

where $M_{L,R}$ are the amplitudes on the left and on the right of the propagator and the sum goes over the helicities h and other quantum numbers. In the poles $M_{R,L}(z_{ij})$ become on-shell physical amplitudes. Then

$$M(0) = -\sum_{ij} \frac{c_{ij}}{z_{ij}} = \sum_{ij}\sum_h \frac{M_L^h(z_{ij})M_R^{-h}(z_{ij})}{P_{ij}^2}. \quad \text{(P4.174)}$$

Since for all helicities of the same kind the amplitude vanishes, we take the gluons n and k with the helicities $(+-)$. For them one takes

$$q^\mu = \frac{1}{2}\langle k^-|\gamma^\mu|n^-\rangle = \frac{1}{2}\langle n^+|\gamma^\mu|k^+\rangle, \quad qp_n = qp_k = 0, \quad \text{(P4.175)}$$

$$4q^2 = \langle k^-|\gamma^\mu|n^-\rangle\langle k^-|\gamma_\mu|n^-\rangle = \langle k^-|2(|k^+\rangle\langle n^+| + |n^-\rangle\langle k^-|)|n^-\rangle = 0, \quad \text{(P4.176)}$$

$$p_n^\mu(z) = \frac{1}{2}\langle n^+|\gamma^\mu|n^+\rangle - \frac{z}{2}\langle n^+|\gamma^\mu|k^+\rangle = \frac{1}{2}\langle n^+|\gamma^\mu(|n^+\rangle - z|k^+\rangle)$$

$$= \frac{1}{2}((\langle n^-| - z\langle k^-|)\gamma^\mu|n^-\rangle; \quad \text{(P4.177)}$$

$$p_k^\mu(z) = \frac{1}{2}\langle k^+|\gamma^\mu|k^+\rangle + \frac{z}{2}\langle n^+|\gamma^\mu|k^+\rangle = \frac{1}{2}((\langle k^+| + z\langle n^+|)\gamma^\mu|k^+\rangle$$

$$= \frac{1}{2}\langle k^-|\gamma^\mu(|k^-\rangle + z|n^-\rangle). \quad \text{(P4.178)}$$

Therefore for this q

$$\langle n^-| \to \langle n^-| - z\langle k^-|, \quad \langle n^+| \to \langle n^+|, \quad \text{(P4.179)}$$

$$|n^+\rangle \to |n^+\rangle - z|k^+\rangle, \quad |n^-\rangle \to |n^-\rangle; \quad \text{(P4.180)}$$

$$\langle k^+| \to \langle k^+| + z\langle n^+|, \quad \langle k^-| \to \langle k^-|, \quad \text{(P4.181)}$$

$$|k^-\rangle \to |k^-\rangle + z|n^-\rangle, \quad |k^+\rangle \to |k^+\rangle. \quad \text{(P4.182)}$$

Is $M(z) \to 0$ as $z \to \infty$ for this q?

4.19 Show that the ghosts are necessary to get the unitarity relation in $qq \to \tilde{q}\bar{\tilde{q}}$ in the 4-th order of perturbation theory.

SOLUTION. There are diagrams with gluons, ghosts and quarks in the intermediate state in the 4-th order for $qq \to \tilde{q}\bar{\tilde{q}}$. They are shown in

Figure 4.6. The diagram with the 4-gluon vertex does not have an imaginary part and we do not draw it. The -1's show that the fermion loops have (-1) according to the Feynman rules. We work in the gauge with $\xi = 1$ and have $-ig_{\mu\nu}$ in the numerator of the gluon propagator. According to the Cutkosky rule, the discontinuity of the diagram comes from the substitution

$$\frac{i}{q^2 + i0}\frac{i}{k^2 + i0} \to (-2\pi i)i\delta(k^2)(-2\pi i)i\delta(q^2). \tag{P4.183}$$

Therefore

$$-i\, discM = 2\operatorname{Im} M = \int \frac{d^4 k}{(2\pi)^2}\,\delta(q^2)\,\delta(k^2)\left[\frac{1}{2}G^{\mu\nu}G^*_{\mu\nu} + Q^*Q - S^*S\right]. \tag{P4.184}$$

Here G, Q and S are the anplitudes of $q\bar{q} \to gg$, $q\bar{q} \to q'\bar{q}'$, and $q\bar{q} \to ghost + antighost$ processes shown in Figure 4.6. The $\frac{1}{2}$ appears because the convolution $G^{\mu\nu}G^*_{\mu\nu}$ takes all the diagrams twice (recall that the diagram with the gluon loop has a symmetry factor $\frac{1}{2}$ itself), Q^*Q comes with the sign $+$ since the numerators of the quark propagators on the mass shell

$$\hat{q} + m = \sum_s u^s_q \bar{u}^s_q, \quad -\hat{k} + m = -\sum_s v^s_k \bar{v}^s_k, \tag{P4.185}$$

and the minus sign in front of the antiquark polarization sum compensates the (-1) from the diagram. According to the unitarity relation

$$2\operatorname{Im} M = \int (2\pi)^4 \delta(p - q - k)\frac{d^3 k}{(2\pi)^3 2E_k}\frac{d^3 q}{(2\pi)^3 2E_q}$$
$$\times [\frac{1}{2}\sum_{ij=1,2} G_{\mu\nu}\varepsilon^{*\nu}_i(q)\varepsilon^{*\mu}_j(k)G^*_{\mu'\nu'}\varepsilon^{\nu'}_i(q)\varepsilon^{\mu'}_j(k) + QQ^*]. \tag{P4.186}$$

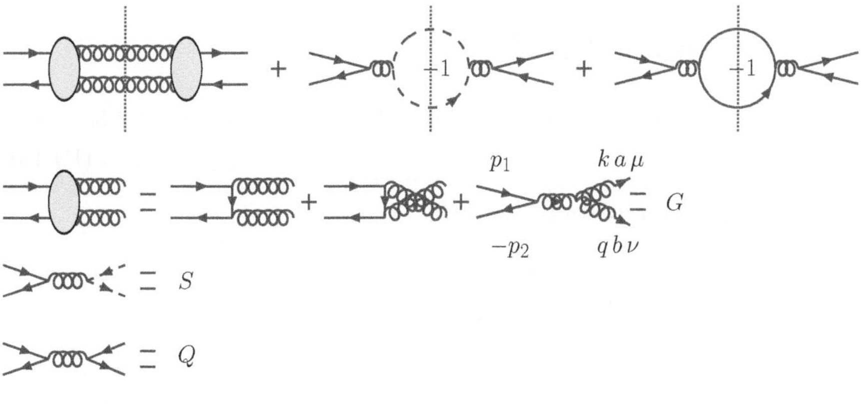

Figure 4.6 Diagrams for calculation of the imaginary part of $q\bar{q} \to \tilde{q}\bar{\tilde{q}}$.

Here $\frac{1}{2}$ takes into account the identity of the produced gluons. We sum only over the physical polarization vectors and do not have ghosts. We denote the vectors of the outgoing particles in these amplitudes as k and q and choose $\varepsilon_{1,2}$ as the transverse polarization vectors for both gluons in G with the momenta k and q. Since $k^2 = q^2 = 0$, four vectors k, q, ε_1 and ε_2 comprise the basis in the 4 dimensional space and

$$g^{\nu\nu'} = \frac{1}{kq}(k^\nu q^{\nu'} + k^{\nu'} q^\nu) - \varepsilon_1^{*\nu}\varepsilon_1^{\nu'} - \varepsilon_2^{*\nu}\varepsilon_2^{\nu'}. \qquad \text{(P4.187)}$$

Then

$$2\,\mathrm{Im}\,M = \int \frac{d^4k}{(2\pi)^2}\delta(q^2)\,\delta(k^2)[QQ^* + \frac{1}{2}G_{\mu\nu}G^*_{\mu'\nu'}$$

$$\times \{\frac{k^\nu q^{\nu'} + k^{\nu'} q^\nu}{kq} - g^{\nu\nu'}\}\{\frac{k^\mu q^{\mu'} + k^{\mu'} q^\mu}{kq} - g^{\mu\mu'}\}]|_{q=p-k}.$$

$$\text{(P4.188)}$$

Therefore to check consistency of the calculation, we have to show that on the mass shell

$$-S^* S = \frac{1}{2}G_{\mu\nu}G^*_{\mu'\nu'}\{\frac{1}{(kq)^2}(k^\nu q^{\nu'} + k^{\nu'} q^\nu)(k^\mu q^{\mu'} + k^{\mu'} q^\mu)$$

$$- \frac{g^{\mu\mu'}}{kq}(k^\nu q^{\nu'} + k^{\nu'} q^\nu) - \frac{g^{\nu\nu'}}{kq}(k^\mu q^{\mu'} + k^{\mu'} q^\mu)\}. \qquad \text{(P4.189)}$$

We have

$$S = -ig^2\bar{v}_2 i\gamma_\rho t^c u_1 \frac{-ig^{\rho\nu}}{(k+q)^2}(-f^{bca}q_\nu) = ig^2\bar{v}_2\hat{q}t^c u_1 \frac{f^{bca}}{2kq}, \qquad \text{(P4.190)}$$

$$G_{\mu\nu} = -ig^2\bar{v}_2 \left[\frac{i\gamma_\nu i(\hat{p}_1 - \hat{k} + m)i\gamma_\mu}{(p_1 - k)^2 - m^2}t^b t^a + \frac{i\gamma_\mu i(\hat{k} - \hat{p}_2 + m)i\gamma_\nu}{(p_2 + k)^2 - m^2}t^a t^b\right.$$

$$+ i\gamma_\alpha t^c \frac{-ig^{\rho\alpha}}{(k+q)^2}f^{abc}$$

$$\left.\times (g_{\mu\nu}(-k+q)_\rho + g_{\nu\rho}(-q - (k+q))_\mu + g_{\rho\mu}(k+q+k)_\nu)\right] u_1.$$

$$\text{(P4.191)}$$

The convolutions read

$$G_{\mu\nu}k^\mu = g^2\bar{v}_2 \left[\gamma_\nu \frac{\hat{k}(-\hat{p}_1 + m) + 2(kp_1)}{2(p_1 k)}t^b t^a - \frac{(\hat{p}_2 + m)\hat{k} + 2kp_2}{2(p_2 k)}\gamma_\nu t^a t^b\right.$$

$$\left.+ \gamma_\rho \frac{-if^{abc}t^c}{2(kq)}(k_\nu(-k+q)_\rho - 2g_{\nu\rho}(qk) + k_\rho(2k+q)_\nu)\right] u_1$$

$$= g^2 \bar{v}_2 \left[\gamma_\nu [t^b t^a] + \gamma_\nu t^c i f^{abc} + \frac{-i f^{abc}}{2(kq)} (k_\nu q_\rho + k_\rho (k+q)_\nu) \right] u_1$$

$$= g^2 \bar{v}_2 \gamma_\rho \left[\frac{-i f^{abc} t^c}{2(kq)} (k_\nu (q+k)_\rho + k_\rho q_\nu) \right] u_1$$

$$= g^2 \frac{-i f^{abc}}{2(kq)} q_\nu \bar{v}_2 t^c \hat{k} u_1 = -S q_\nu, \tag{P4.192}$$

$$G_{\mu\nu} k^\mu q^\nu = 0, \tag{P4.193}$$

$$G_{\mu\nu} q^\nu = -S k_\mu. \tag{P4.194}$$

Here we commute the gamma matrices and use $k^2 = \hat{k}^2 = 0$, $k+q = p_1 + p_2$, and the Dirac equation

$$(\hat{p}_1 - m)u_1 = 0, \quad \bar{v}_2(\hat{p}_2 + m) = 0. \tag{P4.195}$$

Therefore (P4.189) reads

$$\frac{1}{2} G_{\mu\nu} G^*_{\mu'\nu'} \{ \frac{k^\nu q^{\nu'} k^\mu q^{\mu'} + k^{\nu'} q^\nu k^{\mu'} q^\mu}{(kq)^2}$$

$$- \frac{g^{\mu\mu'}}{kq} (k^\nu q^{\nu'} + k^{\nu'} q^\nu) - \frac{g^{\nu\nu'}}{kq} (k^\mu q^{\mu'} + k^{\mu'} q^\mu) \}$$

$$= \frac{1}{2} [2SS^* - 2SS^* - 2S^*S] = -S^*S. \tag{P4.196}$$

One can see that the ghost loops are necessary to keep unitarity. They cancel contributions from the nontransverse polarization vectors of gluons.

4.20 Show that in the covariant gauge the QCD generating functional can be written with the **auxiliary field** B^a:

$$\mathcal{Z}[0] = const \int \mathcal{D}A \mathcal{D}c \mathcal{D}\bar{c} \mathcal{D}B e^{iS_{QCD} + iS_{ghost} + \frac{i}{2} \int dx \{\xi (B^a)^2 + 2B^a (\partial^\mu A^a_\mu)\}}, \tag{P4.197}$$

where S_{QCD} is given in (4.23) and S_{ghost} is given in (4.25).

SOLUTION.

$$e^{iS_{fix}} = e^{-\frac{i}{2\xi} \int dx (\partial^\mu A^a_\mu)^2}$$

$$= const \int \mathcal{D}B^a e^{\frac{i}{2\xi} \int dx \{(\xi B^a)^2 + 2\xi B^a (\partial^\mu A^a_\mu) + (\partial^\mu A^a_\mu)^2\} - \frac{i}{2\xi} \int dx (\partial^\mu A^a_\mu)^2}$$

$$= const \int \mathcal{D}B^a e^{-\frac{i}{2} \int dx \{-\xi (B^a)^2 - 2B^a (\partial^\mu A^a_\mu)\}}. \tag{P4.198}$$

We will also need the integral

$$const \int \mathcal{D}BB_x^a e^{-\frac{i}{2}\int dx\{-\xi(B^a)^2-2B^a(\partial^\mu A_\mu^a)\}}$$

$$= -\frac{1}{\xi}(\partial^\mu A_{x\mu}^a)e^{-\frac{i}{2\xi}\int dx(\partial^\mu A_\mu^a)^2}, \qquad (P4.199)$$

meaning that after integration linear terms in $B_x^a \to -\frac{1}{\xi}(\partial^\mu A_\mu^a)$. Therefore the QCD lagrangian with the auxiliary field B reads

$$\mathcal{L} = -\frac{1}{4}F^2 + \bar\psi(i\hat\partial - m)\psi + \frac{1}{2}\xi(B^a)^2 + B^a(\partial^\mu A_\mu^a) + \bar{c}_x^a(-\partial^\mu D_\mu^{ac}c_x^c).$$
$$(P4.200)$$

4.21 Show that after gauge fixing QCD Lagrangian (P4.200) **is symmetric under the Becchi-Rouet-Stora-Tyutin (BRST) transformation** with a constant infinitesimal Grassmann ε

$$\delta A_\mu^a = \varepsilon D_\mu^{ac}c^c, \quad \delta\psi = ig\varepsilon c^a t^a\psi, \quad \delta\bar\psi = ig\varepsilon\bar\psi t^a c^a, \qquad (P4.201)$$

$$\delta c^a = -\frac{1}{2}g\varepsilon f^{abc}c^b c^c, \quad \delta\bar{c}^a = \varepsilon B^a, \quad \delta B^a = 0. \qquad (P4.202)$$

One can write it as

$$\phi \to \phi' = \phi + \delta\phi, \quad \delta\phi = \varepsilon\hat{Q}\phi, \quad \phi = \{A_\mu^a, \psi, c^a, \bar{c}^a, B^a\} \qquad (P4.203)$$

with the **nonlinear BRST operator $\hat{Q}$**.

SOLUTION. Comparing the BRST transformation with the infinitesimal gauge transformations for A^θ (2.11), ψ (2.3), and $\bar\psi$ (2.4)

$$\delta A_\mu^{\theta a} = \frac{1}{g}D_\mu^{ac}\theta^c = \frac{1}{g}(\partial_\mu\theta^a + gf^{abc}A_\mu^b\theta^c), \qquad (P4.204)$$

$$\delta\psi = i\theta^a t^a, \quad \delta\bar\psi = \bar\psi(-i\theta^a t^a), \qquad (P4.205)$$

one can see that the BRST transformation is a gauge transformation with $\theta_x^a = g\varepsilon c_x^a$ for $\mathcal{L}_{QCD}$. Hence, $\mathcal{L}_{QCD}$ is invariant.

$$\delta(B^a(\partial^\mu A_\mu^a)) = B^a(\partial^\mu\varepsilon D_\mu^{ac}c^c), \qquad (P4.206)$$

$$\delta(\bar{c}^a(-\partial^\mu D_\mu^{ac}c^c)) = \delta(\bar{c}^a)(-\partial^\mu D_\mu^{ac}c^c) + \bar{c}^a\delta(-\partial^\mu D_\mu^{ac}c^c). \qquad (P4.207)$$

$$\delta(B^a(\partial^\mu A_\mu^a)) = B^a(\partial^\mu\varepsilon D_\mu^{ac}c^c) = -\delta(\bar{c}^a)(-\partial^\mu D_\mu^{ac}c^c), \qquad (P4.208)$$

$$\bar{c}^a\delta(-\partial^\mu D_\mu^{ac}c^c) = -\frac{1}{2}g\bar{c}^a(-\partial^\mu)D_\mu^{ac}\varepsilon f^{cbf}c^b c^f$$
$$+ \bar{c}^a(-\partial^\mu gf^{abc}\varepsilon(D_\mu^{bd}c^d))c^c. \qquad (P4.209)$$

Using anticommutativity of c,

$$f^{abc}(\partial_\mu c^b c^c) = f^{abc}(\partial_\mu c^b)c^c + f^{abc}c^b(\partial_\mu c^c)$$
$$= f^{abc}(\partial_\mu c^b)c^c - f^{abc}(\partial_\mu c^c)c^b = 2f^{abc}(\partial_\mu c^b)c^c. \quad \text{(P4.210)}$$

Then,

$$\bar{c}^a \delta(-\partial^\mu D_\mu^{ac} c^c) = -\frac{1}{2}gf^{abc}\bar{c}^a(-\partial^2\varepsilon)c^b c^c + \frac{1}{2}gf^{abc}\bar{c}^a(-\partial^2\varepsilon)c^b c^c$$

$$-\frac{1}{2}g^2\bar{c}^a(-\partial^\mu f^{adc}A_\mu^d)\varepsilon f^{cbf}c^b c^f + \bar{c}^a(-\partial^\mu g^2 f^{abc}\varepsilon f^{bsd}A_\mu^s c^d)c^c$$

$$= -\frac{1}{2}g^2\bar{c}^a\varepsilon(-\partial^\mu)(f^{dao}f^{boc} - 2f^{dbo}f^{aoc})A_\mu^d c^b c^c$$

$$= -\frac{1}{2}g^2\bar{c}^a\varepsilon(-\partial^\mu)(f^{dao}f^{boc} + f^{bdo}f^{aoc} + f^{abo}f^{doc})A_\mu^d c^b c^c$$

$$= 0. \quad \text{(P4.211)}$$

Here we changed the indices $b \leftrightarrow c$ and used Jacobi identity (1.11).

4.22 Show that **BRST operator** $\hat{Q}$ (P4.201–P4.203) **is nilpotent,** i.e.

$$\hat{Q}^2 = 0. \quad \text{(P4.212)}$$

Here one defines $\hat{Q}^2$ from

$$\delta(\hat{Q}\phi) = \varepsilon\hat{Q}^2\phi. \quad \text{(P4.213)}$$

SOLUTION. Indeed,

$$\varepsilon\hat{Q}^2 A_\mu^a = \delta(\hat{Q}A_\mu^a) = \delta(D_\mu^{ac}c^c) = 0, \quad \varepsilon\hat{Q}^2\bar{c}^a = \delta(\hat{Q}\bar{c}^a) = \delta(B^a) = 0,$$

$$\varepsilon\hat{Q}^2\psi = \delta(\hat{Q}\psi) = \delta(igc^a t^a\psi) = -g^2 c^a t^a \varepsilon c^c t^c\psi - \frac{1}{2}\varepsilon ig^2 f^{abc}c^b c^c t^a\psi$$

$$= \varepsilon g^2 c^a c^c (t^a t^c - \frac{1}{2}[t^a t^c])\psi = 0, \quad \text{(P4.214)}$$

$$\varepsilon\hat{Q}^2 c^a = \delta(\hat{Q}c^a) = \delta(-\frac{1}{2}gf^{abc}c^b c^c) = \frac{1}{4}\varepsilon g^2 f^{abc}f^{bde}c^d c^e c^c$$

$$+\frac{1}{4}g^2 f^{abc}c^b f^{cde}\varepsilon c^d c^e = \frac{1}{2}g^2\varepsilon f^{abc}f^{bde}c^d c^e c^c = 0. \quad \text{(P4.215)}$$

Here the last equality holds thanks to Jacobi identity (1.11).

4.23 Show that the $\hat{Q}$ splits the Hamiltonian eigenstates into 3 subspaces such that the asymptotic gluon states with physical and nonphysical polarizations belong to different orthogonal subspaces.

SOLUTION. Since $\hat{Q}$ commutes with the Hamiltonian, it splits its eigenstates into 3 subspaces $\mathcal{H}_0$, $\mathcal{H}_1$, $\mathcal{H}_2$

$$\hat{Q}^2|\phi\rangle = 0$$

$$\hat{Q}|\phi\rangle = 0 \qquad \hat{Q}|\phi\rangle \neq 0 \implies |\phi\rangle = |\phi_1\rangle \in \mathcal{H}_1 \quad \text{(P4.216)}$$

$$|\phi\rangle = \hat{Q}|\phi_1\rangle \qquad |\phi\rangle \neq \hat{Q}|\phi_1\rangle$$
$$|\phi\rangle = |\phi_2\rangle \in \mathcal{H}_2 \qquad |\phi\rangle = |\phi_0\rangle \in \mathcal{H}_0$$

Thanks to (P4.199) for the free ($g = 0$) theory BRST transformation goes into

$$\delta A_\mu^a = \varepsilon \partial_\mu c^a, \quad \delta \bar{c}^a = \varepsilon B^a = -\frac{\varepsilon}{\xi}(\partial^\mu A_\mu^a), \quad \delta\psi = \delta c^a = \delta B^a = 0.$$
$$\text{(P4.217)}$$

Or, after Fourier transform

$$\delta A_{k\mu}^a = \varepsilon i k_\mu c_k^a, \quad \delta \bar{c}_k^a = \varepsilon B_k^a = -\frac{\varepsilon}{\xi} i(k^\mu A_\mu^a), \quad \delta\psi = \delta c^a = \delta B^a = 0.$$
$$\text{(P4.218)}$$

Therefore using

$$k^\mu = k n^{+\mu} = k(1,0,0,1), \quad n^- = \frac{1}{2}(1,0,0,-1), \quad \text{(P4.219)}$$

$$g_{\mu\nu} = g_{\mu\nu\perp} + n_\mu^+ n_\nu^- + n_\mu^- n_\nu^+ \implies A_{k\mu}^a = A_{k\mu\perp}^a + n_\mu^+(n^- A) + n_\mu^-(n^+ A),$$
$$\text{(P4.220)}$$

one has for a one-particle state

$$\mathcal{H}_1 = \{\bar{c}, (n^- A)\}, \quad \mathcal{H}_2 = \{c, (n^+ A)\}, \quad \mathcal{H}_0 = \{A_\perp\}. \quad \text{(P4.221)}$$

Hence, the physical transverse gluons live in $\mathcal{H}_0$, while $\mathcal{H}_{1,2}$ are the nonphysical states. One can see that the asymptotic states from different subspaces $\mathcal{H}_i$ are orthogonal.

4.24 Show that the BRST transformation does not change the measure in the functional integral for QCD generating functional (P4.197).

SOLUTION. Let us change the integration variables to the BRST-transformed fields, which we denote with ~, i.e. $A \rightarrow \tilde{A} = A + \varepsilon \hat{Q} A$, etc. The Jacobian of the BRST transformation comes from these derivatives

$$\frac{\delta \tilde{A}^a_\mu(x)}{\delta A^s_\nu(y)} = \frac{\delta(A^a_\mu(x) + \varepsilon(\partial_\mu \delta^{ac} + gf^{abc}A^b_\mu)c^c_x)}{\delta A^s_\nu(y)}$$

$$= \delta^\nu_\mu \delta(x - y)(\delta^{as} + \varepsilon g f^{asc}c^c_x), \qquad (P4.222)$$

$$\frac{\delta \tilde{c}^a_x}{\delta c^s_y} = \frac{\delta(c^a_x - \frac{1}{2}g\varepsilon f^{abc}c^b_x c^c_x)}{\delta c^s_y}$$

$$= \delta(x - y)(\delta^{as} + \frac{1}{2}g\varepsilon f^{asc}c^c_x - \frac{1}{2}g\varepsilon f^{abs}c^b_x)$$

$$= \delta(x - y)(\delta^{as} + g\varepsilon f^{asc}c^c_x), \qquad (P4.223)$$

$$\frac{\delta \tilde{\bar{c}}^a_x}{\delta \bar{c}^s_y} = \frac{\delta(\bar{c}^a_x + \varepsilon B^a_x)}{\delta \bar{c}^s_y} = \delta(x - y)\delta^{as}, \qquad (P4.224)$$

$$\frac{\delta \tilde{\psi}_x}{\delta \psi_y} = \frac{\delta(\psi_x + ig\varepsilon c^a_x t^a \psi_x)}{\delta \psi_y} = \delta(x - y)(1 + ig\varepsilon c^a_x t^a), \qquad (P4.225)$$

$$\frac{\delta \tilde{\bar{\psi}}_x}{\delta \bar{\psi}_y} = \frac{\delta(\bar{\psi}_x - ig\bar{\psi}_x \varepsilon t^a c^a_x)}{\delta \bar{\psi}_y} = \delta(x - y)(1 - ig\varepsilon c^a_x t^a). \qquad (P4.226)$$

Let us first transform the measure in the gluon integral

$$D\tilde{A} = \det \frac{\delta \tilde{A}}{\delta A} DA = DA \det \left[\delta^\nu_\mu \delta(x - y)(\delta^{as} + \varepsilon g f^{asc}c^c_x) \right]. \qquad (P4.227)$$

All other variables are Grassmann. Therefore

$$D\tilde{c}D\tilde{\bar{c}}D\tilde{\psi}D\tilde{\bar{\psi}} = DcD\bar{c}D\psi D\bar{\psi} \det \frac{\delta(\tilde{c}, \tilde{\bar{c}}, \tilde{\psi}, \tilde{\bar{\psi}})}{\delta(c, \bar{c}, \psi, \bar{\psi})}^{-1}$$

$$= DcD\bar{c}D\psi D\bar{\psi} \det \left\{ \delta(x - y) \right.$$

$$\times \left(\begin{array}{cccc} (\delta^{as} + g\varepsilon f^{asc}c^c_x) & 0 & \cdots & \cdots \\ 0 & \delta^{as} & 0 & 0 \\ 0 & 0 & (1 + ig\varepsilon c^a_x t^a) & 0 \\ 0 & 0 & 0 & (1 - ig\varepsilon c^a_x t^a) \end{array} \right)^{-1} \left. \right\}$$

$$= DcD\bar{c}D\psi D\bar{\psi} \left(\det \frac{\delta \tilde{A}}{\delta A} \right)^{-1}. \qquad (P4.228)$$

Hence, using the Grassmann nature of ε one can see that the measure does not change.

4.25 Derive the Schwinger-Dyson equations for Green functions, which follow from the BRST transformation of the fields in the functional integral. They are called the **generalized Ward or Slavnov - Taylor (ST) identities**.

SOLUTION. The BRST symmetry of the Lagrangian leads to the symmetry of the Green functions revealed through the ST identities. One can consider any Green function as an average of a set of fields $\phi = \{A^a_\mu, \psi, c^a, \bar{c}^a\}$

$$\langle \phi_{x_1}...\phi_{x_n}\rangle = \frac{1}{Z[0]} \int \mathcal{D}A\mathcal{D}c\mathcal{D}\bar{c}\mathcal{D}\psi\mathcal{D}\bar\psi \phi_{x_1}...\phi_{x_n} e^{iS_{QCD}+iS_{fix}+iS_{ghost}}.$$

(P4.229)

Changing the integration variables $\phi \to \phi' + \varepsilon\hat{Q}\phi'$ via the BRST transformation and recalling that the measure and the action are BRST invariant, we get the ST identity

$$\langle\hat{Q}(\phi_{x_1}...\phi_{x_n})\rangle = 0.$$

(P4.230)

4.26 Consider the QCD Lagrangian (P4.200) where the fields B are integrated over:

$$\mathcal{L} = -\frac{1}{4}F^2 + \bar{\psi}(i\hat{\partial} - m)\psi + (\partial^\mu \bar{c}^a_x)D^{ac}_\mu c^c_x - \frac{1}{2\xi}(\partial^\mu A^a_\mu)^2.$$

(P4.231)

a) Demonstrate that Lagrangian (P4.231) is invariant under the BRST transformation of the following form: $\delta\phi_i = \varepsilon Q\phi_i$

$$\delta\psi = ig\varepsilon\, t^a c^a \psi,$$
$$\delta\bar\psi = ig\varepsilon\bar\psi t^a c^a,$$
$$\delta A^a_\mu = \varepsilon D^{ab}_\mu c^b,$$
$$\delta c^a = -\frac{\varepsilon}{2}g f^{abe} c^b c^e,$$
$$\delta\bar{c}^a = -\frac{\varepsilon}{\xi}\left(\partial_\mu A^{\mu a}\right),$$

(P4.232)

where ε is a Grassmann parameter.

b) Demonstrate the nilpotent property, i.e. $\delta Q = 0$, of this transformation.

c) Introduce the real and imaginary parts of the ghost fields

$$c^a = \frac{1}{\sqrt{2}}(\rho^a + i\sigma^a), \qquad \bar{c}^a = \frac{1}{\sqrt{2}}(\rho^a - i\sigma^a),$$

(P4.233)

and demonstrate that Lagrangian (P4.231) is invariant under the BRST transformation of the following form:

$$\delta\psi = ig\varepsilon t^a \sigma^a \psi,$$
$$\delta\bar\psi = ig\varepsilon\bar\psi t^a \sigma^a,$$
$$\delta A^a_\mu = \varepsilon D^{ab}_\mu c^b,$$
$$\delta\rho^a = -i\frac{\varepsilon}{\xi}\left(\partial_\mu A^{\mu a}\right),$$
$$\delta\sigma^a = -g\frac{\varepsilon}{2}f^{abe} c^b c^e.$$

(P4.234)

4.27   Derive the Noether current for the BRST transformation.

SOLUTION. Since Lagrangian (P4.231) does not change under the BRST transformation, the Noether current has the form

$$J^\mu_{BRST} = \frac{\partial \mathcal{L}}{\partial \partial_\mu \phi} \delta \phi. \tag{P4.235}$$

Using

$$\frac{\partial \mathcal{L}}{\partial \partial_\mu A^a_\nu} = -F^{a\mu\nu} - \frac{1}{\xi} g^{\mu\nu} \partial_\alpha A^{\alpha a}, \qquad \frac{\partial \mathcal{L}}{\partial \partial_\mu c^a} = \partial^\mu \bar{c}^a, \tag{P4.236}$$

$$\frac{\partial \mathcal{L}}{\partial \partial_\mu \bar{c}^a} = D^{\mu ac} c^c, \qquad \frac{\partial \mathcal{L}}{\partial \partial_\mu \bar{\psi}} = 0, \qquad \frac{\partial \mathcal{L}}{\partial \partial_\mu \psi^i} = \bar{\psi}_i i \gamma^\mu \tag{P4.237}$$

and (P4.232) we have

$$\begin{aligned}
J^\mu_{BRST} &= \frac{\partial \mathcal{L}}{\partial \partial_\mu \phi} \delta \phi = \left(-F^{a\mu\nu} - \frac{1}{\xi} g^{\mu\nu} (\partial_\alpha A^{\alpha a}) \right) \varepsilon D^{ac}_\nu c^c \\
&\quad - (D^{\mu ac} c^c) \frac{\varepsilon}{\xi} \partial^\gamma A^a_\gamma + (\partial^\mu \bar{c}^a) \left(-\frac{1}{2} g \varepsilon f^{abc} c^b c^c \right) + \bar{\psi}_i i \gamma^\mu i g \varepsilon c^a (t^a \psi)^i \\
&= -\varepsilon F^{a\mu\nu} D^{ac}_\nu c^c - \varepsilon \frac{1}{\xi} (\partial_\alpha A^{\alpha a}) D^{\mu ac} c^c + \varepsilon (D^{\mu ac} c^c) \frac{1}{\xi} \partial^\gamma A^a_\gamma \\
&\quad + \varepsilon \frac{1}{2} g f^{abc} (\partial^\mu \bar{c}^a) c^b c^c + g \varepsilon \bar{\psi} \gamma^\mu c^a t^a \psi \\
&= \varepsilon \left(-F^{a\mu\nu} D^{ac}_\nu c^c + \frac{1}{2} g f^{abc} (\partial^\mu \bar{c}^a) c^b c^c + g \bar{\psi} \gamma^\mu c^a t^a \psi \right).
\end{aligned} \tag{P4.238}$$

4.28   Show how

a) the Slavnov-Taylor identity for $\langle 0 | \mathcal{T}(A^a_{\mu x} \bar{c}^b_y) | 0 \rangle$ relates the ghost and the gluon propagators,

b) the Slavnov-Taylor identity for $\langle 0 | \mathcal{T}(\bar{c}^a_{x_1} A^b_{\mu x_2} \bar{\psi}_{x_3} \psi_{x_4}) | 0 \rangle$ relates the Green functions for $q\bar{q} \to gg$ and $q\bar{q} \to c\bar{c}$ in the Born order.

SOLUTION. a) Under BRST transformation (P4.232)

$$\begin{aligned}
0 &= \delta \langle 0 | \mathcal{T}(A^a_{\mu x} \bar{c}^b_y) | 0 \rangle \\
&= \langle 0 | \mathcal{T}(\varepsilon (D^{ad}_{\mu x} c^d_x) \bar{c}^a_y - \frac{\varepsilon}{\xi} A^a_{\mu x} \partial_{y\gamma} A^{\gamma b}_y) | 0 \rangle.
\end{aligned} \tag{P4.239}$$

Therefore

$$\frac{\partial}{\partial x^\mu} \langle 0 | \mathcal{T}(c^a_x \bar{c}^a_y) | 0 \rangle - \frac{\partial}{\partial y^\gamma} \frac{1}{\xi} \langle 0 | \mathcal{T}(A^a_{\mu x} A^{b\gamma}_y) | 0 \rangle + g f^{aed} \langle 0 | \mathcal{T}(A^e_{\mu x} c^d_x \bar{c}^a_y) | 0 \rangle = 0. \tag{P4.240}$$

For Born propagators, we have

$$\frac{\partial}{\partial x^\mu}\langle 0|\mathcal{T}(c_x^a \bar{c}_y^a)|0\rangle = \frac{\partial}{\partial y^\gamma}\frac{1}{\xi}\langle 0|\mathcal{T}(A_{\mu x}^a A_y^{b\gamma})|0\rangle, \tag{P4.241}$$

or in the momentum space

$$-ik_\mu D(k)^{ab} = \frac{1}{\xi}ik_\gamma D_\mu{}^\gamma(k)^{ab} = \frac{k_\mu}{k^2 + i0}, \tag{P4.242}$$

as one can see from (4.26) and (4.27).

b) Under BRST transformation (P4.232)

$$0 = \delta\langle 0|\mathcal{T}(\bar{c}_{x_1}^a A_{\mu x_2}^b \bar{\psi}_{x_3}\psi_{x_4})|0\rangle$$

$$= \langle 0|\mathcal{T}(-\frac{\varepsilon}{\xi}\left(\partial_{x_1\gamma}A_{x_1}^{\gamma a}\right)A_{\mu x_2}^b \bar{\psi}_{x_3}\psi_{x_4} + \bar{c}_{x_1}^a \varepsilon(D_\mu^{bd}c_{x_2}^d)\bar{\psi}_{x_3}\psi_{x_4})|0\rangle + O(g)$$

$$= -\varepsilon\langle 0|\mathcal{T}(\frac{1}{\xi}\left(\partial_{x_1\gamma}A_{x_1}^{\gamma a}\right)A_{\mu x_2}^b \bar{\psi}_{x_3}\psi_{x_4} + \bar{c}_{x_1}^a (\partial_{x_2\mu}c_{x_2}^b)\bar{\psi}_{x_3}\psi_{x_4})|0\rangle + O(g). \tag{P4.243}$$

Therefore in the Born order

$$\frac{1}{\xi}\partial_{x_1\gamma}\langle 0|\mathcal{T}(A_{x_1}^{\gamma a} A_{\mu x_2}^b \bar{\psi}_{x_3}\psi_{x_4})|0\rangle = -\partial_{x_2\mu}\langle 0|\mathcal{T}(\bar{c}_{x_1}^a c_{x_2}^b \bar{\psi}_{x_3}\psi_{x_4})|0\rangle. \tag{P4.244}$$

For $\xi = 1$ we met the corresponding identity for the amplitudes of these processes (P4.192) in problem 4.19. It can be derived from this one via the LSZ reduction.

4.29 Add the Lagrangian $\mathcal{L}_\omega$ of the free scalar field ω to the QED Lagrangian $\mathcal{L}_{QED}^\xi$: $\mathcal{L} = \mathcal{L}_{QED} + \mathcal{L}_\omega$, where

$$\mathcal{L}_{QED}^\xi = \bar{\psi}\left(i\hat{D} - m\right)\psi - \frac{1}{4}F_{\mu\nu}F^{\mu\nu} - \frac{1}{2\xi}\left(\partial_\mu A^\mu\right)^2, \tag{P4.245}$$

$$\mathcal{L}_\omega = \frac{1}{2}\partial_\mu\omega\partial^\mu\omega. \tag{P4.246}$$

a) Demonstrate that the action $S = \int d^4x\mathcal{L}$ is invariant (up to ϵ^2 terms) under the following infinitesimal transformation, where the first three equations look like the infinitesimal gauge transformation in QED with the parameter $\theta(x) = e\epsilon\,\omega(x)$:

$$\delta_0\psi(x) = ie\epsilon\,\omega(x)\psi(x), \quad \delta_0\bar{\psi}(x) = -ie\bar{\psi}(x)\epsilon\,\omega(x), \tag{P4.247}$$

$$\delta_0 A_\mu(x) = -\epsilon\,\partial_\mu\omega(x), \quad \delta_0\omega(x) = \frac{\epsilon}{\xi}\partial_\mu A^\mu(x), \tag{P4.248}$$

where ϵ is an infinitesimal c-number (not Grassmann!) parameter and e is the electron charge.

b) Demonstrate that the functional measure $\mathcal{D}\omega\mathcal{D}A\mathcal{D}\psi\mathcal{D}\bar{\psi}$ does not change (up to ϵ^2 terms) under this transformation.

c) Show that for an arbitrary correlator of the Heisenberg fields $G[A, \omega, \psi, \bar{\psi}] = \langle 0|\mathcal{T}(g[A, \omega, \psi, \bar{\psi}])|0\rangle$ one has $\delta_0 G = 0$.

d) Use the correlator $\langle 0|\mathcal{T}(A_{\mu x}\omega_y)|0\rangle$ to show that the exact Green function of the photon $\mathcal{D}_{\mu\nu}(q) = \int d^4x e^{iqx}\langle 0|\mathcal{T}(A_\mu(x)A_\nu(0))|0\rangle$ satisfies the relation

$$\mathcal{D}_{\mu\nu}(q)q^\nu = -i\xi\frac{q_\mu}{q^2 + i0}, \qquad (\text{P4.249})$$

i.e. the longitudinal component of QED propagator does not renormalize.

4.30 Derive the Slavnov-Taylor identity for the Green function with the connected Wilson line $U_C(x, x)$

$$tr[\langle(\frac{1}{\xi}\frac{\partial\partial A_z}{\partial z^\nu} + ig\{\frac{\partial\bar{c}_z}{\partial z^\nu}c_x\})U_C(x, x)\rangle] = 0. \qquad (\text{P4.250})$$

SOLUTION. The Wilson line changes via the gauge transform $V = e^{i\theta}$ in the following way

$$U_C(y, x) \rightarrow U'_C(y, x) = V_y U_C(y, x)V_x^{-1}$$
$$\simeq U_C(y, x) + i\theta_y U_C(y, x) - iU_C(y, x)\theta_x. \qquad (\text{P4.251})$$

For gluons the BRST transform is the infinitesimal gauge transformation with $\theta = g\varepsilon c$. Therefore

$$\delta U_C(y, x) = i\varepsilon(c_y U_C(y, x) - U_C(y, x)c_x). \qquad (\text{P4.252})$$

Hence

$$0 = \langle\delta(\bar{c}_z U_C(y, x))\rangle$$
$$= -\frac{\varepsilon}{\xi}\langle(\frac{\partial A_z^\mu}{\partial z^\mu})U_C(y, x))\rangle + g\langle\bar{c}_z i\varepsilon(c_y U_C(y, x) - U_C(y, x)c_x)\rangle$$
$$= -\frac{\varepsilon}{\xi}\langle(\frac{\partial A_z^\mu}{\partial z^\mu})U_C(y, x)\rangle - gi\varepsilon\langle\bar{c}_z(c_y U_C(y, x) - U_C(y, x)c_x)\rangle.$$
$$(\text{P4.253})$$

Finally, we get

$$\frac{1}{\xi}\langle(\frac{\partial A_z^\mu}{\partial z^\mu})U_C(y, x)\rangle = -gi\langle\bar{c}_z(c_y U_C(y, x) - U_C(y, x)c_x)\rangle. \qquad (\text{P4.254})$$

For $y = x$ taking $\frac{\partial}{\partial z^\nu}$, we have (P4.250).

4.31 Derive the **QCD loop** or **Migdal-Makeenko equations** (Mak05)

$$\frac{\partial}{\partial x^\mu}\frac{\delta}{\delta\sigma_{\mu\nu}}trU_C(x,x) = -ig^2\int dx^\nu\delta^{(4)}(x-y)tr\langle U_{C_x^y}(y,x)t^aU_{C_y^x}(x,y)t^a\rangle,$$

(P4.255)

where the area and ordinary derivative were defined on Wilson loops in (P2.138) and (P2.139).

SOLUTION. Here we follow (BGSN82). Schwinger-Dyson equation (3.51) for the functional $F = U_C(x,x)$ reads

$$\langle i\frac{\delta U_C(x,x)}{\delta A_\alpha^a(x)}\rangle = \langle U_C(x,x)\frac{\delta S}{\delta A_\alpha^a(x)}\rangle,$$

(P4.256)

where S is the action with the gauge fixing and ghost terms in the Lorentz gauge (P4.231). We have

$$-\frac{1}{4}\frac{\delta}{\delta A_\alpha^a(x)}\int dx(\partial_\mu A_\nu^a - \partial_\nu A_\mu^a + gf^{abc}A_\mu^b A_\nu^c)F_{\mu\nu}^a = D_\nu^{ad}F^{d\nu\alpha},$$

(P4.257)

$$-\frac{1}{2\xi}\frac{\delta}{\delta A_\alpha^a(x)}\int dx(\partial^\nu A_\nu^b)^2 = \frac{1}{\xi}\partial_\alpha(\partial A^a),$$

(P4.258)

$$\frac{\delta}{\delta A_\alpha^a(x)}\int dx(\partial^\mu \bar{c}_x^a)D_\mu^{ac}c_x^c = gf^{dac}(\partial^\mu \bar{c}_x^d)c_x^c.$$

(P4.259)

For the point $z \in C$

$$\frac{\delta U_C(x,x)}{\delta A_\alpha^a(z)} = ig\int_C dy^\alpha U_{C_y^x}(x,y)t^aU_{C_x^y}(y,x)\delta^{(4)}(y-z).$$

(P4.260)

Multiplying by t^a, taking the trace and the $\mathcal{T}$-product, one gets the SD equation

$$-g\int_C dy^\alpha\delta^{(4)}(y-x)\langle tr[U_{C_y^x}(x,y)t^aU_{C_x^y}(y,x)t^a]\rangle$$

$$= \langle tr[U_C(x,x)t^a(D_\nu^{ad}F^{d\nu\alpha} + \frac{1}{\xi}\partial_\alpha(\partial A^a) + gf^{dac}(\partial^\mu \bar{c}_x^d)c_x^c)]\rangle$$

$$= \langle tr[(t^a D_\nu^{ad}F^{d\nu\alpha} + \frac{1}{\xi}\partial_\alpha(\partial A) + ig\{(\partial^\mu \bar{c}_x)c_x\})U_C(x,x)]\rangle$$

$$= \langle tr[(t^a D_{x\nu}^{ad}F_x^{d\nu\alpha})U_C(x,x)]\rangle = \langle tr[(\partial_{x\nu}F_x^{\nu\alpha} - ig[A_{x\nu},F_x^{\nu\alpha}])U_C(x,x)]\rangle$$

$$= \langle tr[(D_{x\nu}F_x^{\nu\alpha})U_C(x,x)]\rangle.$$

(P4.261)

One can see that thanks to Slavnov-Taylor identity (P4.250) ghost and gauge fixing contributions cancel. Using the representation of this trace via the area and ordinary derivative of the Wilson loop (P2.144), one gets Migdal-Makeenko equations (P4.255).

BRST symmetry and Slavnov-Taylor identities are clearly explained in (Tay79; Fra08; Ram81). Asymptotic behavior of amplitudes in Regge limit one can find in (FR11). Spinor-helicity techniques are given in details in (MP91). Recursive techniques are discussed in (BCFW05; Dix96). For loop equations the reader is referred to (Mak05; BGSN82).

FURTHER READING

[BCFW05] Ruth Britto, Freddy Cachazo, Bo Feng, and Edward Witten. Direct proof of tree-level recursion relation in Yang-Mills theory. *Phys. Rev. Lett.*, 94:181602, 2005.

[BGSN82] Richard A. Brandt, A. Gocksch, M. A. Sato, and F. Neri. LOOP SPACE. *Phys. Rev. D*, 26:3611, 1982.

[CL84] T. P. Cheng and L. F. Li. *Gauge theory of elementary particle physics*, chapter 9.1, 9.2. Oxford University Press, Oxford, UK, 1984.

[DGH14] J. F. Donoghue, E. Golowich, and Barry R. Holstein. *Dynamics of the standard model*, volume 2, chapter Appendix A: Functional Integration. CUP, 2014.

[Dix96] Lance J. Dixon. Calculating scattering amplitudes efficiently. In *Theoretical Advanced Study Institute in Elementary Particle Physics (TASI 95): QCD and Beyond*, pages 539–584, 1, 1996.

[FR11] Jeffrey R. Forshaw and D. A. Ross. *Quantum chromodynamics and the pomeron*, volume 9. Cambridge University Press, 1, 2011.

[Fra08] Paul. H. Frampton. *Gauge field theories: Third revised and improved edition*, chapter 3.4, Becchi–Rouet–Stora–Tyutin Transformation. 2008.

[Mak05] Yu. Makeenko. *Methods of contemporary gauge theory*, chapter 12. Cambridge Monographs on Mathematical Physics. Cambridge University Press, 11 2005.

[Man98] Michelangelo L. Mangano. Introduction to QCD. In *1998 European School of High-Energy Physics*, pages 53–97, 1998.

[MP91] Michelangelo L. Mangano and Stephen J. Parke. Multiparton amplitudes in gauge theories. *Phys. Rept.*, 200:301–367, 1991.

[PS95] Michael E. Peskin and Daniel V. Schroeder. *An introduction to quantum field theory*, chapter 9.4, 16.2, 16.6. Addison-Wesley, Reading, USA, 1995.

[Ram81] Pierre Ramond. *Field Theory. A Modern Primer*, volume 51 of *Frontiers in Physics*. 1981.

[Tay79] J. C. Taylor. *Gauge theories of weak interactions*, chapter 12.1–12.5 Ward-Takahashi identities. Cambridge Monographs on Mathematical Physics. Cambridge University Press, Cambridge, UK, 2, 1979.

[Ynd13] Francisco J Ynduráin. *Quantum chromodynamics: an introduction to the theory of quarks and gluons*, chapter 6. Springer Science & Business Media, 2013.

CHAPTER 5

Strong Coupling

T HIS chapter discusses the situation when the small charge perturbation theory does not work. It introduces the imaginary time, goes on with the construction of the Wilson lattice action and strong coupling expansion for the potential between two static quarks. Then it discusses the Regge trajectories and the string model of hadrons relating the string tension from the strong coupling expansion, Nambu-Goto string, and the slope of the Regge trajectories.

1. **Euclidean time.** The functional integral for a scalar field theory in Minkowski space-time reads

$$Z = \int \mathcal{D}\phi e^{iS[\phi]} = \int \mathcal{D}\phi e^{i\int \frac{1}{2}((\partial_\mu \phi)^2 - m^2\phi^2)d^4x}$$

$$= \int \mathcal{D}\phi e^{i\int \frac{1}{2}\phi(-\partial^2 - m^2 + i\varepsilon)\phi d^4x}. \qquad (5.1)$$

Here we introduced $i\varepsilon$ which ensures convergence of the functional integral and reveals in propagators as Feynman prescription. For numerical calculations the functional integral is commonly regulated by the Wick rotation to the Euclidean time, i.e.

$$x^0 = -i\tau. \qquad (5.2)$$

This rotation is consistent with the $i\varepsilon$ prescription since in the coordinate space the propagator

$$D(x) = \frac{-i}{4\pi^2(x^2 - i\varepsilon)} \qquad (5.3)$$

has the poles

$$x_i^0 = \pm(|\vec{x}| + i\varepsilon). \qquad (5.4)$$

DOI: 10.1201/9781003272403-5

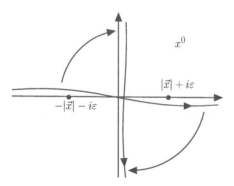

Figure 5.1 The propagator poles and contour rotation in x^0 plane.

This Wick rotation gives

$$iS = i \int_{-\infty}^{+\infty} dt \int \frac{1}{2} \left[\phi_t^2 - (\vec{\nabla}\phi)^2 - m^2\phi^2 \right] d\vec{x}$$

$$= i \int_{+i\infty}^{-i\infty} dt \int \frac{d\vec{x}}{2} \left[\phi_t^2 - (\vec{\nabla}\phi)^2 - m^2\phi^2 \right]$$

$$= i \int_{+i\infty}^{-i\infty} d(-i\tau) \int \frac{d\vec{x}}{2} \left[\left(\frac{\partial\phi}{\partial(-i\tau)} \right)^2 - (\vec{\nabla}\phi)^2 - m^2\phi^2 \right]$$

$$= - \int_{-\infty}^{+\infty} d\tau \int \frac{d\vec{x}}{2} \left[\phi_\tau^2 + (\vec{\nabla}\phi)^2 + m^2\phi^2 \right] = -S_E[\phi] \quad \Longrightarrow$$

$$Z = \int \mathcal{D}\phi \, e^{-S_E[\phi]}. \tag{5.5}$$

2. **Wilson action.** We work in the nonperturbative gauge field normalization and deal with the pure Euclidean Yang-Mills theory. One can **discretize space-time with a unit step** a and **attrubute the gauge field to links of the lattice via the Wilson line**

$$U_\mu(x) = U_{-\mu}^\dagger(x + a\hat{\mu}) = P e^{i \int_x^{x+a\hat{\mu}} dz^\rho A_\rho(z)} = e^{iA_\mu(x)a} + O(a^2)$$

$$= 1 + iA_\mu(x)a + O(a^2), \tag{5.6}$$

which may be called the **link variable**. Then the functional integral w.r.t. the gauge fields goes into the functional integral over these link variables

$$\int \mathcal{D}A_\mu^a(x) \to \int \mathcal{D}U_\mu(x). \tag{5.7}$$

The **plaquette variable** can be built from the link variables as

$$U(\partial p) = U_\nu^\dagger(x)U_\mu^\dagger(x + a\hat{\nu})U_\nu(x + a\hat{\mu})U_\mu(x). \tag{5.8}$$

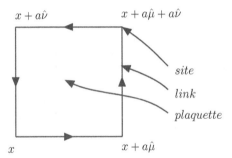

$x + a\hat{\nu}$ $x + a\hat{\mu} + a\hat{\nu}$

site

link

plaquette

x $x + a\hat{\mu}$

Figure 5.2 A plaquette.

It is shown in Figure 5.2. Since this is the Wilson line over the closed contour its trace is gauge invariant. Hence, one can build an action from these variables. The **Wilson action** is

$$S_W[U] = \beta \sum_p \left[1 - \frac{1}{N_c} \operatorname{Re} tr\, U(\partial p) \right]. \tag{5.9}$$

Here β is to be fixed in the continuum limit and the sum goes over all plaquettes, i.e. over all points x and directions μ and ν. $\operatorname{Re} tr\, U(\partial p)$ does not depend on the orientation of the Wilson loop and one can rewrite it as

$$\operatorname{Re} tr\, U(\partial p) = \frac{1}{2}(tr\, U(\partial p) + tr\, U(\partial p)^\dagger). \tag{5.10}$$

Since changing the orientation of the Wilson loop results in Hermitian conjugation, one can rewrite the Wilson action as

$$S_W[U] = \frac{1}{2}\beta \sum_{oriented\ p} \left[1 - \frac{1}{N_c} tr\, U(\partial p) \right]. \tag{5.11}$$

In the latter action, each plaquette is counted twice: with clockwise and anticlockwise orientation. In the continuum limit

$$S_W[U] = \frac{\beta}{2} \sum_{oriented\ p} \left[1 - \frac{1}{N_c} tr\, U(\partial p) \right] \simeq \frac{\beta}{2} \sum_{x\mu\nu} \left[1 - (1 - \frac{a^4}{4N_c} F^a_{\mu\nu} F^a_{\mu\nu}) \right]$$

$$= \frac{\beta}{2} \sum_{x\mu\nu} \left[\frac{a^4}{4N_c} F^a_{\mu\nu} F^a_{\mu\nu} \right] \simeq \frac{\beta g^2}{2N_c} \int d^4 x_E \frac{1}{4g^2} F^a_{\mu\nu} F^a_{\mu\nu}$$

$$\implies \quad \beta = \frac{2N_c}{g^2}, \quad Z = \int \mathcal{D}U e^{-S_W[U]}. \tag{5.12}$$

3. **Strong coupling expansion.** One can calculate the Wilson loop average in the pure Yang-Mills theory in the limit of large g or small β.

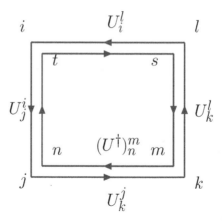

Figure 5.3 Integration over the link variables in the single plaquette.

Using integrals (P1.205–P1.209) from problem 1.63 in Chapter 1, we have for a single plaquette average (see Figure 5.3)

$$
\frac{1}{N_c}\langle trU(\partial p)\rangle = \frac{\int \mathcal{D}U \; \frac{1}{N_c}trU(\partial p) \; e^{-S_W[U]}}{\int \mathcal{D}U e^{-S_W[U]}}
$$

$$
= \frac{\int \mathcal{D}U \; \frac{1}{N_c}trU(\partial p) \; e^{\frac{\beta}{2N_c}\sum\limits_{oriented\,p} tr\,U(\partial p)}}{\int \mathcal{D}U \; e^{\frac{\beta}{2N_c}\sum\limits_{oriented\,p} tr\,U(\partial p)}}
$$

$$
\simeq \frac{\int \mathcal{D}U \; \frac{1}{N_c}trU(\partial p)\left\{1 + \frac{\beta}{2N_c}\sum\limits_{oriented\,p'}[tr\,U(\partial p')]\right\}}{\int \mathcal{D}U \left\{1 + \frac{\beta}{2N_c}\sum\limits_{oriented\,p'}[tr\,U(\partial p')]\right\}}
$$

$$
= \frac{\beta}{2N_c^2}\frac{\int \mathcal{D}U \; trU(\partial p)\, trU(\partial p)^\dagger}{\int \mathcal{D}U}
$$

$$
= \frac{\beta}{2N_c^2}\frac{1}{N_c}\delta^m_k\delta^j_n \times \frac{1}{N_c}\delta^k_m\delta^s_l \times \frac{1}{N_c}\delta^t_i\delta^l_s \times \frac{1}{N_c}\delta^i_t\delta^n_j
$$

$$
= \frac{\beta}{2N_c^2}\frac{1}{N_c^4}\delta^i_i\delta^k_k\delta^m_m\delta^s_s = \frac{\beta}{2N_c^2}. \tag{5.13}
$$

Then for the Wilson loop with the time length T and space length R, $T \gg R$ one tiles the rectangle area of the loop with the plaquettes to

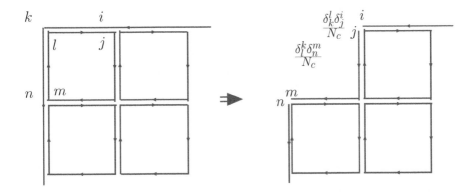

Figure 5.4 Integration over the link variables adjacent to the corner of the $R \times T$ loop.

get (see Figure 5.4)

$$\frac{1}{N_c}\langle trU_{\square T}\rangle_R \simeq \frac{1}{\frac{A_{\min}}{a^2}!}\int \mathcal{D}U \frac{1}{N_c}trU_{\square T}\Big|_R \left\{\sum_{oriented\ p'}\frac{\beta}{2N_c}tr\, U(\partial p')\right\}^{\frac{A_{\min}}{a^2}}$$

$$= \left(\frac{\beta}{2N_c}\right)^{\frac{A_{\min}}{a^2}}\frac{1}{N_c}N_c^{\#sites-\#links}$$

$$= \left|\#sites = (\frac{R}{a}+1)(\frac{T}{a}+1),\quad \#links = (\frac{R}{a}+1)\frac{T}{a}+\frac{R}{a}(\frac{T}{a}+1)\right|$$

$$= \left(\frac{\beta}{2N_c^2}\right)^{\frac{TR}{a^2}} = \left(\frac{1}{N_c g^2}\right)^{\frac{TR}{a^2}} = e^{-KA_{\min}},\quad K = \frac{1}{a^2}\ln\left(N_c g^2\right).$$

$$(5.14)$$

Here $A_{\min}$ is the **minimal area of the surface spanned by the Wilson loop** and K is called the **string tension.** Surfaces of non-minimal area give β in higher powers. They are subdominant since in the strong coupling regime $\beta \to 0$. Therefore one can extract the interaction potential of 2 static quarks as (2.27)

$$V(R) = \lim_{T\to\infty}\frac{-1}{T}\ln\langle\frac{1}{N_c}trU_{\square T}\rangle_R = \lim_{T\to\infty}\frac{-1}{T}\ln\left(\frac{1}{N_c g^2}\right)^{\frac{TR}{a^2}} = KR.$$

$$(5.15)$$

Hence, in the strong coupling regime the matrix element of the $T \times R$ Wilson loop $\sim e^{-KA_{\min}}$ and the potential is linear $\sim R$. Such behavior is called the **area law.** It is a sign of **confinement.** The following terms in the strong coupling (small β) expansion will not change the leading term, i.e. confinement is seen in lattice Yang-Mills theory in all orders of

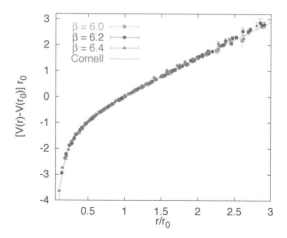

Figure 5.5 The quenched Wilson action $SU(3)$ potential, normalised to $V(r_0) = 0$. Figure from (Bal01). Reprinted from Physics Reports, Vol. 343, issues 1-2, pages 1-136, Gunnar S. Bali, QCD forces and heavy quark bound states, Page No. 42, Copyright 2001, with permission from Elsevier.

the strong coupling expansion. Experimentally, $K \simeq 1 GeV/fm$, which can be related to the slope of Regge trajectories of hadrons. In contrast, in the weak coupling expansion we have the **perimeter law** (P2.112)

$$\langle \frac{1}{N_c} tr U_\square \rangle \simeq e^{-const \frac{L}{a} + O(a^0)}, \quad L = 2T + 2R. \tag{5.16}$$

which gives the QCD Coulomb potential and cannot confine quarks in any order of the small g expansion. At moderate g the static quark potential is calculated numerically by Monte-Carlo integration. The results are well fitted to

$$V(R) \simeq -A\frac{1}{R} + KR + const, \tag{5.17}$$

see Figure 5.5. For $const = 0$ this form is called the **Cornell potential** (EGK[+]75). **Area law for the static quark potential is the Wilson criterion of confinement.** Similarly, for a baryon one has

$$V(\vec{r}_1, \vec{r}_2, \vec{r}_3) \simeq -\frac{A}{2}(\frac{1}{|\vec{r}_1 - \vec{r}_2|} + \frac{1}{|\vec{r}_2 - \vec{r}_3|} + \frac{1}{|\vec{r}_1 - \vec{r}_3|}) + KL_{\min} + const, \tag{5.18}$$

where $L_{\min}$ is the minimal length of a line connecting 3 quarks. These results are obtained taking into account only gluon fields. It means that the functional integral over quark fields is taken as a functional determinant and this determinant is not expanded but taken as a constant.

It is the so-called **quenched approximation**, which amounts to suppressing all quark loops.

4. **Continuum limit.** After the Wick rotation the Eucedian action has the form of an action for a statistical system of set of links with the temperature $T = \frac{1}{\beta} = \frac{g2}{2N_c}$. We are interested in the continuum limit of this system, i.e. in the limit $a \to 0$. We expect that in this limit the physical quantities we calculate (like the string tension or K) remain stable and equal to their physical values. However, for the dimensionless action there is only one length parameter a. It means that any physical length ξ (or mass $m = \frac{1}{\xi}$, or $K = \frac{1}{\xi^2}$) measured in this scale a diverges, i.e. $\frac{\xi}{a} \to \infty$. In statistical physics infinite correlation length is a sign of a second order phase transition. Therefore the continuum limit is defined in the fixed points $T = T^*$ (or $g = g^*$) of the second order phase transition of the lattice system. As we will see in the next chapter, perturbative QCD tells us that

$$\frac{dg^2}{d \ln \mu^2} = -\beta_0 \frac{g^4}{(4\pi)^2}, \quad \beta_0 > 0, \tag{5.19}$$

where $\mu \sim \frac{1}{a}$ is the mass scale. It means that $g^* \to 0$ is the fixed point at $\frac{1}{a} \to 0$. Lattice theory seems to indicate that no other fixed points exist in the $SU(N)$ gauge theory. Therefore in the continuous limit ratios of two physical correlation lengths or masses should go to the constant values independent of a, which is called the scaling. In gluodynamics one can express masses via the ratios with the string tension known from the Regge trajectories which we discuss below or via the ratio with the hadronic scale $r_0 \simeq 0.5 fm$ defined through the static potential between quarks

$$r^2 \frac{dV(r)}{dr}\Big|_{r_0} = 1.65, \tag{5.20}$$

known from the bottomonium phenomenology (Som94). Figures 5.6 and 5.7 give examples of extrapolation of such ratios to the continuum limit. Note that this phase transition is not to be confused with the phase transition in QCD at nonzero temperature.

5. **String model of hadrons** is one of many models describing confining potential and Regge trajectories. One should consider it as a model demonstrating specific features of QCD bound states rather than a fundamental theory. However, string theory began from such discussions. Here we follow (Fel81). By analogy with a point particle whose action is proportional to the length of its word-line

$$S = -m \int_C \sqrt{|g_{zz}|} dz = -m \int \sqrt{\left|g_{\mu\nu} \frac{dx^\mu}{dz} \frac{dx^\nu}{dz}\right|} dz, \tag{5.21}$$

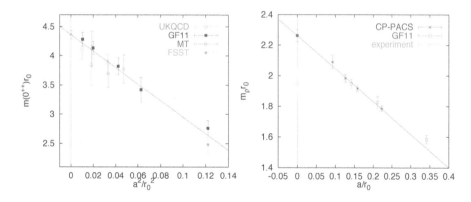

Figure 5.6 The scalar glueball mass in units of r_0 as a function of a^2 (Bal01). Reprinted from Physics Reports, Vol. 343, issues 1-2, pages 1–136, Gunnar S. Bali, QCD forces and heavy quark bound states, Page No. 23, Copyright 2001, with permission from Elsevier.

Figure 5.7 The ratio, $m_\rho r_0$, in quenched QCD, extrapolated to the continuum limit (Bal01). Reprinted from Physics Reports, Vol. 343, issues 1-2, pages 1–136, Gunnar S. Bali, QCD forces and heavy quark bound states, Page No. 22, Copyright 2001, with permission from Elsevier.

one builds the action of a string parameterized by $\vec{x} = \vec{x}(\sigma)$ at a fixed time (see Figure 5.8). In space-time the string sweeps out a time-like sheet called the world-sheet parameterized by

$$x^\mu = x^\mu(\sigma, \tau). \tag{5.22}$$

We will consider a finite string with $\sigma \in [0, \pi]$ and time-like variable τ. Then one defines the action for a **Nambu-Goto string** proportional to the area of its worldsheet

$$S = -\frac{1}{2\pi\alpha} \int_0^\pi d\sigma \int_{\tau_1}^{\tau_2} d\tau \sqrt{|h|}, \quad h = \det \begin{pmatrix} h_{\sigma\sigma} & h_{\sigma\tau} \\ h_{\tau\sigma} & h_{\tau\tau} \end{pmatrix} = h_{\sigma\sigma}h_{\tau\tau} - h_{\sigma\tau}^2, \tag{5.23}$$

where h is the determinant of the metric on the word-sheet

$$h_{\sigma\sigma} = g_{\mu\nu}\frac{\partial x^\mu}{\partial \sigma}\frac{\partial x^\nu}{\partial \sigma}, \quad h_{\sigma\tau} = g_{\mu\nu}\frac{\partial x^\mu}{\partial \sigma}\frac{\partial x^\nu}{\partial \tau}, \quad h_{\tau\tau} = g_{\mu\nu}\frac{\partial x^\mu}{\partial \tau}\frac{\partial x^\nu}{\partial \tau}. \tag{5.24}$$

We will work in the inertial frame, i.e. $g^{\mu\nu} = diag(1, -1, -1, -1)$ and denote $\frac{\partial x^\mu}{\partial \sigma} = x'^\mu$, $\frac{\partial x^\nu}{\partial \tau} = \dot{x}^\nu$. Demanding the extremality of the action under variations of the wordsheet with

$$x^\mu(\sigma, \tau) \to x^\mu(\sigma, \tau) + \varepsilon y^\mu(\sigma, \tau), \quad y^\mu(\sigma, \tau_1) = y^\mu(\sigma, \tau_2) = 0, \tag{5.25}$$

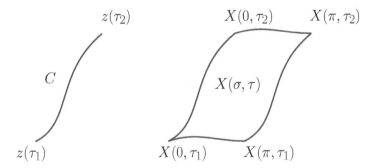

Figure 5.8 Particle world-line and string world-sheet.

we get the following e.o.m. and boundary conditions

$$\frac{\partial S}{\partial \varepsilon}|_{\varepsilon=0} = \int_0^\pi d\sigma \int_{\tau_1}^{\tau_2} d\tau \left[\frac{\partial L}{\partial x'^\mu} \frac{\partial y^\mu}{\partial \sigma} + \frac{\partial L}{\partial \dot{x}^\mu} \frac{\partial y^\mu}{\partial \tau} \right]$$

$$= \int_{\tau_1}^{\tau_2} d\tau y^\mu \frac{\partial L}{\partial x'^\mu}|_{\sigma=0}^{\sigma=\pi} - \int_0^\pi d\sigma \int_{\tau_1}^{\tau_2} d\tau y^\mu \left[\frac{\partial}{\partial \sigma} \frac{\partial L}{\partial x'^\mu} + \frac{\partial}{\partial \tau} \frac{\partial L}{\partial \dot{x}^\mu} \right]$$

$$\implies \frac{\partial P_\sigma^\mu}{\partial \sigma} + \frac{\partial P_\tau^\mu}{\partial \tau} = 0, \quad P_\sigma^\mu|_{\sigma=0,\pi} = 0, \tag{5.26}$$

$$P_\tau^\mu(\sigma,\tau) = \frac{\partial L}{\partial \dot{x}^\mu}, \quad P_\sigma^\mu(\sigma,\tau) = \frac{\partial L}{\partial x'^\mu}. \tag{5.27}$$

The action is invariant w.r.t. space-time translations $x \to x + a$. Therefore the 4-momentum and angular momentum of the string are conserved. They reads

$$P^\mu = - \int_0^\pi d\sigma P_\tau^\mu(\sigma,\tau), \quad J^{\mu\nu} = \int d\sigma[x^\mu P_\tau^\nu - x^\nu P_\tau^\mu]. \tag{5.28}$$

For a rotating string (see Figure 5.9)

$$x^0 = \tau, \quad x^1 = A(\sigma - \frac{\pi}{2}) \cos\omega\tau, \quad x^2 = A(\sigma - \frac{\pi}{2}) \sin\omega\tau, \quad x^3 = 0 \tag{5.29}$$

one gets

$$P^0 = \frac{A\pi}{4\alpha}, \quad \vec{P} = 0, \quad J^{21} = \frac{A^2\pi^2}{16\alpha}, \quad J^{23} = J^{31} = 0. \tag{5.30}$$

Therefore the spinning string behaves like a particle with mass M and spin J :

$$M = \frac{A\pi}{4\alpha}, \quad J = \frac{A^2\pi^2}{16\alpha} = \alpha M^2. \tag{5.31}$$

In other words for a Nambu-Goto string **spin grows linearly with the square of the mass**. Such behavior is experimentally observed for

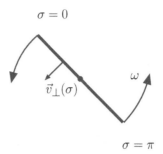

$$\sigma = 0$$

$$\vec{v}_\perp(\sigma)$$

$$\omega$$

$$\sigma = \pi$$

Figure 5.9 The rotating string.

hadrons. They lie on approximately straight lines in $J - M^2$ coordinates. These lines are known as **Regge trajectories**, see Figure 5.10. In the inertial frame with time variable t, one can identify $\tau = t$. Then the arc-length ds of the string at fixed time t and the velocity of a point on the string $\vec{v}$ are

$$ds = \sqrt{\frac{\partial \vec{x}}{\partial \sigma}\frac{\partial \vec{x}}{\partial \sigma}}d\sigma, \quad \left|\frac{\partial \vec{x}}{\partial s}\right| = 1, \quad \vec{v} = \frac{\partial \vec{x}}{\partial t}. \tag{5.32}$$

In these variables the Nambu string action reads

$$S = -\frac{1}{2\pi\alpha}\int_{t_1}^{t_2} dt \int_0^L ds \sqrt{1 - \vec{v}_\perp^2}, \quad \vec{v}_\perp = \vec{v} - \frac{\partial \vec{x}}{\partial s}(\vec{v}\frac{\partial \vec{x}}{\partial s}). \tag{5.33}$$

On the other hand in the rest frame of a small portion of the string we can introduce the coordinates (t^0, x^0, y^0, z^0) with $dz^0 = ds$. In this frame the string is static and

$$L = -H \quad \Longrightarrow \quad L(t^0, x^0, y^0, z^0) = -K\delta(x^0)\delta(y^0), \tag{5.34}$$

where we assumed that the string is infinitely thin and its energy per unit length is K as the strong coupling expansion for the static quark potential teaches us. Since the rest frame moves with the velocity $\vec{v}_\perp$ perpendicular to the string, the action reads

$$S = \int L(t^0, x^0, y^0, z^0)dt^0 dx^0 dy^0 dz^0 = -K\int_0^L ds \int_{t_1}^{t_2} dt \sqrt{1 - \vec{v}_\perp^2}, \tag{5.35}$$

since $dt^0 = dt\sqrt{1 - \vec{v}_\perp^2}$. As a result one can express the slope of the Regge trajectories α through the string tension K

$$\alpha = \frac{1}{2\pi K}. \tag{5.36}$$

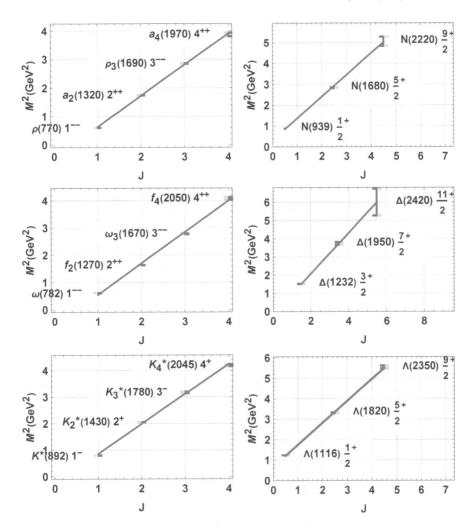

Figure 5.10 Typical Regge trajectories for mesons (left) and baryons (right).

Regge trajectories have the slope $\alpha \simeq 1 GeV^{-2}$, see Figure 5.10. Therefore the string tension

$$K \simeq \frac{1}{2\pi\alpha} \simeq \frac{GeV^2}{2\pi} \simeq \frac{1}{2\pi \times 0.2 GeV\ fm} GeV^2 \simeq 1\frac{GeV}{fm}. \qquad (5.37)$$

EXERCISES

5.1 Derive the QCD action in the Euclidean space.

SOLUTION. We have the metric tensor in the Minkowski space $(1, -1, -1, -1)$ and in the Euclidean space $(+1, +1, +1, +1)$. To go to the Euclidean space we make the following coordinate change (5.2)

$$x_{iE} = x^i = -x_i, \quad x_{4E} = ix^0 = ix_0, \tag{P5.1}$$

$$\partial_{iE} = \frac{\partial}{\partial x_{iE}} = \frac{\partial}{\partial x^i} = \partial_i = -\frac{\partial}{\partial x_i} = -\partial^i, \tag{P5.2}$$

$$\partial_{4E} = \frac{\partial}{\partial x_{4E}} = -i\frac{\partial}{\partial x^0} = -i\frac{\partial}{\partial x_0} = -i\partial_0. \tag{P5.3}$$

We also denote

$$A_{iE} = A_i = -A^i, \quad A_{4E} = -iA^0 = -iA_0. \tag{P5.4}$$

Then, we get for the scalar product and the strength tensor

$$\partial_\mu A^\mu = \partial_0 A^0 + \partial_i A^i = i\partial_{4E} i A_{4E} + \partial_{iE}(-A_{iE}) = -\partial_{\mu E} A_{\mu E}, \tag{P5.5}$$

$$F_{ij} = \partial_i A_j - \partial_j A_i - ig[A_i A_j] = F^{ij} = F_{ijE}, \tag{P5.6}$$

$$F_{i0} = -F^{i0} = iF_{i4E}. \tag{P5.7}$$

Therefore

$$\begin{aligned}
-iS_{YM} &= -i \int (-i) dx_{4E} d\vec{x}_E \left(-\frac{1}{4} F^a_{\mu\nu} F^{a\mu\nu}\right) \\
&= \frac{1}{4} \int dx_{4E} d\vec{x}_E \left(F^a_{ijE} F^a_{ijE} - 2i F^a_{i4E} i F^a_{i4E}\right) \\
&= \frac{1}{4} \int dx_{4E} d\vec{x}_E \, F^a_{\mu\nu E} F^a_{\mu\nu E} = S_{YM\,E} \geq 0. \tag{P5.8}
\end{aligned}$$

To get the anticommutation relation for the γ-matrices in the Euclidean space $\{\gamma_{\mu E}, \gamma_{\nu E}\} = 2\delta_{\mu\nu}$, one can take

$$\gamma^i = -\gamma_i = i\gamma_{iE}, \quad \gamma_0 = \gamma^0 = \gamma_4. \tag{P5.9}$$

Then the Dirac Lagrangian reads

$$\begin{aligned}
\mathcal{L}_f &= \bar{\psi}(i(\partial_0 - igA_0)\gamma^0 + i(\partial_i - igA_i)\gamma^i - m)\psi \\
&= \bar{\psi}(-(\partial_{4E} - igA_{4E})\gamma_{4E} + i(\partial_{iE} - igA_{iE})i\gamma_{iE} - m)\psi \\
&= \bar{\psi}(-(\partial_{4E} - igA_{4E})\gamma_{4E} - (\partial_{iE} - igA_{iE})\gamma_{iE} - m)\psi \\
&= -\bar{\psi}((\partial_{\mu E} - igA_{\mu E})\gamma_{\mu E} + m)\psi = -\bar{\psi}(D_{\mu E}\gamma_{\mu E} + m)\psi, \tag{P5.10}
\end{aligned}$$

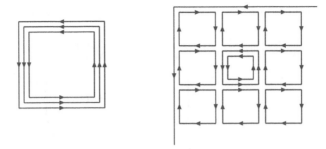

Figure 5.11 Link configuration for the first correction to the RT Wilson loop average. Left: links for integral (P5.13). Right: tiling of the RT loop.

and the fermion action has the form

$$-iS_f = \int d^4 x_E \bar{\psi} (D_{\mu E} \gamma_{\mu E} + m) \psi = S_{f\,E}. \qquad (P5.11)$$

Finally, one arrives at the generating functional in the Euclidean space

$$Z = \int \mathcal{D}A \mathcal{D}\psi \mathcal{D}\bar{\psi} e^{-S_{f\,E} - S_{YM\,E}} = \int \mathcal{D}A \det(D_{\mu E} \gamma_{\mu E} + m) e^{-S_{YM\,E}}. \qquad (P5.12)$$

Keeping the fermion determinant constant results in the quenched approximation.

5.2 Calculate the first correction in β to the RT Wilson loop and the string tension in the strong coupling limit.

SOLUTION. We follow (Cre85). The first correction comes from the minimal surface tiled with the plaquettes as is shown in Figure 5.11-right. In this tiling one of the places is occupied by 2 plaquettes with the opposite orientation w.r.t. the main contribution in Figure 5.4. For this configuration one has to integrate over the links shown in Figure 5.11-left. Via (P1.210) one gets

$$\int \mathcal{D}U \, [tr U(\partial p)]^3 = 1. \qquad (P5.13)$$

There are $\dfrac{(\frac{RT}{a^2}+1)\frac{RT}{a^2}}{2}$ ways to choose 2 plaquettes from $(\frac{RT}{a^2}+1)$, and there are $\frac{RT}{a^2}!$ ways to place $(\frac{RT}{a^2}+1)$ plaquettes in the $\frac{RT}{a^2}$ places in the

RT Wilson loop when we treat the chosen 2 plaquettes as 1. Therefore

$$\frac{1}{N_c}\langle trU_{\square T}\rangle^{(1)}_R \simeq \frac{1}{(\frac{RT}{a^2}+1)!}\int \mathcal{D}U \, \frac{1}{N_c}trU_{\square T}{}_R$$

$$\times \left\{ \sum_{oriented\ p'} \frac{\beta}{2N_c}tr\, U(\partial p') \right\}^{\frac{RT}{a^2}+1}$$

$$= \frac{(\frac{RT}{a^2}+1)\frac{RT}{a^2}}{2}\frac{\frac{RT}{a^2}!}{(\frac{RT}{a^2}+1)!}\left(\frac{\beta}{2N_c}\right)^{\frac{RT}{a^2}+1}$$

$$\times \frac{1}{N_c}N_c^{(\#nodes-4)-(\#links-4)}$$

$$= \left(\frac{\beta}{2N_c^2}\right)^{\frac{TR}{a^2}}\frac{\beta}{2N_c}\frac{RT}{2a^2}, \quad \Longrightarrow \quad \text{(P5.14)}$$

$$e^{-KRT} = \left(\frac{\beta}{2N_c^2}\right)^{\frac{TR}{a^2}}(1+\frac{\beta RT}{4N_ca^2}), \quad \text{(P5.15)}$$

$$a^2K = \ln\left(\frac{18}{\beta}\right) - \frac{\beta}{12} + O(\beta^2). \quad \text{(P5.16)}$$

5.3 Estimate the leading contribution to the **mass gap**, or the energy of a lightest glueball state in the quenched $SU(N_c)$ lattice theory in the strong coupling regime.

SOLUTION. We follow (Cre85). One considers the correlation function of 2 plaquettes separated at a Euclidean time distance $T \to \infty$.

$$\langle 0|trU(\partial p')^\dagger trU(\partial p)|0\rangle = \sum_n \langle 0|trU(\partial p')^\dagger |n\rangle\langle n| trU(\partial p)|0\rangle$$

$$= \sum_n \langle 0|e^{HT}trU(\partial p)e^{-HT}|n\rangle\langle n| trU(\partial p')^\dagger|0\rangle$$

$$\underset{T\to\infty}{\to} |\langle 0|trU(\partial p)|n_1\rangle|^2 e^{-m_1T}. \quad \text{(P5.17)}$$

Here we inserted a complete set of states between the two plaquettes and left only the state with the smallest mass in the limit $T \to \infty$. Therefore one gets the mass as

$$m = -\lim_{T\to\infty}\frac{1}{T}\ln\langle 0|trU(\partial p')^\dagger trU(\partial p)|0\rangle. \quad \text{(P5.18)}$$

In the strong coupling regime, this matrix element corresponds to the tube-like tiling shown in Figure 5.12. It reads

$$\langle 0|trU(\partial p')^\dagger trU(\partial p)|0\rangle \simeq \frac{1}{\frac{A_{\min}}{a^2}!}\int \mathcal{D}U \, trU(\partial p')^\dagger trU(\partial p)$$

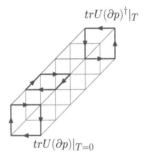

$$trU(\partial p)^\dagger|_T$$

$$trU(\partial p)|_{T=0}$$

Figure 5.12 Link configuration for the mass gap calculation.

$$\times \left\{ \sum_{oriented\ p''} \frac{\beta}{2N_c} tr\, U(\partial p'') \right\}^{\frac{A_{min}}{a^2}}$$

$$= \left(\frac{\beta}{2N_c} \right)^{\frac{A_{min}}{a^2}} N_c^{\#sites-\#links}$$

$$= \left| \#_{sites} = 4(\frac{T}{a}+1), \quad \#_{links} = 4\frac{T}{a}+4(\frac{T}{a}+1) \right|$$

$$= \left(\frac{\beta}{2N_c^2} \right)^{4\frac{T}{a}} \implies m = \frac{4}{a}\ln\left(\frac{2N_c^2}{\beta} \right). \tag{P5.19}$$

5.4 Calculate the potential energy for 3 static quarks in the color singlet state in the strong coupling regime. Get the small β limit of (5.18).

HINT. Use problem 2.27 from Chapter 2.

5.5 Show that one can fix one link without changing the average of a gauge invariant polynomial of link variables.

SOLUTION. For a gauge invariant polynomial $P(U)$

$$\langle P(U) \rangle = \frac{1}{Z} \int \mathcal{D}U e^{-S_W[U]} P(U) = \int dU_\mu(x) F(P, U_\mu(x)), \tag{P5.20}$$

$$F(P, U_\mu(x)) = \frac{1}{Z} \int_{\substack{all\ but \\ U_\mu(x)}} \mathcal{D}U e^{-S_W[U]} P(U)$$

$$= \frac{1}{Z} \int_{\substack{all\ but \\ U_\mu^V(x)}} \mathcal{D}U^V e^{-S_W[U^V]} P(U^V)$$

$$= F(P, V(x)U_\mu(x)V^\dagger(x+a\hat\mu)), \tag{P5.21}$$

where we did the gauge transformation with $V(y)$

$$U_\nu(y) \to U_\nu^V(y) = V(y)U_\nu(y)V^\dagger(y+a\hat\nu), \tag{P5.22}$$

and took into account the gauge invariance of the measure, action and P. Therefore F does not depend on the specific value of $U_\mu(x)$ and one can choose it arbitrarily. In the same manner, one can fix values for all links comprising an arbitrary connected or disconnected tree graph, fixing so the gauge. However, one cannot fix values of the links in a closed loop since the trace over the closed loop is gauge invariant.

5.6 Show that P^μ (5.28) is a conserved charge associated with space-time translational invariance.

SOLUTION. The Nambu-Goto string action

$$S = -\frac{1}{2\pi\alpha} \int_0^\pi d\sigma \int_{\tau_1}^{\tau_2} d\tau \sqrt{(x'\dot{x})^2 - x'^2 \dot{x}^2} \qquad \text{(P5.23)}$$

depends only on space-time coordinates' derivatives $\dot{x}$ and x'. Therefore it is invariant under the constant space-time translations

$$x^\mu \to x^\mu + \delta x^\mu, \quad \delta x^\mu = \varepsilon^\mu = const. \qquad \text{(P5.24)}$$

The conservation law for the 4 Noether currents ($\mu = 0, 1, 2, 3$)

$$J_a^\mu = \frac{\partial \mathcal{L}}{\partial(\partial_a x^\mu)}, \quad a = (\tau, \sigma) \qquad \text{(P5.25)}$$

is

$$\frac{\partial}{\partial a} J_a^\mu = \frac{\partial}{\partial \sigma}\left(\frac{\partial \mathcal{L}}{\partial x'^\mu}\right) + \frac{\partial}{\partial \tau}\left(\frac{\partial \mathcal{L}}{\partial \dot{x}^\mu}\right) = \frac{\partial}{\partial \sigma} P_\sigma^\mu + \frac{\partial}{\partial \tau} P_\tau^\mu = 0. \qquad \text{(P5.26)}$$

It coinsides with the e.o.m. The conserved charges are then the integrals over the worldsheet of the τ-components of these currents

$$P^\mu = -\int_0^\pi d\sigma\, P_\tau^\mu. \qquad \text{(P5.27)}$$

5.7 Show that the 4-momentum of the string is conserved.

SOLUTION. Using the e.o.m. and boundary conditions

$$\frac{\partial}{\partial \sigma}\frac{\partial L}{\partial x'^\mu} + \frac{\partial}{\partial \tau}\frac{\partial L}{\partial \dot{x}^\mu} = 0, \quad \frac{\partial L}{\partial x'^\mu}\Big|_{\sigma=0} = \frac{\partial L}{\partial x'^\mu}\Big|_{\sigma=\pi} = 0 \qquad \text{(P5.28)}$$

one gets

$$\frac{d}{d\tau}P^\mu = -\frac{d}{d\tau}\int_0^\pi d\sigma \frac{\partial L}{\partial \dot{x}^\mu} = \int_0^\pi d\sigma \frac{d}{d\sigma}\frac{\partial L}{\partial x'^\mu} = -\frac{\partial L}{\partial x'^\mu}\Big|_{\sigma=0}^{\sigma=\pi} = 0. \qquad \text{(P5.29)}$$

5.8 Show that the angular momentum of the string is conserved.

SOLUTION.

$$\frac{d}{d\tau} J^{\mu\nu} = \frac{d}{d\tau} \int_0^\pi d\sigma [x^\mu P_\tau^\nu - x^\nu P_\tau^\mu]$$

$$= \int_0^\pi d\sigma [\dot{x}^\mu P_\tau^\nu + x^\mu \dot{P}_\tau^\nu - \dot{x}^\nu P_\tau^\mu - x^\nu \dot{P}_\tau^\mu]$$

$$= \int_0^\pi d\sigma [\dot{x}^\mu P_\tau^\nu - x^\mu P_\sigma^{\prime\nu} - \dot{x}^\nu P_\tau^\mu + x^\nu P_\sigma^{\prime\mu}]$$

$$= [-x^\mu P_\sigma^\nu + x^\nu P_\sigma^\mu]|_{\sigma=0}^{\sigma=\pi}$$

$$+ \int_0^\pi d\sigma [\dot{x}^\mu P_\tau^\nu + x^{\prime\mu} P_\sigma^\nu - \dot{x}^\nu P_\tau^\mu - x^{\prime\nu} P_\sigma^\mu]$$

$$= \frac{1}{2\pi\alpha} \int_0^\pi d\sigma \left[\frac{\dot{x}^\mu \left(x^{\prime 2} \dot{x}^\nu - (x'\dot{x}) x^{\prime\nu} \right) + x^{\prime\mu} (\dot{x}^2 x^{\prime\nu} - (x'\dot{x}) \dot{x}^\nu)}{\sqrt{(x'\dot{x})^2 - x^{\prime 2} \dot{x}^2}} \right.$$

$$\left. - (\mu \leftrightarrow \nu) \right] = 0. \tag{P5.30}$$

Here we used

$$S = -\frac{1}{2\pi\alpha} \int_0^\pi d\sigma \int_{\tau_1}^{\tau_2} d\tau \sqrt{(x'\dot{x})^2 - x^{\prime 2} \dot{x}^2} \tag{P5.31}$$

and

$$P_\tau^\mu(\sigma, \tau) = \frac{\partial L}{\partial \dot{x}^\mu} = \frac{x^{\prime 2} \dot{x}^\mu - (x'\dot{x}) x^{\prime\mu}}{2\pi\alpha \sqrt{(x'\dot{x})^2 - x^{\prime 2} \dot{x}^2}}, \tag{P5.32}$$

$$P_\sigma^\mu(\sigma, \tau) = \frac{\partial L}{\partial x^{\prime\mu}} = \frac{\dot{x}^2 x^{\prime\mu} - (x'\dot{x}) \dot{x}^\mu}{2\pi\alpha \sqrt{(x'\dot{x})^2 - x^{\prime 2} \dot{x}^2}}, \tag{P5.33}$$

$$P_\sigma^{\prime\mu} + \dot{P}_\tau^\mu = 0, \quad P_\sigma^\mu|_{\sigma=\pi} = P_\sigma^\mu|_{\sigma=0} = 0. \tag{P5.34}$$

5.9 Show that the end points of the string move with the speed of light.

SOLUTION. Explicit expressions for P_τ^μ and P_σ^μ (P5.33) give

$$P_\sigma^2 = \frac{x^{\prime 2} \dot{x}^4 - (x'\dot{x})^2 \dot{x}^2}{4\pi^2\alpha^2 \left((x'\dot{x})^2 - x^{\prime 2} \dot{x}^2 \right)} = -\frac{\dot{x}^2}{4\pi^2\alpha^2}. \tag{P5.35}$$

Then at the endpoints $P_\sigma^\mu|_{\sigma=\pi} = P_\sigma^\mu|_{\sigma=0} = 0$. Hence

$$\dot{x}^2|_{\sigma=0,\pi} = 0, \tag{P5.36}$$

i.e. the velocity 4-vector of the endpoints is light-like. Also, we need

$$0 = P_\tau^\mu x_\mu' = P_\sigma^\mu \dot{x}_\mu, \quad P_\tau^2 = \frac{x^{\prime 4} \dot{x}^2 - (x'\dot{x})^2 x^{\prime 2}}{4\pi^2\alpha^2 \left((x'\dot{x})^2 - x^{\prime 2} \dot{x}^2 \right)} = -\frac{x^{\prime 2}}{4\pi^2\alpha^2}. \tag{P5.37}$$

5.10 Show that the rotating string

$$x^0 = \tau, \quad x^1 = A(\sigma - \frac{\pi}{2})\cos\omega\tau, \quad x^2 = A(\sigma - \frac{\pi}{2})\sin\omega\tau, \quad x^3 = 0$$
$$\text{(P5.38)}$$

satisfies the equations of motion and boundary conditions $P_\sigma'^\mu + \dot{P}_\tau^\mu = 0$, $P_\sigma^\mu|_{\sigma=\pi} = P_\sigma^\mu|_{\sigma=0} = 0$ for $\omega A\frac{\pi}{2} = 1$.

SOLUTION.

$$\dot{x}^\mu = (1, -\omega A(\sigma - \frac{\pi}{2})\sin\omega\tau, \omega A(\sigma - \frac{\pi}{2})\cos\omega\tau, 0), \quad \text{(P5.39)}$$

$$\dot{x}^2 = 1 - \omega^2 A^2(\sigma - \frac{\pi}{2})^2, \quad \text{(P5.40)}$$

$$x'^\mu = (0, A\cos\omega\tau, A\sin\omega\tau, 0), \quad x'^2 = -A^2, \quad (x'\dot{x}) = 0. \quad \text{(P5.41)}$$

$$P_\sigma^\mu(\sigma, t) = \frac{\dot{x}^2 x'^\mu}{2\pi\alpha\sqrt{-x'^2\dot{x}^2}} = \sqrt{1 - \omega^2 A^2(\sigma - \frac{\pi}{2})^2}\frac{(0, \cos\omega\tau, \sin\omega\tau, 0)}{2\pi\alpha},$$
$$\text{(P5.42)}$$

$$P_\sigma^\mu(0, t) = 0 \quad \Longrightarrow \quad 1 = \omega A\frac{\pi}{2} \quad \Longrightarrow \quad P_\sigma^\mu(\pi, t) = 0, \quad \text{(P5.43)}$$

$$P_\sigma'^\mu(\sigma, t) = \frac{-\omega^2 A^2(\sigma - \frac{\pi}{2})}{\sqrt{1 - \omega^2 A^2(\sigma - \frac{\pi}{2})^2}}\frac{(0, \cos\omega\tau, \sin\omega\tau, 0)}{2\pi\alpha}. \quad \text{(P5.44)}$$

$$P_\tau^\mu(\sigma, t) = \frac{x'^2 \dot{x}^\mu}{2\pi\alpha\sqrt{-x'^2\dot{x}^2}}$$
$$= -\frac{A(1, -\omega A(\sigma - \frac{\pi}{2})\sin\omega\tau, \omega A(\sigma - \frac{\pi}{2})\cos\omega\tau, 0)}{2\pi\alpha\sqrt{1 - \omega^2 A^2(\sigma - \frac{\pi}{2})^2}}, \quad \text{(P5.45)}$$

$$\dot{P}_\tau^\mu = \frac{\omega^2 A^2(\sigma - \frac{\pi}{2})(0, \cos\omega\tau, \sin\omega\tau, 0)}{2\pi\alpha\sqrt{1 - \omega^2 A^2(\sigma - \frac{\pi}{2})^2}} \quad \Longrightarrow \quad P_\sigma'^\mu + \dot{P}_\tau^\mu = 0.$$
$$\text{(P5.46)}$$

5.11 Calculate the 4-momentum for the rotating string

$$x^0 = \tau, \quad x^1 = A(\sigma - \frac{\pi}{2})\cos\omega\tau, \quad x^2 = A(\sigma - \frac{\pi}{2})\sin\omega\tau, \quad x^3 = 0$$
$$\text{(P5.47)}$$

with $\omega A\frac{\pi}{2} = 1$.

SOLUTION. We have

$$\int_0^\pi d\sigma \vec{P}_\tau(\sigma, t) = 0, \quad \text{(P5.48)}$$

$$E = -\int_0^\pi d\sigma P_\tau^0(\sigma, t) = \int_0^\pi \frac{A d\sigma}{2\pi\alpha\sqrt{1 - \omega^2 A^2(\sigma - \frac{\pi}{2})^2}}$$
$$= \int_0^\pi \frac{A d\sigma}{2\pi\alpha\sqrt{1 - (\frac{2\sigma}{\pi} - 1)^2}} = \frac{A}{2\pi\alpha}\frac{\pi^2}{2}. \quad \text{(P5.49)}$$

5.12 Calculate the angular momentum of the rotating string.

SOLUTION.

$$J^{\mu\nu} = \int_0^\pi d\sigma [x^\mu P_\tau^\nu - x^\nu P_\tau^\mu] = \int_0^\pi d\sigma \frac{x'^2(x^\mu \dot{x}^\nu - x^\nu \dot{x}^\mu)}{2\pi\alpha\sqrt{-x'^2 \dot{x}^2}} \implies$$

$$J^{10} = J^{02} = 0 = \dots$$

$$J^{12} = \int_0^\pi d\sigma \frac{-\omega A^3(\sigma - \frac{\pi}{2})^2}{2\pi\alpha\sqrt{1 - \omega^2 A^2(\sigma - \frac{\pi}{2})^2}} = \int_0^\pi d\sigma \frac{-\omega A^3(\sigma - \frac{\pi}{2})^2}{2\pi\alpha\sqrt{1 - (\frac{2\sigma}{\pi} - 1)^2}}$$

$$= \frac{-\omega A^3}{2\pi\alpha} \frac{\pi^4}{16} = -\frac{A^2 \pi^2}{16\alpha}. \tag{P5.50}$$

There are many excellent introductory books on lattice QCD, e. g. (Cre85; Mak05; Rot12). Comprehensive surveys on bound states include (Bal01; FH12; KZ07). Basic introduction to string theory one can find (Zwi06).

FURTHER READING

[Bal01] Gunnar S. Bali. QCD forces and heavy quark bound states. *Phys. Rept.*, 343:1–136, 2001.

[Cre85] Michael Creutz. *Quarks, gluons and lattices.* Cambridge Monographs on Mathematical Physics. Cambridge University Press, Cambridge, UK, 6, 1985.

[EFG09] D. Ebert, R. N. Faustov, and V. O. Galkin. Mass spectra and Regge trajectories of light mesons in the relativistic quark model. *Phys. Rev. D*, 79:114029, 2009.

[EGK+75] E. Eichten, K. Gottfried, T. Kinoshita, J. Kogut, K. D. Lane, and T. M. Yan. Spectrum of charmed quark-antiquark bound states. *Phys. Rev. Lett.*, 34:369–372, Feb 1975.

[Fel81] B. Felsager. *Geometry, particles and fields*, chapter 8.8. Graduate Texts in Contemporary Physics. Univ.Pr., Odense, 1981.

[FH12] Zoltan Fodor and Christian Hoelbling. Light Hadron Masses from Lattice QCD. *Rev. Mod. Phys.*, 84:449, 2012.

[KZ07] Eberhard Klempt and Alexander Zaitsev. Glueballs, Hybrids, Multiquarks. Experimental facts versus QCD inspired concepts. *Phys. Rept.*, 454:1–202, 2007.

[Mak05] Yu. Makeenko. *Methods of contemporary gauge theory*, chapter 6, 12. Cambridge Monographs on Mathematical Physics. Cambridge University Press, 11, 2005.

[Rot12] Heinz J. Rothe. *Lattice gauge theories: An introduction (fourth edition)*, volume 43. World Scientific Publishing Company, 2012.

[Som94] R. Sommer. A New way to set the energy scale in lattice gauge theories and its applications to the static force and alpha-s in SU(2) Yang-Mills theory. *Nucl. Phys. B*, 411:839–854, 1994.

[TMNS01] Toru T. Takahashi, H. Matsufuru, Y. Nemoto, and H. Suganuma. The three quark potential in the SU(3) lattice QCD. *Phys. Rev. Lett.*, 86:18–21, 2001.

[Zwi06] B. Zwiebach. *A first course in string theory*, chapter 6.4, 8.4. Cambridge University Press, 7, 2006.

Effective Action

THIS chapter describes construction of the effective action in general and for QCD in the external gluon, and quark and gluon fields. We calculate the quadratic in the external fields one loop corrections to the action as functional determinants and show that one can absorb the singularities from these corrections into the bare charge and mass. Problems touch upon questions of scale setting and scheme changing.

1. Consider a real scalar field ϕ with the generating functional

$$Z[J] = \int \mathcal{D}\phi\, e^{i\int dx\left(\frac{1}{2}(\partial\phi)^2 - \frac{1}{2}m^2\phi^2 - V(\phi) + J\phi\right)}$$

$$= \int \mathcal{D}\phi\, e^{iS + i\int dx\, J\phi} = e^{-iE[J]}, \qquad (6.1)$$

where for convenience we introduced the energy functional E. In the past, we looked at source fields as at proxies to get the correlation functions. Now we want to understand **how the dynamics changes in the presence of a fixed real physical source field** J. First, the vacuum changes and as a result the field averages

$$\frac{\delta E[J]}{\delta J_x} = i\frac{\delta \ln Z[J]}{\delta J_x} = -\frac{1}{Z[J]}\int \mathcal{D}\phi\, \phi_x e^{iS + i\int dx\, J\phi} = -\langle\phi_x\rangle_J = -\phi_x^{cl}.$$

$$(6.2)$$

The above equation defines the **classical field** ϕ_x^{cl} as **the average field in the presence of a fixed source** J. Hence, the classical field is a function of J and vice versa.

$$\frac{\delta^2 E[J]}{\delta J_x \delta J_y} = -\frac{\delta \phi_x^{cl}}{\delta J_y} = i\frac{\delta \ln Z[J]}{\delta J_x \delta J_y} = -\frac{\delta}{\delta J_y}\frac{1}{Z[J]}\int \mathcal{D}\phi\, \phi_x e^{iS + i\int dx\, J\phi}$$

$$= \frac{i}{Z[J]^2}\int \mathcal{D}\phi\, \phi_y e^{iS + i\int dx\, J\phi}\int \mathcal{D}\phi\, \phi_x e^{iS + i\int dx\, J\phi}$$

$$- \frac{i}{Z[J]}\int \mathcal{D}\phi\, \phi_y \phi_x e^{iS + i\int dx\, J\phi} = -i(\langle\phi_x\phi_y\rangle_J - \langle\phi_x\rangle_J\langle\phi_y\rangle_J)$$

$$= -i\langle\phi_x\phi_y\rangle_{Jconn} = -iD(x,y). \qquad (6.3)$$

DOI: 10.1201/9781003272403-6

Here one defines the **propagator D as the connected part of the average of ϕ_x and ϕ_y** in the presence of J. Along the same line one gets

$$\frac{\delta^n E[J]}{\delta J_{x_1} ... \delta J_{x_n}} = i^{n+1} \langle \phi_{x_1} ... \phi_{x_n} \rangle_{J conn}. \tag{6.4}$$

Therefore $-E[J]$ **is often called the generating functional for connected Green functions**.

2. One defines the **effective action** $\Gamma[\phi^{cl}]$ as

$$\Gamma[\phi^{cl}] = -E[J] - \int J_z \phi_z^{cl} dz. \tag{6.5}$$

It has the following properties

$$\frac{\delta \Gamma[\phi^{cl}]}{\delta \phi_x^{cl}} = -\frac{\delta E[J]}{\delta \phi_x^{cl}} - \frac{\delta}{\delta \phi_x^{cl}} \int J_z \phi_z^{cl} dz$$

$$= -\int dz \frac{\delta E[J]}{\delta J_z} \frac{\delta J_z}{\delta \phi_x^{cl}} - J_x - \int \frac{\delta J_z}{\delta \phi_x^{cl}} \phi_z^{cl} dz$$

$$= \int dz \phi_z^{cl} \frac{\delta J_z}{\delta \phi_x^{cl}} - \int \frac{\delta J_z}{\delta \phi_x^{cl}} \phi_z^{cl} dz - J_x = -J_x. \tag{6.6}$$

Therefore

$$\frac{\delta}{\delta J_y} \frac{\delta \Gamma}{\delta \phi_x^{cl}} = -\delta(x - y). \tag{6.7}$$

This expression can be rewritten as

$$-\delta(x - y) = \int dz \frac{\delta \phi_z^{cl}}{\delta J_y} \frac{\delta^2 \Gamma}{\delta \phi_z^{cl} \delta \phi_x^{cl}} = -\int dz \frac{\delta^2 E[J]}{\delta J_z \delta J_y} \frac{\delta^2 \Gamma}{\delta \phi_z^{cl} \delta \phi_x^{cl}} \tag{6.8}$$

$$= i \int dz D(z, y) \frac{\delta^2 \Gamma}{\delta \phi_z^{cl} \delta \phi_x^{cl}} \implies \frac{\delta^2 \Gamma}{\delta \phi_z^{cl} \delta \phi_x^{cl}} = i D^{-1}(z, x). \tag{6.9}$$

So the **second functional derivative of the effective action is the inverse propagator**.

3. **Calculation of the effective action.** Defining

$$\phi_x = \phi_x^{cl} + \eta_x, \tag{6.10}$$

$$e^{i\Gamma[\phi^{cl}]} = \int D\phi e^{iS + i \int dx J\eta}$$

$$= e^{iS[\phi^{cl}]} \int D\eta e^{i \int dx \eta_x \left(\frac{\delta S}{\delta \phi_x} |_{\phi^{cl}} + J_x \right) + \frac{i}{2} \int dx dy \frac{\delta^2 S[\phi_{cl}]}{\delta \phi_x \delta \phi_y} \eta_x \eta_y + \cdots}. \tag{6.11}$$

Since by construction

$$\frac{\delta S}{\delta \phi_x}\Big|_{\phi^{cl}} + J_x = 0, \tag{6.12}$$

one has

$$e^{i\Gamma[\phi^{cl}]} = e^{iS[\phi^{cl}]} \int \mathcal{D}\eta \, e^{\frac{i}{2} \int dx dy \frac{\delta^2 S[\phi_{cl}]}{\delta \phi_x \delta \phi_y} \eta_x \eta_y + \dots}$$

$$\simeq const \, e^{iS[\phi^{cl}]} \det \left(\frac{\delta^2 S[\phi_{cl}]}{\delta \phi_x \delta \phi_y} \right)^{-\frac{1}{2}}$$

$$= const \, e^{iS[\phi^{cl}] - \frac{1}{2} tr \ln \frac{\delta^2 S[\phi_{cl}]}{\delta \phi_x \delta \phi_y}} \implies$$

$$\Gamma[\phi^{cl}] = S[\phi^{cl}] + \frac{i}{2} tr \ln \frac{\delta^2 S[\phi_{cl}]}{\delta \phi_x \delta \phi_y} + \dots \tag{6.13}$$

4. **Effective action for QCD.** The Faddev-Popov action in the external classical field b was derived in problem 4.7 in Chapter 4. We consider $A^\mu = a^\mu + b^\mu$, where a^μ **is a quantum field and** b^μ **is an external classical field.** The action reads

$$\mathcal{L}_{FP}^b = \mathcal{L}_{QCD}^b - \frac{1}{2\xi g^2}(D^{ab\mu}a_\mu^b)^2 + \bar{c}^a(-(D^2)^{ac} - (D^{ab\mu}a_\mu^d)f^{bdc})c^c, \tag{6.14}$$

where

$$D^\mu = \partial^\mu - ib^\mu, \quad F_{\mu\nu}^a = \partial_\mu b_\nu^a - \partial_\nu b_\mu^a + f^{abc}b_\mu^b b_\nu^c, \tag{6.15}$$

$$\mathcal{L}_{QCD}^b = -\frac{1}{4g^2}(F_{\mu\nu}^a + D_\mu^{ab}a_\nu^b - D_\nu^{ab}a_\mu^b + f^{abc}a_\mu^b a_\nu^c)^2 + \bar{\psi}(i\hat{D} + \hat{a} - m)\psi. \tag{6.16}$$

Using $\xi = 1$ and **keeping only constant and quadratic in the quantum fields terms,** we have

$$\mathcal{L}_{FP}^b = -\frac{1}{4g^2}(F_{\mu\nu}^a)^2 - \frac{1}{2g^2}F^{\mu\nu a}f^{abc}a_\mu^b a_\nu^c - \frac{1}{4g^2}(D_\mu^{ab}a_\nu^b - D_\nu^{ab}a_\mu^b)^2$$

$$- \frac{1}{2g^2}(D^{ab\mu}a_\mu^b)^2 + \bar{\psi}(i\hat{D} - m)\psi + \bar{c}^a(-(D^2)^{ac})c^c$$

$$+ \text{ (linear in fields terms and terms with fields}$$

$$\text{in 3-rd and higher power).} \tag{6.17}$$

Taking into account that via integration by parts

$$(D_\mu^{ab}a_\nu^b)(\dots)^a = ((\delta^{ab}\partial_\mu - ib_\mu^c T_{ab}^c)a_\nu^b)(\dots)^a$$

$$\rightarrow a_\nu^b(-\delta_\mu^{ba}\partial + ib_\mu^c T_{ba}^c)(\dots)^a = -a_\nu^b D_\mu^{ba}(\dots)^a, \tag{6.18}$$

and recalling $[D^\mu D^\nu]^{bc} = -iF^{a\mu\nu}T^a_{bc}$ and $iT^a_{bc} = -f^{bac}$, one can rewrite

$$-\frac{1}{2}(D^{ab}_\mu a^b_\nu - D^{ab}_\nu a^b_\mu)^2 - (D^{ab\mu}a^b_\mu)^2 = -D^{ab}_\mu a^b_\nu D^{\mu ac}a^{\nu c}$$
$$+ D^{ab}_\mu a^b_\nu D^{\nu ac}a^{\mu c} - D^{ab\mu}a^b_\mu D^{ac\nu}a^c_\nu$$
$$= a^b_\nu D^{ba}_\mu D^{\mu ac}a^{\nu c} - a^b_\nu D^{ba}_\mu D^{\nu ac}a^{\mu c} + a^b_\mu D^{ba\mu}D^{ac\nu}a^c_\nu$$
$$= a^b_\nu (D^2)^{bc}a^{\nu c} - a^b_\nu(D^{\mu ba}D^{\nu ac} - D^{ba\nu}D^{ac\mu})a^c_\mu$$
$$= a^b_\nu (D^2)^{bc}a^{\nu c} - a^b_\nu[D^\mu D^\nu]^{bc}a^c_\mu = a^b_\nu(D^2)^{bc}a^{\nu c} - a^b_\nu(-iF^{a\mu\nu}T^a_{bc})a^c_\mu$$
$$= a^b_\nu (D^2)^{bc}a^{\nu c} + a^b_\nu F^{a\nu\mu}f^{bac}a^c_\mu. \tag{6.19}$$

As a result,

$$\mathcal{L}^b_{FP} = -\frac{1}{4g^2}(F^a_{\mu\nu})^2 + \frac{1}{2g^2}a^b_\mu((D^2)^{bc}g^{\mu\nu} + 2F^{a\mu\nu}f^{bac})a^c_\nu$$
$$+ \bar\psi(i\hat D - m)\psi + \bar c^a(-(D^2)^{ac})c^c + ... \tag{6.20}$$

Therefore the first correction to the effective action **for n_f light quark flavors** reads

$$e^{i\Gamma[b]} = e^{-\frac{i}{4g^2}\int F^2 dx} \int \mathcal{D}a e^{\frac{i}{2g^2}\int dx a^b_\mu((D^2)^{bc}g^{\mu\nu} + 2F^{a\mu\nu}f^{bac})a^c_\nu}$$
$$\times \int \mathcal{D}\psi \mathcal{D}\bar\psi e^{i\int dx\bar\psi(i\hat D - m)\psi} \int \mathcal{D}c\mathcal{D}\bar c e^{i\int dx\bar c^a(-(D^2)^{ac})c^c} \tag{6.21}$$
$$= e^{-\frac{i}{4g^2}\int F^2 dx} \det\left[((-D^2)^{bc}g^{\mu\nu} - 2F^{a\mu\nu}f^{bac})\right]^{-\frac{1}{2}}$$
$$\times \det\left[i\hat D\right]^{+n_f} \det\left[(-(D^2)^{ac}\right]^{+1}. \tag{6.22}$$

Here we neglected masses of the quarks writing

$$\det\left[i\hat D - m\right] \to \det\left[i\hat D\right], \tag{6.23}$$

which is justified when the characteristic energy scale $p \gg m$. In the opposite case when the quarks are too heavy, i.e. $p \ll m$

$$\det\left[i\hat D - m\right] \to \det[m] = const, \tag{6.24}$$

and such quarks give a constant correction to the effective action. This constant does not change equations of motion and can be neglected.

5. **Running coupling.** Using the results of exercise 6.6 for the correction to the classical action proportional to the classical action

$$\ln \det \left[\left((-D^2)^{bc} g^{\mu\nu} - 2F^{a\mu\nu} f^{bac} \right) \right]^{-\frac{1}{2}} = const$$

$$-\frac{1}{2} \int \frac{d^d p}{(2\pi)^d} F^a_{\mu\nu}(-p) F^{a\mu\nu}(p) \frac{i}{(4\pi)^2}$$

$$\times \left(-\frac{5}{3} N_c \right) \left(\frac{1}{\varepsilon} - \gamma - \ln \frac{-p^2}{4\pi\mu^2} + \frac{59}{30} + O(\varepsilon) \right), \qquad (6.25)$$

$$\ln \det \left[i\hat{D} \right]^{+n_f} \simeq const + n_f \int \frac{d^d p}{(2\pi)^d} F^a_{\mu\nu}(-p) F^{a\mu\nu}(p) \frac{i}{(4\pi)^2}$$

$$\times \left(-\frac{1}{6} \right) \left(\frac{1}{\varepsilon} - \gamma - \ln \frac{-p^2}{4\pi\mu^2} + \frac{5}{3} + O(\varepsilon) \right), \qquad (6.26)$$

$$\ln \det \left[(-(D^2)^{ac})^{+1} \right] = const + \int \frac{d^d p}{(2\pi)^d} F^a_{\mu\nu}(-p) F^{a\mu\nu}(p) \frac{i}{(4\pi)^2}$$

$$\times \frac{N_c}{4} \frac{1}{3} \left(\frac{1}{\varepsilon} - \gamma - \ln \frac{-p^2}{4\pi\mu^2} + \frac{8}{3} + O(\varepsilon) \right), \qquad (6.27)$$

we get

$$e^{i\Gamma[b]} = e^{-\frac{i}{4g^2} \int F^2 dx}$$

$$\times e^{const + \frac{i}{(4\pi)^2} \int \frac{d^d p}{(2\pi)^d} F^a_{\mu\nu}(-p) F^{a\mu\nu}(p) \frac{\beta_0}{4} \left(\frac{1}{\varepsilon} - \gamma - \ln \frac{-p^2}{4\pi\mu^2} + C + O(\varepsilon) \right)},$$

$$\beta_0 = \frac{11}{3} N_c - \frac{2}{3} n_f, \quad C = \frac{1}{\beta_0} \left(\frac{67}{9} N_c - \frac{10}{9} n_f \right). \qquad (6.28)$$

Here $d = 4 - 2\varepsilon$, μ is the regularization scale, β_0 **is called (the first coefficient of) the beta function.** We can look at this expression as at action with the effective charge

$$\Gamma[b] = -\frac{1}{4} \int \frac{d^d p}{(2\pi)^d} F^a_{\mu\nu}(-p) F^{a\mu\nu}(p) \frac{1}{g^2(p^2)}, \qquad (6.29)$$

$$\frac{1}{g^2(p^2)} = \frac{1}{g^2} - \frac{\beta_0}{(4\pi)^2} \left(\frac{1}{\varepsilon} - \gamma - \ln \frac{-p^2}{4\pi\mu^2} + C + O(\varepsilon) \right). \qquad (6.30)$$

and assume that the bare charge g absorbs the singularity

$$\frac{1}{g^2} = \frac{1}{g^2(-\mu^2)} + \frac{\beta_0}{(4\pi)^2} \left(\frac{1}{\varepsilon} - \gamma - \ln \frac{1}{4\pi} \right). \qquad (6.31)$$

Note that we could have taken any constant in addition to $\frac{1}{\varepsilon} - \gamma - \ln \frac{1}{4\pi}$ into the bare charge. Taking only these terms is known as $\overline{MS}$ **prescription.** Now one can find the relation between the charges at different scales

$$\frac{1}{g^2(q^2)} - \frac{1}{g^2(p^2)} = \frac{\beta_0}{(4\pi)^2} \ln \frac{q^2}{p^2} \quad \Longrightarrow \quad g^2(q^2) = \frac{g^2(p^2)}{1 + \frac{g^2(p^2)}{(4\pi)^2}\beta_0 \ln \frac{q^2}{p^2}}. \tag{6.32}$$

Since $\beta_0 > 0$, $g^2(q^2)$ decreases as q^2 increases, i.e. at high energy QCD is a weakly coupled theory. This behavior is called the **asymptotic freedom.** In the differential form (6.32) is known as **Gell-Mann – Low** equation

$$\frac{g^2(p^2) - g^2(q^2)}{\ln \frac{p^2}{q^2}} \quad \rightarrow \quad \frac{dg^2(p^2)}{d\ln(-p^2)} = -\frac{\beta_0}{(4\pi)^2}g^4(p^2). \tag{6.33}$$

Regularization scale μ can be traded for the **QCD mass scale** Λ_{QCD}

$$g^2(\Lambda_{QCD}^2) = \infty \quad \Longrightarrow \quad \Lambda_{QCD}^2 = -\mu^2 e^{-\frac{(4\pi)^2}{\beta_0 g^2(-\mu^2)}}, \tag{6.34}$$

$$g^2(q^2) = \frac{g^2(p^2)}{1 + \frac{g^2(p^2)}{(4\pi)^2}\beta_0 \ln \frac{\Lambda_{QCD}^2}{p^2} + \frac{g^2(p^2)}{(4\pi)^2}\beta_0 \ln \frac{q^2}{\Lambda_{QCD}^2}} = \frac{(4\pi)^2}{\beta_0 \ln \frac{q^2}{\Lambda_{QCD}^2}}. \tag{6.35}$$

In terms of $\alpha_s = \frac{g^2}{4\pi}$

$$\alpha_s(q^2) = \frac{\alpha_s(p^2)}{1 + \frac{\alpha_s(p^2)}{4\pi}\beta_0 \ln \frac{q^2}{p^2}}, \quad \frac{d\alpha_s(p^2)}{d\ln p^2} = -\frac{\alpha_s^2(p^2)}{4\pi}\beta_0. \tag{6.36}$$

Experimentally

$$\Lambda_{QCD} \simeq 250 MeV \quad \Longrightarrow \quad \alpha_s(1 GeV^2) \simeq 0.4. \tag{6.37}$$

Therefore **QCD becomes a strongly coupled theory at the scale** $\lesssim \Lambda_{QCD}$ **and the perturbation theory is not applicable at low energy.** Now the β-function $\beta(\alpha_s(p^2))$

$$\frac{d\alpha_s(p^2)}{d\ln p^2} = \beta(\alpha_s(p^2)) = -\frac{\alpha_s^2(p^2)}{4\pi}(\beta_0 + \beta_1\alpha_s + \beta_2\alpha_s^2 + ...). \tag{6.38}$$

is known up to 5 loops and it is easier to fix the argument of α_s at some physically known scale like m_Z and recalculate the Λ_{QCD} from it

$$\alpha_s(m_Z) = 0.1179... \tag{6.39}$$

The scale dependence of the strong coupling is shown in Figure 6.1.

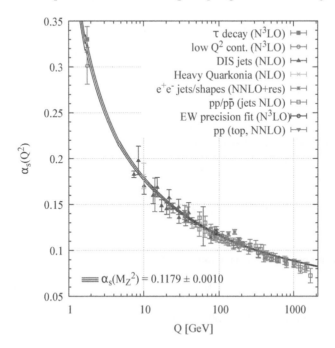

Figure 6.1 Strong coupling constant as a function of energy scale. Picture from (Z$^+$20).

6. **Effective action in both gluon and quark external fields.** One can also treat the quark field as **a sum of external field ϕ and quantum field χ**

$$\psi = \phi + \chi, \tag{6.40}$$

Then the Lagrangian in the presence of both external fields reads

$$\mathcal{L}_{FP}^{b,\phi} = -\frac{1}{4g^2}(F_{\mu\nu}^a)^2 + \bar{\phi}(i\hat{D} - m)\phi + \frac{1}{2g^2}a_\mu^b((D^2)^{bc}g^{\mu\nu} + 2F^{a\mu\nu}f^{bac})a_\nu^c$$
$$+ \bar{c}^a(-(D^2)^{ac})c^c + \bar{\phi}\hat{a}\chi + \bar{\chi}\hat{a}\phi + \bar{\chi}(i\hat{D} - m)\chi + \dots \tag{6.41}$$

Again, we kept only the terms in the second power of the external fields. The ghost sector does not change, while the quark and gluon fields are now entangled. To disentangle them one can do the following substitution in the functional integral

$$\chi = \chi - \frac{1}{(i\hat{D} - m)}\hat{a}\phi, \quad \bar{\chi} = \bar{\chi} - \bar{\phi}\hat{a}\frac{1}{(i\hat{D} - m)}, \tag{6.42}$$

which does not change the integration measure. Then,

$$
\mathcal{L}_{FP}^{b,\phi} = -\frac{1}{4g^2}(F_{\mu\nu}^a)^2 + \bar{\phi}(i\hat{D} - m)\phi + \frac{1}{2g^2}a_\mu^b((D^2)^{bc}g^{\mu\nu} + 2F^{a\mu\nu}f^{bac})a_\nu^c
$$
$$
+ \bar{c}^a(-(D^2)^{ac})c^c + \bar{\phi}\hat{a}\chi + \bar{\chi}\hat{a}\phi + \bar{\chi}(i\hat{D} - m)\chi
$$
$$
- \bar{\phi}\hat{a}\frac{1}{(i\hat{D} - m)}\hat{a}\phi - \bar{\phi}\hat{a}\frac{1}{(i\hat{D} - m)}\hat{a}\phi - \bar{\chi}(i\hat{D} - m)\frac{1}{(i\hat{D} - m)}\hat{a}\phi
$$
$$
- \bar{\phi}\hat{a}\frac{1}{(i\hat{D} - m)}(i\hat{D} - m)\chi + \bar{\phi}\hat{a}\frac{1}{(i\hat{D} - m)}(i\hat{D} - m)\frac{1}{(i\hat{D} - m)}\hat{a}\phi + \dots
$$

$$(6.43)$$

and we get a diagonal Lagrangian

$$
\mathcal{L}_{FP}^{b,\phi} = -\frac{1}{4g^2}(F_{\mu\nu}^a)^2 + \bar{\phi}(i\hat{D} - m)\phi + \bar{c}^a(-(D^2)^{ac})c^c + \bar{\chi}(i\hat{D} - m)\chi
$$
$$
+ \frac{1}{2g^2}a_\mu^b\left[((D^2)^{bc}g^{\mu\nu} + 2F^{a\mu\nu}f^{bac}) - \bar{\phi}t^b\gamma^\mu\frac{2g^2}{(i\hat{D} - m)}t^c\gamma^\nu\phi\right]a_\nu^c + \dots
$$

$$(6.44)$$

As a result, the effective action reads

$$
e^{i\Gamma[b,\phi]} = e^{i\int dx\left[-\frac{F^2}{4g^2} + \bar{\phi}(i\hat{D} - m)\phi\right]} \det\left[i\hat{D} - m\right]^{+n_f} \det\left[(-(D^2)^{ac}\right]^{+1}
$$
$$
\times \det\left[((-D^2)^{bc}g^{\mu\nu} - 2F^{a\mu\nu}f^{bac})\right.
$$
$$
\times \left(1 + \frac{1}{((-D^2)^{bc}g^{\mu\nu} - 2F^{a\mu\nu}f^{bac})}\bar{\phi}t^b\gamma^\mu\frac{2g^2}{(i\hat{D} - m)}t^c\gamma^\nu\phi\right)\right]^{-\frac{1}{2}}
$$
$$
= e^{-\frac{i}{4g^2}\int F^2 dx} \det\left[i\hat{D} - m\right]^{+n_f} \det\left[(-(D^2)^{ac}\right]^{+1}
$$
$$
\times \det\left[1 + \frac{1}{((-D^2)^{bc}g^{\mu\nu} - 2F^{a\mu\nu}f^{bac})}\bar{\phi}t^b\gamma^\mu\frac{2g^2}{(i\hat{D} - m)}t^c\gamma^\nu\phi\right]^{-\frac{1}{2}}
$$
$$
\times \det\left[((-D^2)^{bc}g^{\mu\nu} - 2F^{a\mu\nu}f^{bac})\right]^{-\frac{1}{2}}.
$$

$$(6.45)$$

7. **Running mass.** Using the result of exersise 6.6,

$$
\ln\det\left[1 + \frac{1}{((-D^2)^{bc}g^{\mu\nu} - 2F^{a\mu\nu}f^{bac})}\bar{\phi}t^b\gamma^\mu\frac{2g^2}{(i\hat{D} - m)}t^c\gamma^\nu\phi\right]^{-\frac{1}{2}}
$$
$$
= -\frac{1}{2}i\frac{2g^2 C_F}{(4\pi)^2}\int\frac{d^d p}{(2\pi)^d}\bar{\phi}(-p)[X\hat{p} + Ym]\phi(p) + O(\varepsilon) + O(b), \quad (6.46)
$$

$$
X \underset{-p^2 \gg m^2}{=} -\left(\frac{1}{\varepsilon} - \gamma - \ln\left(\frac{-p^2}{4\pi\mu^2}\right) + 1\right), \tag{6.47}
$$

$$
Y \underset{-p^2 \gg m^2}{=} 4\left(\frac{1}{\varepsilon} - \gamma - \ln\left(\frac{-p^2}{4\pi\mu^2}\right) + \frac{3}{2}\right). \tag{6.48}
$$

Therefore, the quadratic in the fields fermion part of the effective action will have the form

$$\Gamma[\phi] = \int \frac{d^d p}{(2\pi)^d} \tilde{\phi}(-p) \left(1 - \frac{g^2 C_F}{(4\pi)^2} X\right) \left[\hat{p} - m \frac{\left(1 + \frac{g^2 C_F}{(4\pi)^2} Y\right)}{\left(1 - \frac{g^2 C_F}{(4\pi)^2} X\right)}\right] \tilde{\phi}(p)$$

(6.49)

$$= \int \frac{d^d p}{(2\pi)^d} \tilde{\phi}_R(-p) \left[\hat{p} - m_R(p^2)\right] \tilde{\phi}_R(p),$$

(6.50)

where

$$\tilde{\phi}_R(p) = \left(1 - \frac{g^2 C_F}{(4\pi)^2} X\right)^{\frac{1}{2}} \tilde{\phi}(p), \quad \tilde{\phi}_R(-p) = \tilde{\phi}(-p) \left(1 - \frac{g^2 C_F}{(4\pi)^2} X\right)^{\frac{1}{2}},$$

(6.51)

$$m_R(p^2) = m \frac{\left(1 + \frac{g^2 C_F}{(4\pi)^2} Y\right)}{\left(1 - \frac{g^2 C_F}{(4\pi)^2} X\right)} \simeq m \left(1 + \frac{g^2 C_F}{(4\pi)^2} (Y + X)\right)$$

$$= m \left(1 + \frac{g^2 C_F}{(4\pi)^2} 3 \left(\frac{1}{\varepsilon} - \gamma - \ln\left(\frac{-p^2}{4\pi\mu^2}\right) + \frac{5}{3}\right)\right)$$

$$= m_R(-\mu^2) \left(1 - \frac{g^2 C_F}{(4\pi)^2} 3 \ln\left(\frac{-p^2}{\mu^2}\right)\right) + O(g^4).$$

(6.52)

Here one may absorb $\left(1 - \frac{g^2 C_F}{(4\pi)^2} X\right)^{\frac{1}{2}}$ into the definition of the renormalized field ϕ_R and rewrite the Lagrangian with effective mass $m_R(p^2)$, assuming that the bare mass absorbs the singularity through the $\overline{MS}$ prescription

$$m_R(-\mu^2) = m \left(1 + \frac{g^2 C_F}{(4\pi)^2} 3 \left(\frac{1}{\varepsilon} - \gamma - \ln\left(\frac{1}{4\pi}\right)\right)\right)$$

(6.53)

$$= m \left(1 + \frac{g^2(-\mu^2)}{1 + \frac{\beta_0 g^2(-\mu^2)}{(4\pi)^2} \left(\frac{1}{\varepsilon} - \gamma - \ln\frac{1}{4\pi} + O(\varepsilon)\right)}\right.$$

$$\times \left. \frac{C_F}{(4\pi)^2} 3 \left(\frac{1}{\varepsilon} - \gamma - \ln\left(\frac{1}{4\pi}\right)\right)\right)$$

(6.54)

$$= m \left(1 + \frac{C_F g^2(-\mu^2)}{(4\pi)^2} 3 \left(\frac{1}{\varepsilon} - \gamma - \ln\left(\frac{1}{4\pi}\right)\right) + O(g^4)\right).$$

(6.55)

For close p^2 and q^2

$$\frac{m_R(p^2)}{m_R(q^2)} \simeq 1 - \frac{g^2\left(p^2\right) C_F}{(4\pi)^2} 3 \ln\left(\frac{p^2}{q^2}\right) \implies \frac{d \ln m\left(p^2\right)}{d \ln(-p^2)} = -\frac{3 C_F}{(4\pi)^2} g^2\left(p^2\right).$$

(6.56)

Using Gell-Mann - Low equation (6.33)

$$\ln \frac{m\left(p^2\right)}{m(q^2)} = -\frac{3C_F}{(4\pi)^2} \int g^2\left(p^2\right) d\ln(-p^2)$$

$$= -\frac{3C_F}{(4\pi)^2} \frac{(4\pi)^2}{-\beta_0} \int \frac{dg^2(p^2)}{g^2(p^2)} = \frac{3C_F}{\beta_0} \ln \frac{g^2(p^2)}{g^2(q^2)}. \qquad (6.57)$$

Finally,

$$m\left(p^2\right) = m(q^2) \left(\frac{g^2(p^2)}{g^2(q^2)}\right)^{\frac{3C_F}{\beta_0}} = m(q^2) \left(\frac{\alpha_s(p^2)}{\alpha_s(q^2)}\right)^{\frac{3C_F}{\beta_0}}. \qquad (6.58)$$

These masses calculated via perturbation theory at large energy scale are called **current quark masses**. They do not take into account the energy of nonperturbative gluon clouds interacting with quarks at low energy in hadrons. As the energy increases the current masses decrease, e.g.

$$\frac{m\left(m_Z^2\right)}{m(1GeV^2)} \sim \left(\frac{\alpha_s(m_Z^2)}{\alpha_s(1GeV^2)}\right)^{\frac{3C_F}{\beta_0}} \sim \left(\frac{0.1}{0.4}\right)^{\frac{4}{11-\frac{2}{3}(\sim 4)}} \sim \frac{1}{2}. \qquad (6.59)$$

EXERCISES

6.1 Show that ϕ^{cl} obeys the classical Lagrange equation, i.e. minimizes the action

$$\frac{\delta \int dx \left(\mathcal{L} + J\phi\right)}{\delta \phi_x}\Big|_{\phi^{cl}} = 0. \qquad (P6.1)$$

SOLUTION. Indeed, in the discretized truncation

$$\phi_z^{cl} = \frac{1}{Z[J]} \int \mathcal{D}\phi \, \phi_z e^{i \int dx(\mathcal{L}+J\phi)}$$

$$\simeq \frac{1}{Z[J]} \left(\Pi \int d\phi_x\right) \phi_x e^{i\left(\frac{1}{2}\phi_x A_{xy}\phi_y + J_x\phi_x\right)}. \qquad (P6.2)$$

$$0 = \frac{\partial}{\partial \phi_z} \left(\frac{1}{2}\phi_x A_{xy}\phi_y + J_x\phi_x\right)\Big|_{\phi^*} = A_{zy}\phi_y + J_z\Big|_{\phi^*}$$

$$\implies \phi_y^* = -A_{yz}^{-1} J_z, \quad \phi_x = \eta_x - A_{xz}^{-1} J_z$$

$$\implies \frac{1}{2}\phi_x A_{xy}\phi_y + J_x\phi_x = \frac{1}{2}(\eta_x - A_{xz}^{-1} J_z)A_{xy}(\eta_y - A_{yz'}^{-1} J_{z'})$$

$$+ J_x(\eta_x - A_{xz}^{-1} J_z)$$

$$= \frac{1}{2}\eta_x A_{xy}\eta_y + \frac{1}{2}A_{xz'}^{-1} J_{z'} A_{xy} A_{yz}^{-1} J_z - \frac{1}{2}A_{xz}^{-1} J_z A_{xy}\eta_y$$

$$- \frac{1}{2}\eta_x A_{xy} A_{yz}^{-1} J_z + J_x\eta_x - J_x A_{xz}^{-1} J_z$$

$$= \frac{1}{2}\eta_x A_{xy}\eta_y - \frac{1}{2}J_y A_{yz}^{-1} J_z. \qquad (P6.3)$$

Then,

$$\phi_z^{cl} \simeq \frac{1}{Z[J]} \left(\Pi \int d\eta_x \right) (\eta_x - A_{xz}^{-1} J_z) e^{i\left(\frac{1}{2}\eta_x A_{xy}\eta_y - \frac{1}{2}J_y A_{yz}^{-1} J_z\right)}$$
$$= -A_{xz}^{-1} J_z = \phi_y^*. \tag{P6.4}$$

6.2 Show that for a nonsingular matrix M

$$\frac{\partial M^{-1}}{\partial \alpha} = -M^{-1} \frac{\partial M}{\partial \alpha} M^{-1}. \tag{P6.5}$$

6.3 Show that

$$\frac{\delta^n \Gamma}{\delta \phi_{x_1}^{cl} ... \delta \phi_{x_n}^{cl}} = -i \langle \phi_{x_1} ... \phi_{x_n} \rangle_{1PI}. \tag{P6.6}$$

SOLUTION. Indeed,

$$\frac{\delta}{\delta J_y} = \int dt \frac{\delta \phi_t^{cl}}{\delta J_y} \frac{\delta}{\delta \phi_t^{cl}} = i \int dt D(y, t) \frac{\delta}{\delta \phi_t^{cl}}. \tag{P6.7}$$

Therefore

$$\frac{\delta^2 E[J]}{\delta J_x \delta J_y \delta J_z} = i \int dt D(t, y) \frac{\delta}{\delta \phi_t^{cl}} \left[\frac{\delta^2 \Gamma}{\delta \phi_z^{cl} \delta \phi_x^{cl}} \right]^{-1}$$
$$= -i \int dt dr do D(t, y) \left[\frac{\delta^2 \Gamma}{\delta \phi_z^{cl} \delta \phi_r^{cl}} \right]^{-1} \frac{\delta}{\delta \phi_t^{cl}} \frac{\delta^2 \Gamma}{\delta \phi_r^{cl} \delta \phi_o^{cl}} \left[\frac{\delta^2 \Gamma}{\delta \phi_o^{cl} \delta \phi_x^{cl}} \right]^{-1}$$
$$= i \int dt dr do D(t, y) D(r, z) D(o, x) \frac{\delta^3 \Gamma}{\delta \phi_t^{cl} \delta \phi_r^{cl} \delta \phi_o^{cl}}, \tag{P6.8}$$
$$\implies \frac{\delta^3 \Gamma}{\delta \phi_t^{cl} \delta \phi_r^{cl} \delta \phi_o^{cl}} = -i \langle \phi_t \phi_r \phi_o \rangle_{1PI}. \tag{P6.9}$$

Hence, one can see that the third functional derivative of the effective action is the "vertex", i.e. the **1-particle irreducible part of the connected Green function**. Along the same line

$$\frac{\delta^n \Gamma}{\delta \phi_{x_1}^{cl} ... \delta \phi_{x_n}^{cl}} = -i \langle \phi_{x_1} ... \phi_{x_n} \rangle_{1PI}. \tag{P6.10}$$

6.4 Show that in the background field method for QCD renormalization constants Z_A and Z_g are related

$$Z_A^{-\frac{1}{2}} = Z_g, \quad \text{where} \quad A = Z_A^{\frac{1}{2}} A_R, \quad g = Z_g g_R. \tag{P6.11}$$

SOLUTION. Indeed, thanks to explicit gauge invariance of the effective action the correction should take the form

$$\sim const(\varepsilon) F_{\mu\nu R}^a F_R^{a\mu\nu}. \tag{P6.12}$$

In the usual field normalization

$$F_{\mu\nu}^a = Z_A^{\frac{1}{2}}(\partial_\mu A_{R\nu}^a - \partial_\nu A_{R\mu}^a + Z_g Z_A^{\frac{1}{2}} g_R f^{abc} A_{R\mu}^b A_{R\nu}^c), \qquad (P6.13)$$

and gauge invariance demands

$$Z_g Z_A^{\frac{1}{2}} = 1. \qquad (P6.14)$$

In our normalization

$$b = gA, \qquad (P6.15)$$

and the field b does not change in the renormalization.

6.5 Use the bracket notation in d-dimensional space

$$\langle x|y\rangle = \delta(x-y), \quad \langle q|p\rangle = (2\pi)^d \delta(p-q), \quad \langle x|p\rangle = e^{-ipx}, \qquad (P6.16)$$

$$\hat{1} = \int |x\rangle\langle x| d^d x = \int |p\rangle\langle p| \frac{d^d p}{(2\pi)^d}, \qquad (P6.17)$$

$$F(p) = \langle p|F\rangle = \langle p|x\rangle\langle x|F\rangle = \int d^d x e^{ipx} F(x). \qquad (P6.18)$$

We will often omit writing integration explicitly when summing repeated indices, i.e.

$$\int |x\rangle\langle x| d^d x \rightarrow |x\rangle\langle x|, \quad \int |p\rangle\langle p| \frac{d^d p}{(2\pi)^d} \rightarrow |p\rangle\langle p|. \qquad (P6.19)$$

Understand the following equalities

$$\langle p|b(x)|q\rangle = \langle p|y\rangle\langle y|b|x\rangle\langle x|q\rangle = \langle p|x\rangle b(x)\langle x|q\rangle$$
$$= \int d^d x b(x) e^{i(p-q)x} = \tilde{b}(p-q), \qquad (P6.20)$$

$$F^{a\mu\nu}(p) = \int d^d x e^{ipx}(\partial^\mu b^{a\nu} - \partial^\nu b^{a\mu}) + O(b^2)$$
$$= -(ip^\mu \tilde{b}(p)^{a\nu} - ip^\nu \tilde{b}^{a\mu}(p)) + O(b^2), \qquad (P6.21)$$

$$i\hat{\partial}_\mu|p\rangle = i\partial_\mu|x\rangle\langle x|p\rangle = i|x\rangle \frac{\partial}{\partial x^\mu} e^{-ipx} = p_\mu|p\rangle, \qquad (P6.22)$$

$$\langle p|i\overrightarrow{\partial}_\mu = -\langle p|i\overleftarrow{\partial}_\mu = -\langle p|x\rangle\langle x|i\overleftarrow{\partial}_\mu$$
$$= -i\langle x|\frac{\partial}{\partial x^\mu} e^{ipx} = \langle p|p_\mu, \qquad (P6.23)$$

$$\langle \bar{\phi}|p\rangle = \langle p|\phi\rangle^\dagger \gamma^0 = (\langle p|x\rangle\langle x|\phi\rangle)^\dagger \gamma^0 = (\int d^d x e^{ipx} \phi_x)^\dagger \gamma^0$$
$$= (\tilde{\phi}(p))^\dagger \gamma^0 = \tilde{\bar{\phi}}(-p), \qquad (P6.24)$$

$$\int dx \bar{\phi}_x (i\hat{\partial}_x - m)\phi_x = \langle \bar{\phi}|x\rangle (i\hat{\partial}_x - m)\langle x|\phi\rangle = \langle \bar{\phi}|x\rangle (i\hat{\partial}_x - m)\langle x|p\rangle \langle p|\phi\rangle$$

$$= \langle \bar{\phi}|p\rangle (\hat{p} - m)\langle p|\phi\rangle = \int \frac{d^d p}{(2\pi)^d} \tilde{\bar{\phi}}(-p)(\hat{p} - m)\tilde{\phi}(p),$$

$$(P6.25)$$

$$\int d^d x (F^a_{\mu\nu}(x))^2 = \langle F|x\rangle \langle x|F\rangle = \langle F|p\rangle \langle p|F\rangle = \int \frac{d^d p}{(2\pi)^d} F^a_{\mu\nu}(-p) F^{a\mu\nu}(p)$$

$$= -\int \frac{d^d p}{(2\pi)^d} (ip_\mu \tilde{b}(-p)^a_\nu - ip_\nu \tilde{b}(-p)^a_\mu)$$

$$\times (ip^\mu \tilde{b}(p)^{a\nu} - ip^\nu \tilde{b}^{a\mu}(p)) + O(b^3)$$

$$= 2 \int \frac{d^d p}{(2\pi)^d} \tilde{b}^{a\mu}(-p)(p^2 g_{\mu\nu} - p_\mu p_\nu)\tilde{b}(p)^{a\nu} + O(b^3).$$

$$(P6.26)$$

$$tr(\sigma^{\mu\nu}\sigma^{\alpha\beta}) = 4(g^{\mu\alpha}g^{\nu\beta} - g^{\mu\beta}g^{\nu\alpha}). \qquad (P6.27)$$

6.6 Calculate the following determinants up to terms quadratic in classical fields:

a. the ghost determinant $\det\left[(-(D^2)^{ac}\right]$,

b. the quark determinant $\det\left[i\hat{D}\right]$,

c. the gluon determinant $\det\left[((-D^2)^{bc}g^{\mu\nu} - 2F^{a\mu\nu}f^{bac})\right]$,

d. the mixed determinant

$$\det\left[1 + \frac{1}{((-D^2)^{bc}g^{\mu\nu} - 2F^{a\mu\nu}f^{bac})}\bar{\phi}t^b\gamma^\mu \frac{2g^2}{(i\hat{D}-m)}t^c\gamma^\nu\phi\right].$$

SOLUTION.

a. We need

$$\ln\det\left[((iD)^2)^{ac}\right] = tr\ln\left[(i\partial^\mu + b^\mu)^{ab}(i\partial_\mu + b_\mu)^{ba}\right]$$

$$= tr\ln\left[(i\partial)^2 + bi\partial + i\partial b + b^2\right]$$

$$= tr\ln\left[(i\partial)^2(1 + \frac{1}{(i\partial)^2}(bi\partial + i\partial b + b^2))\right]$$

$$= const + tr\ln\left[1 + \frac{1}{(i\partial)^2}(bi\partial + i\partial b + b^2)\right]$$

$$= const + tr\left[\frac{1}{(i\partial)^2}b^2\right.$$

$$\left. - \frac{1}{2}\frac{1}{(i\partial)^2}(bi\partial + i\partial b)\frac{1}{(i\partial)^2}(bi\partial + i\partial b)\right]. \quad (P6.28)$$

Here we used

$$\ln\left[1 + x\right] = x - \frac{x^2}{2} + \dots \tag{P6.29}$$

and the tracelessness of T

$$tr\left[\frac{1}{(i\partial)^2}(bi\partial + i\partial b)\right] = tr\left[\frac{1}{(i\partial)^2}(b^a i\partial + i\partial b^a)\right] T_{bb}^a = 0. \tag{P6.30}$$

Next,

$$tr\left[\frac{1}{(i\partial)^2}b^2\right] = \langle p|\frac{1}{(i\partial)^2}|q\rangle\langle q|b_\mu^a|l\rangle\langle l|b^{\mu c}|p\rangle(T_{on}^a T_{no}^c)$$

$$= N_c \int \frac{d^dp}{(2\pi)^d}\frac{d^dq}{(2\pi)^d}\frac{d^dl}{(2\pi)^d}\langle p|q\rangle\frac{\delta^{ac}}{q^2}\tilde{b}_\mu^a(q-l)\tilde{b}^{\mu c}(l-p)$$

$$= N_c \int \frac{d^dp}{(2\pi)^d}\tilde{b}_\mu^a(p)\tilde{b}^{\mu a}(-p)\int\frac{d^dl}{(2\pi)^d}\frac{1}{(p+l)^2}. \tag{P6.31}$$

$$tr\left[-\frac{1}{2}\frac{1}{(i\partial)^2}(bi\partial + i\partial b)\frac{1}{(i\partial)^2}(bi\partial + i\partial b)\right]$$

$$= -\frac{N_c}{2}\langle p|\frac{1}{(i\partial)^2}|q\rangle\langle q|b^a i\partial + i\partial b^a|l\rangle\langle l|\frac{1}{(i\partial)^2}|n\rangle\langle n|b^a i\partial + i\partial b^a|p\rangle$$

$$= -\frac{N_c}{2}\frac{1}{p^2l^2}\langle p|b_\mu^a l^\mu + p^\mu b_\mu^a|l\rangle\langle l|b_\nu^a p^\nu + l^\nu b_\nu^a|p\rangle$$

$$= -\frac{N_c}{2}\int\frac{d^dp}{(2\pi)^d}\frac{d^dl}{(2\pi)^d}\tilde{b}_\mu^a(p-l)\tilde{b}_\nu^a(l-p)\frac{(l^\mu + p^\mu)(p^\nu + l^\nu)}{p^2l^2}$$

$$= -\frac{N_c}{2}\int\frac{d^dp}{(2\pi)^d}\tilde{b}_\mu^a(p)\tilde{b}_\nu^a(-p)\int\frac{d^dl}{(2\pi)^d}\frac{(2l^\mu + p^\mu)(p^\nu + 2l^\nu)}{(p+l)^2l^2}. \tag{P6.32}$$

Therefore,

$$\ln\det\left[(iD)^2\right] = const + \frac{N_c}{2}\int\frac{d^dp}{(2\pi)^d}\tilde{b}_\mu^a(p)\tilde{b}_\nu^a(-p)$$

$$\times \int\frac{d^dl}{(2\pi)^d}\frac{2g^{\mu\nu}l^2 - (2l^\mu + p^\mu)(p^\nu + 2l^\nu)}{(p+l)^2l^2}. \tag{P6.33}$$

Here, recovering the regularization scale μ,

$$\int\frac{\mu^{4-d}d^dl}{(2\pi)^d}\frac{2g^{\mu\nu}l^2 - (2l^\mu + p^\mu)(p^\nu + 2l^\nu)}{(p+l)^2l^2}$$

$$= \int_0^1 dx \int\frac{\mu^{4-d}d^dl}{(2\pi)^d}\frac{2g^{\mu\nu}l^2 - (2l^\mu + p^\mu)(p^\nu + 2l^\nu)}{((px+l)^2 + p^2x(1-x))^2} \tag{P6.34}$$

$$= \int_0^1 dx \int\frac{\mu^{4-d}d^dk}{(2\pi)^d}\left[\frac{2g^{\mu\nu}(k-px)^2}{(k^2 + p^2x(1-x))^2}\right.$$

$$\left. -\frac{(2(k-px)^\mu + p^\mu)(p^\nu + 2(k-px)^\nu)}{(k^2 + p^2x(1-x))^2}\right] \tag{P6.35}$$

$$
= \int_0^1 dx \int \frac{\mu^{4-d} d^d k}{(2\pi)^d} \frac{2g^{\mu\nu}(k^2 + p^2 x^2) - 4k^\mu k^\nu - (1-2x)^2 p^\mu p^\nu)}{(k^2 + p^2 x(1-x))^2}
$$
(P6.36)

$$
= \int_0^1 dx \int \frac{\mu^{4-d} d^d k}{(2\pi)^d} \frac{2g^{\mu\nu} k^2 (1-\frac{2}{d}) + 2g^{\mu\nu} p^2 x^2 - (1-2x)^2 p^\mu p^\nu)}{(k^2 + p^2 x(1-x))^2}
$$
(P6.37)

$$
= \frac{i\mu^{4-d}}{(4\pi)^{\frac{d}{2}}} \int_0^1 dx \left\{ -\frac{d}{2}\Gamma(1-\frac{d}{2}) \left(-p^2 x(1-x)\right)^{\frac{d}{2}-1} 2g^{\mu\nu}(1-\frac{2}{d}) \right.
$$

$$
\left. + \Gamma(2-\frac{d}{2}) \left(-p^2 x(1-x)\right)^{\frac{d}{2}-2} (2g^{\mu\nu} p^2 x^2 - (1-2x)^2 p^\mu p^\nu) \right\}
$$
(P6.38)

$$
= \frac{i\mu^{4-d}\Gamma(2-\frac{d}{2})}{(4\pi)^{\frac{d}{2}}} \int_0^1 dx \left(-p^2 x(1-x)\right)^{\frac{d}{2}-2}
$$

$$
\times \left\{ 2g^{\mu\nu} p^2 \left(x^2 - x(1-x)\right) - (1-2x)^2 p^\mu p^\nu) \right\}
$$
(P6.39)

$$
= \frac{i\mu^{4-d}\Gamma(2-\frac{d}{2})}{(4\pi)^{\frac{d}{2}}} \int_0^1 dx \left(-p^2 x(1-x)\right)^{\frac{d}{2}-2}
$$

$$
\times \left\{ g^{\mu\nu} p^2 \left(x^2 - 2x(1-x) + (1-x)^2\right) - (1-2x)^2 p^\mu p^\nu) \right\}
$$
(P6.40)

$$
= \left\{ g^{\mu\nu} p^2 - p^\mu p^\nu \right\} \frac{i\Gamma(2-\frac{d}{2})}{(4\pi)^{\frac{d}{2}}} \left(\frac{-p^2}{\mu^2}\right)^{\frac{d}{2}-2}
$$

$$
\times \int_0^1 dx \, (x(1-x))^{\frac{d}{2}-2} (1-2x)^2
$$
(P6.41)

$$
= \left\{ g^{\mu\nu} p^2 - p^\mu p^\nu \right\} \frac{i\Gamma(2-\frac{d}{2})}{(4\pi)^{\frac{d}{2}}} \left(\frac{-p^2}{\mu^2}\right)^{\frac{d}{2}-2} \frac{2^{2-d}\sqrt{\pi}\Gamma(\frac{d}{2}-1)}{\Gamma(\frac{d+1}{2})}
$$
(P6.42)

$$
= |d = 4 - 2\varepsilon|
$$

$$
\simeq \frac{i\left\{ g^{\mu\nu} p^2 - p^\mu p^\nu \right\}}{(4\pi)^2} \frac{1}{3} \left(\frac{1}{\varepsilon} - \gamma - \ln\frac{-p^2}{4\pi\mu^2} + \frac{8}{3} + O(\varepsilon)\right).
$$
(P6.43)

Finally, we have

$$\ln \det \left[(iD)^2 \right] = const + \frac{N_c}{2} \int \frac{d^d p}{(2\pi)^d} \tilde{b}^a_\mu(p) \tilde{b}^a_\nu(-p) \frac{\{ g^{\mu\nu} p^2 - p^\mu p^\nu \}}{(4\pi)^2}$$

$$\times \frac{i}{3} \left(\frac{1}{\varepsilon} - \gamma - \ln \frac{-p^2}{4\pi\mu^2} + \frac{8}{3} + O(\varepsilon) \right)$$

$$= const + \int \frac{d^d p}{(2\pi)^d} F^a_{\mu\nu}(-p) F^{a\mu\nu}(p)$$

$$\times \frac{i}{(4\pi)^2} \frac{N_c}{4} \frac{1}{3} \left(\frac{1}{\varepsilon} - \gamma - \ln \frac{-p^2}{4\pi\mu^2} + \frac{8}{3} + O(\varepsilon, b^3) \right).$$

$$(P6.44)$$

b. We need using (P3.85),

$$\det \left[i\hat{D} \right] = \det \left[(i\hat{D})^2 \right]^{\frac{1}{2}} = \det \left[(iD)^2 + \frac{1}{2}\sigma F \right]^{\frac{1}{2}}. \qquad (P6.45)$$

$$\ln \det \left[i\hat{D} \right] = \frac{1}{2} tr \ln \left[(iD)^2 + \frac{1}{2}\sigma F \right]$$

$$= \frac{1}{2} tr \ln \left[(i\partial)^2 + bi\partial + i\partial b + b^2 + \frac{1}{2}\sigma F \right]$$

$$= \frac{1}{2} tr \ln \left[(i\partial)^2 (1 + \frac{1}{(i\partial)^2}(bi\partial + i\partial b + b^2) + \frac{1}{(i\partial)^2}\frac{1}{2}\sigma F) \right]$$

$$= 2 \ln \det \left[(iD)^2 \right] + \frac{1}{2} tr \left[-\frac{1}{2}\frac{1}{(i\partial)^2}\frac{1}{2}\sigma F \frac{1}{(i\partial)^2}\frac{1}{2}\sigma F) \right].$$

$$(P6.46)$$

Here we used the convention (Col86) that trace of the unit Dirac matrix in d-dimensional space is 4, $tr1 = 4$. Next,

$$\frac{1}{2} tr \left[-\frac{1}{2}\frac{1}{(i\partial)^2}\frac{1}{2}\sigma F \frac{1}{(i\partial)^2}\frac{1}{2}\sigma F) \right]$$

$$= -\frac{1}{2^4} \langle p | \frac{1}{(i\partial)^2} | q \rangle \langle q | \sigma F | l \rangle \langle l | \frac{1}{(i\partial)^2} | n \rangle \langle n | \sigma F | p \rangle$$

$$= -\frac{1}{2^4} tr(\sigma^{\mu\nu}\sigma^{\alpha\beta}) tr(t^a t^b) \frac{1}{l^2 p^2} \langle p | F^a_{\mu\nu} | l \rangle \langle l | F^b_{\alpha\beta} | p \rangle$$

$$= -\frac{1}{2^2} \frac{1}{l^2 p^2} \langle p | F^a_{\mu\nu} | l \rangle \langle l | F^{a\mu\nu} | p \rangle$$

$$= -\frac{1}{2^2} \int \frac{d^d p}{(2\pi)^d} \frac{d^d l}{(2\pi)^d} \frac{1}{l^2 p^2} F^a_{\mu\nu}(p - l) F^{a\mu\nu}(l - p)$$

$$= -\frac{1}{2^2} \int \frac{d^d p}{(2\pi)^d} F^a_{\mu\nu}(p) F^{a\mu\nu}(-p) \int \frac{d^d l}{(2\pi)^d} \frac{1}{l^2 (p + l)^2}. \qquad (P6.47)$$

Then,

$$\int \frac{\mu^{4-d}d^d l}{(2\pi)^d} \frac{1}{(p+l)^2 l^2} = \int_0^1 dx \int \frac{\mu^{4-d}d^d l}{(2\pi)^d} \frac{1}{((px+l)^2 + p^2 x(1-x))^2}$$

$$= \int_0^1 dx \int \frac{\mu^{4-d}d^d k}{(2\pi)^d} \frac{1}{(k^2 + p^2 x(1-x))^2}$$

$$= \frac{i\Gamma(2-\frac{d}{2})}{(4\pi)^{\frac{d}{2}}} \left(\frac{-p^2}{\mu^2}\right)^{\frac{d}{2}-2} \int_0^1 dx \, (x(1-x))^{\frac{d}{2}-2}$$

$$= \frac{i\Gamma(2-\frac{d}{2})}{(4\pi)^{\frac{d}{2}}} \left(\frac{-p^2}{\mu^2}\right)^{\frac{d}{2}-2} \frac{\sqrt{\pi} 2^{3-d} \Gamma(\frac{d}{2}-1)}{\Gamma(\frac{d-1}{2})} = |d = 4 - 2\varepsilon|$$

$$= \frac{i}{(4\pi)^2} \left(\frac{1}{\varepsilon} - \gamma - \ln \frac{-p^2}{4\pi\mu^2} + 2 + O(\varepsilon)\right). \tag{P6.48}$$

Recalling that in the previous problem we calculated $\det[(iD)^2]$ in the adjoint representation, while here we have the fundumental one, we get a different color factor

$$(T^a_{on} T^c_{no}) = N\delta^{ac} \to tr(t^a t^c) = \frac{1}{2}\delta^{ac}. \tag{P6.49}$$

Finally,

$$\ln\det\left[i\hat{D}\right] = const + \int \frac{d^d p}{(2\pi)^d} F^a_{\mu\nu}(-p) F^{a\mu\nu}(p) \frac{i}{(4\pi)^2}$$

$$\times \left[2\frac{1}{4}\frac{1}{2}\frac{1}{3}\left(\frac{1}{\varepsilon} - \gamma - \ln \frac{-p^2}{4\pi\mu^2} + \frac{8}{3} + O(\varepsilon)\right)\right.$$

$$\left. + \left(-\frac{1}{2^2}\right)\left(\frac{1}{\varepsilon} - \gamma - \ln \frac{-p^2}{4\pi\mu^2} + 2 + O(\varepsilon)\right)\right] + O(b^3) \tag{P6.50}$$

$$= const + \int \frac{d^d p}{(2\pi)^d} F^a_{\mu\nu}(-p) F^{a\mu\nu}(p) \frac{i}{(4\pi)^2}$$

$$\times \left(-\frac{1}{6}\right)\left(\frac{1}{\varepsilon} - \gamma - \ln \frac{-p^2}{4\pi\mu^2} + \frac{5}{3} + O(\varepsilon)\right) + O(b^3). \tag{P6.51}$$

c. We need

$$\ln\det\left[((iD)^2)^{ac} g^{\mu\nu} - 2F^{b\mu\nu} f^{abc}\right]$$

$$= \ln\det\left[((iD)^2)^{ac} g^{\mu\nu} + 2F^{b\mu\nu} iT^b_{ac}\right]$$

$$= tr\ln\left[(i\partial^\rho + b^\rho)^{ab}(i\partial_\rho + b_\rho)^{bc} g^{\mu\nu} + 2F^{b\mu\nu} iT^b_{ac}\right]$$

$$= tr\ln\left[((i\partial)^2 + bi\partial + i\partial b + b^2) g^{\mu\nu} + 2iF^{\mu\nu}\right]$$

$$= tr\ln\left[(i\partial)^2 g^{\mu\nu}(\delta^\rho_\nu + \frac{1}{(i\partial)^2}(\delta^\rho_\nu(bi\partial + i\partial b + b^2) + 2iF_\nu{}^\rho))\right]$$

$$= d\ln\det\left[(iD)^2\right] + tr\left[-\frac{1}{2}\frac{1}{(i\partial)^2}2iF^{\mu\nu}\frac{1}{(i\partial)^2}2iF_{\nu\mu}\right]. \tag{P6.52}$$

Next,

$$tr\left[-\frac{1}{2}\frac{1}{(i\partial)^2}2iF^{\mu\nu}\frac{1}{(i\partial)^2}2iF_{\nu\mu}\right]$$

$$= 2\langle p|\frac{1}{(i\partial)^2}|q\rangle\langle q|F^{\mu\nu}|l\rangle\langle l|\frac{1}{(i\partial)^2}|n\rangle\langle n|F_{\nu\mu}|p\rangle$$

$$= 2(T^a_{dc}T^b_{cd})\frac{1}{l^2p^2}\langle p|F^{a\mu\nu}|l\rangle\langle l|F^b_{\nu\mu}|p\rangle = -2N_c\frac{1}{l^2p^2}\langle p|F^{a\mu\nu}|l\rangle\langle l|F^a_{\mu\nu}|p\rangle$$

$$= -2N_c\int\frac{d^dp}{(2\pi)^d}F^a_{\mu\nu}(p)F^{a\mu\nu}(-p)\int\frac{d^dl}{(2\pi)^d}\frac{1}{l^2(p+l)^2}. \tag{P6.53}$$

As a result,

$$\ln\det\left[((iD)^2)^{ac}g^{\mu\nu}-2F^{b\mu\nu}f^{abc}\right] =$$

$$= const + \frac{i}{(4\pi)^2}\int\frac{d^dp}{(2\pi)^d}F^a_{\mu\nu}(-p)F^{a\mu\nu}(p)$$

$$\times\left[\frac{N_c}{4}\frac{d}{3}\left(\frac{1}{\varepsilon}-\gamma-\ln\frac{-p^2}{4\pi\mu^2}+\frac{8}{3}+O(\varepsilon)\right)\right.$$

$$\left.-2N_c\left(\frac{1}{\varepsilon}-\gamma-\ln\frac{-p^2}{4\pi\mu^2}+2+O(\varepsilon)\right)\right]+O(b^3) \tag{P6.54}$$

$$= const + \int\frac{d^dp}{(2\pi)^d}F^a_{\mu\nu}(-p)F^{a\mu\nu}(p)\frac{i}{(4\pi)^2}$$

$$\times\left(-\frac{5}{3}N_c\right)\left(\frac{1}{\varepsilon}-\gamma-\ln\frac{-p^2}{4\pi\mu^2}+\frac{59}{30}+O(\varepsilon)\right)+O(b^3). \tag{P6.55}$$

d. We need

$$\ln\det\left[\delta^\rho_\nu\delta^{dc}+\frac{1}{((-D^2)^{de}g^{\nu\sigma}-2F^{a\nu\sigma}f^{dae})}\bar{\phi}t^e\gamma^\sigma\frac{2g^2}{(i\hat{D}-m)}t^c\gamma^\rho\phi\right]$$

$$= tr\ln\left[\delta^\rho_\nu\delta^{dc}+\frac{1}{-\partial^2}\bar{\phi}t^d\gamma_\nu\frac{2g^2}{i\hat{\partial}-m}t^c\gamma_\rho\phi\right]$$

$$= tr\left[\frac{1}{-\partial^2}\bar{\phi}t^b\gamma^\mu\frac{2g^2}{i\hat{\partial}-m}t^b\gamma_\mu\phi\right]$$

$$= 2g^2C_F\langle p|\frac{1}{-\partial^2}|q\rangle\langle q|\bar{\phi}|l\rangle\langle l|\gamma^\mu\frac{1}{i\hat{\partial}-m}\gamma_\mu|n\rangle\langle n|\phi|p\rangle$$

$$= 2g^2C_F\int\frac{d^dp}{(2\pi)^d}\int\frac{d^dl}{(2\pi)^d}\frac{1}{p^2}\tilde{\bar{\phi}}(p-l)\gamma^\mu\frac{\hat{l}+m}{l^2-m^2}\gamma_\mu\tilde{\phi}(l-p)$$

$$= 2g^2C_F\int\frac{d^dp}{(2\pi)^d}\tilde{\bar{\phi}}(-p)_i\tilde{\phi}(p)_j\int\frac{d^dl}{(2\pi)^d}\frac{1}{(l-p)^2}[\gamma^\mu\frac{\hat{l}+m}{l^2-m^2}\gamma_\mu]_{ij}. \tag{P6.56}$$

Here

$$\int \frac{d^d l}{(2\pi)^d} \frac{\mu^{4-d}}{(l-p)^2} [\gamma^\mu \frac{\hat{l}+m}{l^2-m^2} \gamma_\mu]_{ij}$$

$$= \int_0^1 dx \int \frac{d^d l}{(2\pi)^d} \frac{\mu^{4-d}[(2-d)\hat{l} + dm]_{ij}}{((l-xp)^2 - (1-x)(m^2 - xp^2))^2}$$

$$= \int_0^1 dx \int \frac{d^d k}{(2\pi)^d} \frac{\mu^{4-d}[(2-d)x\hat{p} + dm]_{ij}}{(k^2 - (1-x)(m^2 - xp^2))^2}$$

$$= \frac{i\Gamma(2-\frac{d}{2})}{(4\pi)^2} \int_0^1 dx \left((1-x)\frac{(m^2-xp^2)}{4\pi\mu^2}\right)^{\frac{d}{2}-2} [(2-d)x\hat{p} + dm]_{ij}$$

$$= |d = 4 - 2\varepsilon|$$

$$= \frac{i\Gamma(2-\frac{d}{2})}{(4\pi)^2} \int_0^1 dx \left[1 + \left(\frac{d}{2} - 2\right)\ln\left((1-x)\frac{(m^2-xp^2)}{4\pi\mu^2}\right)\right]$$

$$\times [(2-d)x\hat{p} + dm]_{ij} + O(\varepsilon)$$

$$= \frac{i}{(4\pi)^2}[X\hat{p} + Ym]_{ij} + O(\varepsilon), \tag{P6.57}$$

where

$$X = -\left(\frac{1}{\varepsilon} - \gamma - \ln\left(\frac{m^2-p^2}{4\pi\mu^2}\right) + 1 + \frac{m^2}{p^2} + \frac{m^4}{p^4}\ln\left(1 - \frac{p^2}{m^2}\right)\right)$$

$$\underset{-p^2 \gg m^2}{\longrightarrow} -1 * \left(\frac{1}{\varepsilon} - \gamma - \ln\left(\frac{-p^2}{4\pi\mu^2}\right) + 1\right), \tag{P6.58}$$

$$Y = 4\left(\frac{1}{\varepsilon} - \gamma - \ln\left(\frac{m^2-p^2}{4\pi\mu^2}\right) + \frac{3}{2} + \frac{m^2}{p^2}\ln\left(1 - \frac{p^2}{m^2}\right)\right)$$

$$\underset{-p^2 \gg m^2}{\longrightarrow} 4\left(\frac{1}{\varepsilon} - \gamma - \ln\left(\frac{-p^2}{4\pi\mu^2}\right) + \frac{3}{2}\right). \tag{P6.59}$$

As a result,

$$\ln\det\left[\delta_\nu^\rho \delta^{dc} + \frac{1}{((-D^2)^{de}g^{\nu\sigma} - 2F^{a\nu\sigma}f^{dae})}\bar{\phi}t^e\gamma^\sigma \frac{2g^2}{(i\hat{D}-m)}t^c\gamma^\rho\phi\right]$$

$$= i\frac{2g^2 C_F}{(4\pi)^2}\int \frac{d^d p}{(2\pi)^d}\tilde{\bar{\phi}}(-p)[X\hat{p} + Ym]\tilde{\phi}(p) + O(\varepsilon) + O(b). \tag{P6.60}$$

6.7 Find the one loop relations between Λ_{QCD}'s calculated for 3, 4, 5, and 6 active flavors via the continuity of the coupling constant at the quark masses.

ANSWER.

$$\Lambda_3 = \Lambda_4 \left(\frac{m_c}{\Lambda_4}\right)^{\frac{2}{27}}, \quad \Lambda_4 = \Lambda_5 \left(\frac{m_b}{\Lambda_5}\right)^{\frac{2}{25}}, \quad \Lambda_5 = \Lambda_6 \left(\frac{m_t}{\Lambda_5}\right)^{\frac{2}{23}};$$

$$\text{(P6.61)}$$

$$\Lambda_4 = \Lambda_3 \left(\frac{\Lambda_3}{m_c}\right)^{\frac{2}{25}}, \quad \Lambda_5 = \Lambda_4 \left(\frac{\Lambda_4}{m_b}\right)^{\frac{2}{23}}, \quad \Lambda_6 = \Lambda_5 \left(\frac{\Lambda_5}{m_t}\right)^{\frac{2}{21}}.$$

$$\text{(P6.62)}$$

6.8 In $\overline{MS}$-scheme the 2-loop Gell-Mann–Low equation reads (Z$^+$20)

$$\frac{d\alpha_s(p^2)}{d\ln p^2} = -\frac{\alpha_s^2(p^2)}{4\pi}(\beta_0 + \beta_1 \alpha_s(p^2)), \ \beta_0 = 11 - \frac{2}{3}n_f, \ \beta_1 = \frac{153 - 19n_f}{6\pi}.$$

$$\text{(P6.63)}$$

Solve this equation exactly. You will get the implicit result. Find approximate explicit form for $p^2 \gg \Lambda_{QCD}^2$.

SOLUTION.

$$\begin{aligned}
\ln \frac{p^2}{q^2} &= -4\pi \int_{\alpha_s(q^2)}^{\alpha_s(p^2)} \frac{d\alpha_s}{\alpha_s^2(\beta_0 + \beta_1 \alpha_s)} \\
&= -4\pi \int_{\alpha_s(q^2)}^{\alpha_s(p^2)} d\alpha_s \left[\frac{1}{\alpha_s^2 \beta_0} + \frac{\beta_1^2}{\beta_0^2(\beta_0 + \beta_1 \alpha_s)} - \frac{\beta_1}{\alpha_s \beta_0^2}\right] \\
&= -\frac{4\pi}{\beta_0} \left[\frac{1}{\alpha_s(q^2)} - \frac{1}{\alpha_s(p^2)} + \frac{\beta_1}{\beta_0} \ln \frac{(\beta_0 + \beta_1 \alpha_s(p^2))\alpha_s(q^2)}{(\beta_0 + \beta_1 \alpha_s(q^2))\alpha_s(p^2)}\right].
\end{aligned}$$

$$\text{(P6.64)}$$

Using

$$\alpha_s(\Lambda_{QCD}^2) = \infty \implies \frac{\beta_0}{4\pi} \ln \frac{p^2}{\Lambda_{QCD}^2} = \frac{1}{\alpha_s(p^2)} - \frac{\beta_1}{\beta_0} \ln \frac{\beta_0 + \beta_1 \alpha_s(p^2)}{\beta_1 \alpha_s(p^2)},$$

$$\text{(P6.65)}$$

$$\alpha_s(p^2) = \frac{1}{\frac{\beta_0}{4\pi} \ln \frac{p^2}{\Lambda_{QCD}^2} + \frac{\beta_1}{\beta_0} \ln \frac{\beta_0 + \beta_1 \alpha_s(p^2)}{\beta_1 \alpha_s(p^2)}}.$$

$$\text{(P6.66)}$$

Taking as the first approximation the LO result

$$\alpha_s^{(1)}(p^2) = \frac{4\pi}{\beta_0 \ln \frac{p^2}{\Lambda_{QCD}^2}},$$

$$\text{(P6.67)}$$

one gets

$$\alpha_s^{(2)}(p^2) = \cfrac{1}{\frac{\beta_0}{4\pi}\ln\frac{p^2}{\Lambda_{QCD}^2}\left(1 + \cfrac{\frac{\beta_1}{\beta_0}\ln\frac{\beta_0+\beta_1\alpha_s(p^2)}{\beta_1\alpha_s(p^2)}}{\frac{\beta_0}{4\pi}\ln\frac{p^2}{\Lambda_{QCD}^2}}\right)} \tag{P6.68}$$

$$\simeq \cfrac{1}{\frac{\beta_0}{4\pi}\ln\frac{p^2}{\Lambda_{QCD}^2}}\left(1 - \cfrac{\frac{\beta_1}{\beta_0}\ln\frac{\beta_0}{\beta_1\alpha_s(p^2)}}{\frac{\beta_0}{4\pi}\ln\frac{p^2}{\Lambda_{QCD}^2}}\right) \tag{P6.69}$$

$$\simeq \cfrac{4\pi}{\beta_0\ln\frac{p^2}{\Lambda_{QCD}^2}}\left(1 - \cfrac{4\pi\beta_1\ln\ln\frac{p^2}{\Lambda_{QCD}^2}}{\beta_0^2\ln\frac{p^2}{\Lambda_{QCD}^2}}\right). \tag{P6.70}$$

6.9 Show that in the ratio $R(s) = \frac{\sigma_{e^+e^- \to hadrons}(s)}{\sigma_{e^+e^- \to \mu^+\mu^-}(s)}$ at high energy all leading logarithmic terms combine into $\alpha_s(s)$.

SOLUTION. We follow the presentation of (DKS03). Neglecting all quark and lepton masses at high energy one gets $R = N_c\sum_q e_q^2$ in the leading zeroth order in α_s. Suppose we calculated the whole series perturbatively. It reads

$$\tilde{R} = \frac{R}{N_c\sum_q e_q^2} = 1 + \sum_{n=1}^{\infty}\alpha_s^n(\mu^2)r_n(t), \quad t = \ln\frac{\mu^2}{s}. \tag{P6.71}$$

Here μ^2 is an arbitrary renormalization scale and $r_n(t)$ are some functions. Since $\tilde{R}$ is dimensionless these functions depend only on the ratio $\frac{\mu^2}{s}$. $\tilde{R}$ is a measurable quantity. Therefore, it must not depend on the renormalization scale choice μ^2. Hence, the functions $r_n(t)$ cannot be arbitrary. They must obey the equation

$$0 = \mu^2\frac{d}{d\mu^2}\tilde{R} = \frac{d}{dt}\tilde{R} = \left(\frac{\partial}{\partial t} + \beta(\alpha_s)\frac{\partial}{\partial\alpha_s}\right)\tilde{R}(t,\alpha_s)$$

$$= \sum_{n=1}^{\infty}[\alpha_s^n r_n'(t) + \beta(\alpha_s)n\alpha_s^{n-1}r_n] \tag{P6.72}$$

$$= \sum_{n=1}^{\infty}[\alpha_s^n r_n'(t) - \frac{nr_n}{4\pi}\sum_{k=0}^{\infty}\beta_k\alpha_s^{n+k+1}] \tag{P6.73}$$

$$= \alpha_s r_1'(t) + \sum_{n=2}^{\infty}\alpha_s^n[r_n'(t) - \sum_{k=1}^{n-1}\frac{(n-k)r_{n-k}}{4\pi}\beta_{k-1}] \tag{P6.74}$$

$$= \alpha_s r_1' + \alpha_s^2(r_2' - \frac{\beta_0}{4\pi}r_1) + \alpha_s^3(r_3' - \frac{2\beta_0}{4\pi}r_2 - \frac{\beta_1}{4\pi}r_1)$$

$$+ \alpha_s^4(r_4' - \frac{3\beta_0}{4\pi}r_3 - \frac{2\beta_1}{4\pi}r_2 - \frac{\beta_2}{4\pi}r_1) + \dots \tag{P6.75}$$

Here each coefficient of α_s^n vanish independently and we have a set of equations

$$r_1 = c_1, \tag{P6.76}$$

$$r_2 = c_1 \frac{\beta_0 t}{4\pi} + c_2, \tag{P6.77}$$

$$r_3 = c_1 \left(\frac{\beta_0 t}{4\pi} \right)^2 + \left(\frac{2\beta_0}{4\pi} c_2 + \frac{\beta_1}{4\pi} c_1 \right) t + c_3, \tag{P6.78}$$

$$r_4 = c_1 \left(\frac{\beta_0 t}{4\pi} \right)^3 + \frac{\beta_0}{4\pi} \left(\frac{3\beta_0}{4\pi} c_2 + \frac{5}{2} \frac{\beta_1}{4\pi} c_1 \right) t^2 + ...t + c_4, \tag{P6.79}$$

$$...$$

$$r_n = c_1 \left(\frac{\beta_0 t}{4\pi} \right)^{n-1} + ... + c_n, \tag{P6.80}$$

$$...$$

Therefore

$$\tilde{R}(t, \alpha_s) = 1 + c_1 \alpha_s \left(1 + \frac{\alpha_s \beta_0 t}{4\pi} + \left(\frac{\alpha_s \beta_0 t}{4\pi} \right)^2 + \left(\frac{\alpha_s \beta_0 t}{4\pi} \right)^3 + ... \right)$$

$$+ c_2 \alpha_s^2 + \alpha_s^3 \left(c_3 + \left(\frac{2\beta_0}{4\pi} c_2 + \frac{\beta_1}{4\pi} c_1 \right) t \right)$$

$$+ \alpha_s^4 \left(c_4 + ...t + ...t^2 \right) + \alpha_s^n \left(c_n + ... + ...t^{n-2} \right) + ... \tag{P6.81}$$

We grouped the leading terms $\sim \alpha_s (\alpha_s t)^k = \alpha_s (\alpha_s \ln \frac{\mu^2}{s})^k$ in the first line. The other terms have less powers of $\ln \frac{\mu^2}{s}$ for each power of α_s. For $s \gg \mu^2$ the first line gives the dominant contribution which changes sign from order to order. Therefore, if we know the series to a finite order only, we get a huge error from the truncation of the first line. However, the series in the first line can be resummed to give the charge at the physically natural scale s

$$\tilde{R}(t, \alpha_s) = 1 + c_1 \frac{\alpha_s}{1 - \frac{\beta_0 t}{4\pi}} + O(\alpha_s^k t^{k-2}) = 1 + c_1 \frac{\alpha_s(\mu^2)}{1 + \frac{\beta_0}{4\pi} \ln \frac{s}{\mu^2}} + ...$$

$$= 1 + c_1 \alpha_s(s) + ... \tag{P6.82}$$

Such a resummation absorbs all the leading logarithms of the ratio $\frac{\mu^2}{s}$ into the value of the running coupling. Generally speaking, setting the scale of the charge to the value relevant for the process allows one to reduce the radiative corrections.

6.10 Show that the multiscale problem can be converted to a one scale problem up to corrections of α_s^2 using the geometric mean of the scales as a new scale.

SOLUTION. If there is a process with n different scales it has the factor

$$\alpha_s(p_1^2)\alpha_s(p_2^2)...\alpha_s(p_n^2) = \alpha_s(M^2)^n \Pi_{i=1}^n (1 + \beta_0 \frac{\alpha_s(M^2)}{4\pi} \ln \frac{M^2}{p_i^2})$$

$$= \alpha_s(M^2)^n (1 + \beta_0 \frac{\alpha_s(M^2)}{4\pi} \ln \frac{M^2}{p_1^2 p_2^2 ... p_n^2} + O(\alpha_s^2))$$

$$= \alpha_s(M^2)(1 + O(\alpha_s^2))|_{M^2 = (p_1^2 p_2^2 ... p_n^2)^{\frac{1}{n}}}. \quad \text{(P6.83)}$$

6.11 Fix the scale in the 2-loop $\overline{MS}$-scheme ratio (DS79)

$$R(s)^{(2)} = \frac{\sigma_{e^+e^- \to hadrons}(s)}{\sigma_{e^+e^- \to \mu^+\mu^-}(s)}$$

$$= N_c \sum_q e_q^2 \left(1 + \frac{\alpha_s(s)}{\pi} + \left(\frac{\alpha_s}{\pi}\right)^2 (1.98 - 0.115 n_f)\right) \quad \text{(P6.84)}$$

via

a. (Ste81) principle of minimal sensitivity (PMS), i.e. make the truncated series independent of μ;

b. fastest apparent convergence (FAC), i.e. make the first term in the truncated series give the result of the whole series;

c. (BLM83) Brodsky-Lepage-Mackenzie (BLM) procedure, i.e. make c_i independent of n_f;

d. estimate the error.

SOLUTION. From the solution of problem 6.9 we know that at 2 loops

$$\tilde{R}^{(2)}(\mu^2) = \frac{R^{(2)}}{N_c \sum_q e_q^2} = 1 + c_1 \alpha_s(\mu^2)\left(1 - \frac{\alpha_s(\mu^2)}{4\pi}\beta_0 \ln \frac{s}{\mu^2}\right) + c_2 \alpha_s^2(\mu^2)$$

$$\text{(P6.85)}$$

$$= 1 + c_1 \alpha_s(s) + c_2 \alpha_s^2(\mu^2). \quad \text{(P6.86)}$$

Here

$$c_1 = \frac{1}{\pi}, \quad c_2 = \frac{1.98 - 0.115 n_f}{\pi^2}\Big|_{n_f=6} = \frac{1.29}{\pi^2}. \quad \text{(P6.87)}$$

a. In PMS one chooses μ demanding that

$$\frac{\partial \tilde{R}^{(2)}}{\partial \ln \mu^2} = 0 \implies \mu_{PMS}^2 = se^{-\frac{4\pi}{\beta_0}\left(\frac{\beta_1}{2\beta_0} + \frac{c_2}{c_1}\right)} = s \times 0.37, \quad \text{for } n_f = 6;$$

(P6.88)

$$\tilde{R}^{(2)} = 1 + \frac{\alpha_s(s)}{\pi} + 1.29 \left(\frac{\alpha_s(0.37s)}{\pi}\right)^2$$

(P6.89)

$$= 1 + \frac{\alpha_s(0.37s)}{\pi} - 0.45 \left(\frac{\alpha_s(0.37s)}{\pi}\right)^2, \quad \text{for } n_f = 6.$$

(P6.90)

Here one uses the Gell-Mann - Low equation (P6.63).

b. In FAC one chooses μ demanding that

$$\tilde{R}^{(2)}(\mu_{FAC}^2) = \tilde{R}^{(1)}(\mu_{FAC}^2) \implies \mu_{FAC}^2 = se^{-\frac{c_2}{c_1}\frac{4\pi}{\beta_0}} = s \times 0.48,$$
$$\text{for } n_f = 6;$$

(P6.91)

$$\tilde{R}^{(2)} = 1 + \frac{\alpha_s(s)}{\pi} + 1.29 \left(\frac{\alpha_s(0.48s)}{\pi}\right)^2 = 1 + \frac{\alpha_s(0.48s)}{\pi},$$
$$\text{for } n_f = 6.$$

(P6.92)

c. In BLM one chooses μ so that the coefficients c_i do not depend on n_f. We have

$$\beta_0 = \frac{11}{3}N_c - \frac{2}{3}n_f \implies n_f = \frac{3}{2}(\frac{11}{3}N_c - \beta_0) \implies$$

(P6.93)

$$\tilde{R}^{(2)} = 1 + \frac{\alpha_s(s)}{\pi} + (1.98 - 0.115n_f)\left(\frac{\alpha_s}{\pi}\right)^2$$

(P6.94)

$$= 1 + \frac{\alpha_s(s)}{\pi}\left(1 + 0.115\frac{3}{2}\beta_0\frac{\alpha_s}{\pi}\right) + (1.98 - 0.115\frac{3}{2}\frac{11}{3}N_c)\left(\frac{\alpha_s}{\pi}\right)^2$$

(P6.95)

$$= 1 + \frac{\alpha_s(s)}{\pi}\left(1 + \frac{\alpha_s(s)}{4\pi}\beta_0 \ln\frac{s}{se^{-6\times0.115}}\right) + 0.08\left(\frac{\alpha_s}{\pi}\right)^2$$

(P6.96)

$$= 1 + \frac{\alpha_s(0.50s)}{\pi} + 0.08\left(\frac{\alpha_s(0.50s)}{\pi}\right)^2.$$

(P6.97)

Here

$$\mu_{BLM}^2 = se^{-6\times0.115} = s \times 0.50 \quad \text{for } \forall n_f.$$

(P6.98)

d. One usually chooses a number $a \sim 1$ (e.g. 2) and estimates the error of truncation as the interval for $\mu^2 \in [\frac{s}{a}, as]$. BLM usually works well for processes without $gg \to gg$ scattering in the lowest order since effectively it assumes that all n_f terms come from vacuum polarization loops, which change the argument of the coupling.

6.12 Fix the scale in the 1-loop $\overline{MS}$-scheme interaction potential between 2 infinitely heavy quarks (Bil80)

$$V(q^2)^{(2)} = -\frac{4\pi C_F}{q^2} \alpha_s(q^2) \left(1 + \frac{\alpha_s}{\pi} (\frac{5}{12}\beta_0 - 2)\right) \qquad \text{(P6.99)}$$

via

a. principle of minimal sensitivity (PMS), i.e. make the truncated series independent of μ;

b. fastest apparent convergence (FAC), i.e. make the first term in the truncated series give the result of the whole series;

c. Brodsky–Lepage–Mackenzie (BLM) procedure, i.e. make c_i independent of n_f;

d. estimate the error.

6.13 Derive the translation rules between two different renormalization schemes for a_s. Show that the first 2 coefficients of the β-function are universal.

SOLUTION. Suppose that in two different schemes

$$\frac{d\alpha_s}{d\ln q^2} = -\frac{\alpha_s^2}{4\pi}(\beta_0 + \beta_1\alpha_s + \beta_2\alpha_s^2 + \beta_3\alpha_s^3 + ...), \qquad \text{(P6.100)}$$

$$\frac{d\alpha_s'}{d\ln q^2} = -\frac{\alpha_s'^2}{4\pi}(\beta_0' + \beta_1'\alpha_s' + \beta_2'\alpha_s'^2 + \beta_3'\alpha_s'^3 + ...). \qquad \text{(P6.101)}$$

Then one can decompose the coupling constant in one scale via the coupling constant in the other scale as

$$\alpha_s'(q^2) = \alpha_s(q^2) + c_1\alpha_s^2(q^2) + c_2\alpha_s^3(q^2) + c_3\alpha_s^4(q^2) + ... \qquad \text{(P6.102)}$$

Plugging this expansion in the Gell-Mann - Low equation, one gets

$$\frac{d\alpha'_s}{d\ln p^2} = \frac{d\alpha_s}{d\ln p^2}(1 + 2c_1\alpha_s + 3c_2\alpha_s^2 + 4\alpha_s^3 + ...)$$

$$= -\frac{\alpha_s^2}{4\pi}(\beta_0 + \beta_1\alpha_s + \beta_2\alpha_s^2 + \beta_3\alpha_s^3 + ...)$$

$$\times (1 + 2c_1\alpha_s + 3c_2\alpha_s^2 + 4c_3\alpha_s^3 + ...)$$

$$= -\frac{1}{4\pi}(\alpha_s + c_1\alpha_s^2 + c_2\alpha_s^3 + c_3\alpha_s^4 + ...)^2$$

$$\times [\beta'_0 + \beta'_1(\alpha_s + c_1\alpha_s^2 + c_2\alpha_s^3 + c_3\alpha_s^4 + ...)$$

$$+ \beta'_2(\alpha_s + c_1\alpha_s^2 + c_2\alpha_s^3 + c_3\alpha_s^4 + ...)^2$$

$$+ \beta'_3(\alpha_s + c_1\alpha_s^2 + c_2\alpha_s^3 + c_3\alpha_s^4 + ...)^3 + ...]. \qquad \text{(P6.103)}$$

Therefore

$$\beta'_0 = \beta_0, \quad \beta'_1 = \beta_1, \qquad\qquad\qquad\qquad\qquad \text{(P6.104)}$$

$$\beta'_2 = \beta_2 - \beta_1 c_1 + (c_2 - c_1^2)\beta_0, \qquad\qquad\qquad \text{(P6.105)}$$

$$\beta'_3 = \beta_3 - 2c_1\beta_2 + \beta_1 c_1^2 + \beta_0(2c_3 - 6c_1 c_2 + 4c_1^3). \qquad \text{(P6.106)}$$

One can see that the first 2 coefficients of the β-function are universal. In the leading order

$$\Lambda'^2_{QCD} = \Lambda^2_{QCD} e^{\frac{4\pi c_1}{\beta_0}}. \qquad\qquad\qquad \text{(P6.107)}$$

6.14 The momentum subtraction (MOM) scheme in the Landau gauge ($\xi = 0$) can be defined as (CG79)

$$\alpha_{sMOM}(q^2) = \alpha_{s\overline{MS}}(q^2)(1 + \frac{\alpha_{s\overline{MS}}}{\pi}(1.28\beta_0 - 7.47) + ...). \quad \text{(P6.108)}$$

a. Fix the scale in α_{sMOM} according to the BLM prescription.

b. Find the $\Lambda_{QCD\,MOM}$ in terms of $\Lambda_{QCD\overline{MS}}$.

6.15 One may define the R-scheme so that (BLM83)

$$R(s) = \frac{\sigma_{e^+e^-\to hadrons}(s)}{\sigma_{e^+e^-\to\mu^+\mu^-}(s)} \overset{def}{=} N_c \sum_q e_q^2 \left(1 + \frac{\alpha_{sR}(s)}{\pi}\right) \qquad \text{(P6.109)}$$

exactly. Find relation of the $\alpha_{sR}(s)$ and $\alpha_{s\overline{MS}}(s)$.

Effective action is discussed in (PS95; IZ80; Ram81), while QCD in the background field is investigated in (Abb81; Abb82). General theory of renormalization is given in (Col86). In addition to these books renormalization of QCD is presented in (DKS03). Different scale setting techniques were proposed in (Ste81; BLM83).

FURTHER READING

[Abb81] L. F. Abbott. The background field method beyond one loop. *Nucl. Phys. B*, 185:189–203, 1981.

[Abb82] L. F. Abbott. Introduction to the background field method. *Acta Phys. Polon. B*, 13:33, 1982.

[Bil80] A. Billoire. How heavy must be quarks in order to build coulombic q anti-q bound states. *Phys. Lett. B*, 92:343–347, 1980.

[BLM83] Stanley J. Brodsky, G. Peter Lepage, and Paul B. Mackenzie. On the elimination of scale ambiguities in perturbative quantum chromodynamics. *Phys. Rev. D*, 28:228, 1983.

[CG79] William Celmaster and Richard J. Gonsalves. The renormalization prescription dependence of the QCD coupling constant. *Phys. Rev. D*, 20:1420, 1979.

[Col86] John C. Collins. *Renormalization: an introduction to renormalization, the renormalization group, and the operator product expansion*, volume 26 of *Cambridge Monographs on Mathematical Physics*, chapter 4.5. Cambridge University Press, Cambridge, 1986.

[DeW12] Bryce S. DeWitt. *Supermanifolds*. Cambridge Monographs on Mathematical Physics. Cambridge University Press, Cambridge, UK, 5, 2012.

[DKS03] G. Dissertori, I. G. Knowles, and M. Schmelling. *Quantum chromodynamics high energy experiments and theory*, chapter 3.4. Clarendon Press, Oxford, 2003.

[DS79] Michael Dine and J. R. Sapirstein. Higher order QCD corrections in e+ e- annihilation. *Phys. Rev. Lett.*, 43:668, 1979.

[IZ80] C. Itzykson and J. B. Zuber. *Quantum field theory*, chapter Vol.2, 12.3. International Series in Pure and Applied Physics. McGraw-Hill, New York, 1980.

[PS95] Michael E. Peskin and Daniel V. Schroeder. *An introduction to quantum field theory*, chapter 11.3–5, 16.6–7, 18.1. Addison-Wesley, Reading, USA, 1995.

[Ram81] Pierre Ramond. *Field theory. A modern primer*, volume 51 of *Frontiers in Physics*. 1981.

[Ste81] Paul M. Stevenson. Optimized perturbation theory. *Phys. Rev. D*, 23:2916, 1981.

[Z+20] P. A. Zyla et al. Review of particle physics. *PTEP*, 2020(8):083C01, 2020. Chapter 9. Quantum chromodynamics.

CHAPTER 7

Renormalization

T HIS chapter introduces renormalization constants, renormalization scale and discusses renormalization point dependence of Green functions via the Callan-Symanzik equation. Then it gives the idea of the Wilson's approach to the renormalization.

1. **Renormalization scale and counterterms.** The bare Lagrangian after gauge fixing in the covariant gauge reads

$$\mathcal{L} = -\frac{1}{4}(\partial_\mu A^a_{0\nu} - \partial_\nu A^a_{0\mu} + gf^{abc}A^b_{0\mu}A^c_{0\nu})^2 + \bar{\psi}_0(i\hat{\partial} + g_0\hat{A}_0 - m_0)\psi_0$$
$$- \frac{1}{2\xi_0}(\partial_\mu A^{a\mu}_0)^2 + (\partial_\mu \bar{c}^a_0)(\delta^{ac}\partial^\mu + g_0 f^{abc}A^{\mu b}_0)c^c_0, \tag{7.1}$$

where $_0$ denotes bare variables. Calculating loop corrections to propagators one can find the **renormalization constants**

$$G_0(x-y) = \langle\psi_{0x}\bar{\psi}_{0y}\rangle = Z_2\langle\psi_x\bar{\psi}_y\rangle = Z_2 G(x-y), \tag{7.2}$$
$$G^{\mu\nu ab}_{0tr}(x-y) = \langle A^{a\mu}_{0x}A^{b\nu}_{0y}\rangle_{tr} = Z_3\langle A^{a\mu}_x A^{b\nu}_y\rangle_{tr} = Z_3 G^{\mu\nu ab}_{tr}(x-y), \tag{7.3}$$
$$D_0(x-y) = \langle c_{0x}\bar{c}_{0y}\rangle = Z^c_2\langle c_x\bar{c}_y\rangle = Z^c_2 D(x-y), \tag{7.4}$$

and redefine the fields and couplings

$$A_0 = Z_3^{\frac{1}{2}}A, \quad c_0 = Z_2^{c\frac{1}{2}}c, \quad \psi_0 = Z_2^{\frac{1}{2}}\psi, \tag{7.5}$$
$$\xi_0 = Z_3\xi, \quad g_0 = Z_g g, \quad m_0 = Z_m m. \tag{7.6}$$

Here one usually chooses the renormalization constant of ξ so that the gauge fixing term does not change. First, we consider the massless case. Renormalization constants absorb the UV-divergent parts of the Feynman diagrams, i.e. divergent parts at large loop momenta, which in the case of multiplicative renormalization for a propagator, e.g. combine into

$$G(\text{Born+one-loop})_0(g_0, \mu, \varepsilon)$$

$$= G(\text{Born})_0 \times \left[1 + \frac{g_0^2 A}{(4\pi)^2} \left(\frac{1}{\varepsilon} - \gamma - \ln \frac{-p^2}{4\pi\mu^2} + const \right) \right]$$

$$\simeq G(\text{Born})_0 \times Z(\frac{M}{\mu}, \varepsilon) \times \left[1 + \frac{g^2 A}{(4\pi)^2} \left(- \ln \frac{-p^2}{M^2} + const \right) \right]$$

$$= Z(\frac{M}{\mu}, \varepsilon) \times G(\text{Born} + \text{one-loop})(M, g), \tag{7.7}$$

where we took M as the **renormalization scale**, i.e. we defined the corresponding constant Z as $(d = 4 - 2\varepsilon)$

$$Z(\frac{M}{\mu}, \varepsilon) = 1 + \frac{g^2 A}{(4\pi)^2} \left(\frac{1}{\varepsilon} - \gamma - \ln \frac{M^2}{4\pi\mu^2} \right). \tag{7.8}$$

To find Z_g we consider a vertex, e.g. the $\bar{\psi} \hat{A} \psi$ vertex with all incoming momenta $\sim p$. We need the **one particle irreducible Green function** (1PI), i.e. the one which cannot be transformed into two disconnected Green functions by removing one propagator, roughly speaking the vertex without propagator corrections in the external legs. For the operator

$$\bar{\psi}_{0x} \hat{A}_{0x} \psi_{0x} = \bar{\psi}_{i0x} A^a_{\mu0x} \psi^j_{0x} (\Gamma^{\mu a})^i_j, \tag{7.9}$$

$$(\Gamma^{\mu a})^i_j (\text{Born}) = (t^a)^i_j \gamma^\mu, \tag{7.10}$$

the 1PI Green function stripped of the fields reads

$$\Gamma(\text{Born} + \text{one-loop})_0(g_0, \mu, \varepsilon) = \Gamma(\text{Born})_0$$

$$\times \left[1 + \frac{g_0^2 A_1}{(4\pi)^2} \left(\frac{1}{\varepsilon} - \gamma - \ln \frac{-p^2}{4\pi\mu^2} + \text{finite part} \right) \right]$$

$$\simeq \Gamma(\text{Born})_0 \times Z_1^{-1}(\frac{M}{\mu}, \varepsilon) \times \left[1 + \frac{g^2 A_1}{(4\pi)^2} \left(- \ln \frac{-p^2}{M^2} + \text{finite part} \right) \right]$$

$$= Z_1^{-1}(\frac{M}{\mu}, \varepsilon) \times \Gamma(\text{Born} + \text{one-loop})(M, g), \tag{7.11}$$

where

$$Z_1^{-1}(\frac{M}{\mu}, \varepsilon) = 1 + \frac{g^2 A_1}{(4\pi)^2} \left(\frac{1}{\varepsilon} - \gamma - \ln \frac{M^2}{4\pi\mu^2} \right). \tag{7.12}$$

Then the unrenormalized 3-point Green function in one loop order reads

$$\langle \psi_{0x} A^\mu_{0y\,tr} \bar{\psi}_{0z} \rangle = \langle \psi_{0x} A^\mu_{0y\,tr} \bar{\psi}_{0z} e^{iS} \rangle$$

$$= ig_0 \int dr G_0(x-r) \Gamma^\mu_{0r} G_0(r-z) G_{0tr\,\mu\nu}(r-y)$$

$$= ig_0 \frac{Z_2^2 Z_3}{Z_1} \underbrace{\int dr G(x-r) \Gamma^\mu_r G(r-z) G_{tr\,\mu\nu}(r-y)}_{UV-finite}.$$

$$\tag{7.13}$$

Here we took the renormalization constants out of the bare propagators and 1PI vertex. Renormalizability of the theory means that renormalized Green functions, i.e. Green functions of renormalized fields expressed via renormalized charge and mass are finite. Therefore, rewriting this Green function via the renormalized fields, we get the renormalized 3-point Green function

$$\langle \psi_x A_{y\,tr} \bar{\psi}_z \rangle = Z_2^{-1} Z_3^{-\frac{1}{2}} \langle \psi_{0x} A_{0y\,tr} \bar{\psi}_{0z} \rangle$$

$$= \underbrace{ig_0 \frac{Z_2 Z_3^{\frac{1}{2}}}{Z_1}}_{g} \underbrace{\int dr G(x-r)\Gamma_r^\mu G(r-z)G_{tr\,\mu\nu}(r-y)}_{UV-finite}. \quad (7.14)$$

To get the finite result for this renormalized Green function one introduces the renormalized charge $g = g_0 \frac{Z_2 Z_3^{\frac{1}{2}}}{Z_1}$:

$$\frac{g_0}{g} = Z_g(\frac{M}{\mu}, \varepsilon) = \frac{Z_1}{Z_2 Z_3^{\frac{1}{2}}} = 1 + \frac{g^2 A_g}{(4\pi)^2}\left(\frac{1}{\varepsilon} - \gamma - \ln\frac{M^2}{4\pi\mu^2}\right), \quad (7.15)$$

$$A_g = -(A_2 + \frac{1}{2}A_3 + A_1). \quad (7.16)$$

Here and in the final expressions for the propagator we changed $g_0 \to g$ in the coefficient of A since

$$g_0 = gZ_g(\frac{M}{\mu}, \varepsilon) = g\left[1 + \frac{g^2 A_g}{(4\pi)^2}\left(\frac{1}{\varepsilon} - \gamma - \ln\frac{M^2}{4\pi\mu^2}\right)\right] = g + O(g^3),$$
$$(7.17)$$

and we do not keep terms of higher orders. Renormalizability of the theory demands that renormalized physical quantities such as cross-sections, etc. expressed via renormalized charge, fields and masses are finite order by order in perturbation theory. In other words the number of renormalization constants does not grow with order of perturbation theory. Therefore, one can calculate unrenormalized Green functions using the bare Lagrangian, express them via the renormalized charge and mass, and divide by the square roots of the renormalization constants of the external fields to get the renormalized, UV-finite Green functions. This way of calculation is called calculation via the unrenormalized perturbation theory. Another option is to change the bare variables to the renormalized ones right in the lagrangian. In the redefined **renormalized variables**, the lagrangian reads

$$\mathcal{L} = \mathcal{L}_{ren} + \mathcal{L}_{c.t.}$$

$$= -\frac{1}{4}Z_3(\partial_\mu A_\nu^a - \partial_\nu A_\mu^a + gf^{abc}Z_3^{\frac{1}{2}}Z_g A_\mu^b A_\nu^c)^2 - \frac{1}{2\xi}(\partial_\mu A^{a\mu})^2$$

$$+ Z_2\bar{\psi}(i\hat{\partial} + Z_3^{\frac{1}{2}}Z_g g\hat{A} - Z_m m)\psi$$

$$+ Z_2^c(\partial_\mu \bar{c}^a)(\delta^{ac}\partial^\mu + gZ_g Z_3^{\frac{1}{2}}f^{abc}A^{\mu b})c^c, \quad (7.18)$$

where we single out the same structure as the bare Lagrangian in $\mathcal{L}_{ren}$ and keep **counter terms** in $\mathcal{L}_{c.t.}$:

$$\mathcal{L}_{ren} = -\frac{1}{4}(\partial_\mu A_\nu^a - \partial_\nu A_\mu^a + g f^{abc} A_\mu^b A_\nu^c)^2 + \bar\psi(i\hat\partial + g\hat A - m)\psi$$
$$- \frac{1}{2\xi}(\partial_\mu A^{a\mu})^2 + (\partial_\mu \bar c^a)(\delta^{ac}\partial^\mu + g f^{abc} A^{\mu b})c^c, \qquad (7.19)$$

$$\mathcal{L}_{c.t.} = \bar\psi(\delta_2 i\hat\partial + \delta_1 g\hat A - \delta_m)\psi - \frac{1}{4}\delta_3(\partial_\mu A_\nu^a - \partial_\nu A_\mu^a)^2$$
$$- \delta^{3g} g f^{abc}(\partial_\mu A_\nu^a)A_\mu^b A_\nu^c - \frac{1}{4}\delta^{4g} g^2(f^{abc} A_\mu^b A_\nu^c)(f^{ab'c'} A^{\mu b'} A^{\nu c'})$$
$$+ (\partial_\mu \bar c^a)(\delta_2^c \delta^{ac}\partial^\mu + \delta_1^c g f^{abc} A^{\mu b})c^c, \qquad (7.20)$$

$$\delta_3 = Z_3 - 1, \quad \delta_2 = Z_2 - 1, \quad \delta_2^c = Z_2^c - 1,$$
$$\delta^{3g} = Z_3^{\frac{3}{2}} Z_g - 1, \quad \delta^{4g} = Z_3^2 Z_g^2 - 1, \quad \delta_m = m(Z_2 Z_m - 1),$$
$$\delta_1^c = Z_2^c Z_g Z_3^{\frac{1}{2}} - 1, \quad \delta_1 = Z_2 Z_3^{\frac{1}{2}} Z_g - 1. \qquad (7.21)$$

Using this lagrangian one calculates renormalized Green functions directly and the counterterms cancel the singularities in the loop integrals. Therefore, one does not divide the Green functions by the square roots of renormalization constants. This method of calculation is called calculation via the renormalized perturbation theory. One gets the same results as in the first method since one uses the same lagrangian in different variables.

One may ask why after renormalization the coupling constants in different vertices are equal to each other, i.e. why Z_g obtained from different vertices in this manner coincide. The reason for this is the BRST invariance of the action (see problem 4.21 in Chapter 4) which holds only for the equal bare coupling constants.

2. **Callan-Symanzik equation.** Consider a bare Green function G_0 of bare fields ϕ_0 in **different** points x_i. It depends on the bare coupling g_0, bare mass, ε and the regularization scale μ, but is independent of the renormalization scale M. Taking out Z, one can rewrite it via the renormalized Green function, i.e. the function of renormalized fields, which depends on the renormalized coupling, renormalized mass and the renormalization scale M. The whole dependence on ε and μ (more precisely $\frac{M}{\mu}$) is absorbed into Z:

$$G_0(x_1,...,x_n) = \langle\phi_{0x_1}...\phi_{0x_n}\rangle = Z^{\frac{n}{2}}\langle\phi_{x_1}...\phi_{x_n}\rangle = Z^{\frac{n}{2}} G(x_1,...,x_n). \qquad (7.22)$$

Here the fields ϕ stand for A, ψ, c and Z will be the renormalization constants for the corresponding fields. As a result, one gets the **Callan-Symanzik equation**

$$0 = M\frac{dG_0}{dM} = \frac{d(Z^{\frac{n}{2}}G)}{d\ln M} = \frac{n}{2}Z^{\frac{n}{2}-1}\frac{dZ}{d\ln M}G + Z^{\frac{n}{2}}\frac{dG}{d\ln M}$$

$$= Z^{\frac{n}{2}}\left(\frac{n}{2}\frac{d\ln Z}{d\ln M} + \frac{d}{d\ln M}\right)G$$

$$= Z^{\frac{n}{2}}\left(\frac{n}{2}\frac{d\ln Z}{d\ln M} + \frac{\partial}{\partial\ln M} + \frac{dg}{d\ln M}\frac{\partial}{\partial g}\right)G \implies$$

$$\left(\frac{\partial}{\partial\ln M} + \beta(g)\frac{\partial}{\partial g} + n\gamma\right)G = 0, \quad \gamma = \frac{1}{2}\frac{d\ln Z}{d\ln M} = \frac{g^2\gamma^{(2)}}{(4\pi)^2} + O(g^4).$$

$$(7.23)$$

Here γ is called the **anomalous dimension** of field ϕ. For the Z from (7.8) $\gamma^{(2)} = -A$. In the dimensional regularization, $Z(\frac{M}{\mu})$ gives

$$\gamma = \frac{1}{2}\frac{d\ln Z}{d\ln M} = -\frac{1}{2}\frac{d\ln Z}{d\ln \mu}. \tag{7.24}$$

The β-function is defined via Gell-Mann - Low equation (6.33).

$$\beta(g) = \frac{dg}{d\ln M} = -\frac{\beta_0}{(4\pi)^2}g^3 + O(g^4). \tag{7.25}$$

In the massless case

$$\left(\frac{\partial}{\partial\ln M} + \beta(g)\frac{\partial}{\partial g} + n\gamma\right)G = 0. \tag{7.26}$$

This equation has the characteristic system

$$\frac{d\ln M}{1} = \frac{dg}{\beta(g)} = \frac{dG}{-n\gamma(g)G}, \tag{7.27}$$

and 2 first integrals. The first integral

$$\ln\frac{M}{P} = \int_{g(P)}^{g(M)}\frac{dg'}{\beta(g')}, \tag{7.28}$$

gives $g(P)$ as a function of M and $g(M)$. The first integral

$$\ln\frac{G(M)}{G(P)} = \int_{g(P)}^{g(M)}\frac{dg'}{\beta(g')}(-n\gamma(g')) = \int_{\ln P}^{\ln M}(-n\gamma(P'))d\ln P' \tag{7.29}$$

gives $G(P)$ as a function of $G(M)$ and $g(M)$ or $G(M)$ and M. The general solution is the solution of the equation

$$F(G(P), g(P)) = 0 \implies G(P) = f(g^2(P)) \implies \tag{7.30}$$

$$G(M) = e^{\int_{g(P)}^{g(M)}\frac{dg'}{\beta(g')}(-n\gamma(g'))}f(g^2(P)), \tag{7.31}$$

and the arbitrary function f is to be found from the initial conditions at the renormalization scale $P \sim p_i$ where p_i are typical for the momenta of the fields ϕ. In one loop

$$G(M) = \left(\frac{g(M)}{g(P)}\right)^{\frac{n\gamma^{(2)}}{\beta_0}} f(g^2(P))$$

$$= \left(\frac{g(M)}{g(P)}\right)^{\frac{n\gamma^{(2)}}{\beta_0}} f(\frac{g^2(M)}{1 + 2\beta_0 \frac{g^2(M)}{(4\pi)^2} \ln \frac{P}{M}}). \qquad (7.32)$$

Note that the Callan-Symanzic equations for the amputated Green functions, i.e. the Green functions divided by the external propagators will have the same structure with the substitution $Z_i \to Z_i^{-1}$ and $\gamma_i \to -\gamma_i$.

3. **Renormalization of local operators.** For a local operator, i.e. the product of n fields **in one point** x there may be an additional singularity which remains after the renormalization of the fields. Then one defines the renormalization constant for this operator so that it cancels the remaining singular terms at the renormalization point M

$$O_0 = Z_O(M)O(M), \qquad (7.33)$$

i.e. one builds the bare Green function

$$G_0(p_1, ..., p_n, k) = \langle \phi_{0p_1} ... \phi_{0p_n} O_{0k} \rangle = Z^{\frac{n}{2}} Z_O \langle \phi_{p_1} ... \phi_{p_n} O_k \rangle$$

$$= Z^{\frac{n}{2}} Z_O G(p_1, ..., p_n, k), \qquad (7.34)$$

so that there are no singularities in the renormalized Green function G. Defining

$$\gamma_O = \frac{d \ln Z_O}{d \ln M} = \frac{g^2}{(4\pi)^2} \gamma_O^{(2)} + O(g^4), \qquad (7.35)$$

one can repeat the derivation for the Callan-Symanzik equation for this Green function

$$\left(\frac{\partial}{\partial \ln M} + \beta(g)\frac{\partial}{\partial g} + n\gamma + \gamma_O\right) G = 0 \qquad (7.36)$$

and its solution in one loop

$$G(M) = \left(\frac{g(M)}{g(P)}\right)^{\frac{n\gamma^{(2)}+\gamma_O^{(2)}}{\beta_0}} f(g^2(P)). \qquad (7.37)$$

4. **Mass terms.** The Lagrangian contains the quark mass via

$$\mathcal{L}_m = -m\bar{\psi}\psi. \qquad (7.38)$$

One can consider it as a perturbation and then any Green function will have the form

$$G = \sum_k \langle \phi_{x_1}...\phi_{x_n} m^k \bar\psi_{z_1}\psi_{z_1}...\bar\psi_{z_k}\psi_{z_k}\rangle = \sum_k m^k G_k. \qquad (7.39)$$

For each Green function G_k, we have the Callan-Symanzik equation

$$\left(\frac{\partial}{\partial \ln M} + \beta(g)\frac{\partial}{\partial g} + n\gamma + k\gamma_{\bar q q}\right)G_k = 0, \quad \gamma_{\bar q q} = \frac{g^2}{(4\pi)^2}\gamma_{\bar q q}^{(2)} + O(g^4). \qquad (7.40)$$

Summing them all, we get the Callan-Symanzik equation for G

$$\left(\frac{\partial}{\partial \ln M} + \beta(g)\frac{\partial}{\partial g} + n\gamma + \gamma_{\bar q q}\frac{\partial}{\partial \ln m}\right)G = 0. \qquad (7.41)$$

The characteristic system for this equation has one more equation

$$\frac{d\ln M}{1} = \frac{dg}{\beta(g)} = \frac{d\ln m}{\gamma_{\bar q q}(g)} = \frac{dG}{-n\gamma G}, \qquad (7.42)$$

and one more first integral

$$\ln \frac{m(M)}{m(P)} = \int_{g(P)}^{g(M)} \frac{\gamma_{\bar q q}(g')}{\beta(g')}dg' \quad \Longrightarrow$$

$$m(M) = m(P)e^{-\frac{\gamma_{\bar q q}^{(2)}}{\beta_0}\int_{g(P)}^{g(M)}\frac{dg'}{g'}} = m(P)\left(\frac{g(P)}{g(M)}\right)^{\frac{\gamma_{\bar q q}^{(2)}}{\beta_0}}. \qquad (7.43)$$

The general solution then is the solution of the equation

$$F(G(P), g^2(P), m(P)) = 0 \quad \Longrightarrow \quad G(P) = f(g^2(P), m(P)) \quad \Longrightarrow \qquad (7.44)$$

$$G(M) = e^{\int_{g(P)}^{g(M)}\frac{dg'}{\beta(g')}(-n\gamma(g'))}f(g^2(P), m(P)), \qquad (7.45)$$

and again the arbitrary function f is to be found from the initial conditions at the renormalization scale $P \sim p_i$ where p_i are the typical momenta of the fields ϕ. In one loop

$$G(M) = \left(\frac{g(M)}{g(P)}\right)^{\frac{n\gamma^{(2)}}{\beta_0}} f(g^2(P), m(P)). \qquad (7.46)$$

Generalization to the Green function of the local operator is straightforward $n\gamma^{(2)} \to n\gamma^{(2)} + \gamma_O^{(2)}$.

5. **Physical observables do not depend on the renormalization point** M. Therefore, the renormalized Green functions which give them

do not depend on the renormalization scale M. It means that for these quantities total dependence on M coming from all renormalization constants Z relating the renormalized and bare Green functions cancel. In other words the Callan-Symanzik equations for these quantities do not have anomalous dimensions and their solutions depend on renormalization scale only via charge and mass.

6. **Wilson approach to renormalization.** Here we briefly follow paragraph 12.1 of (PS95). We demonstrate the idea on the example of the scalar field with

$$Z = \int \mathcal{D}\phi_x e^{\frac{i}{2} \int d^d x \left((\partial_\mu \phi)^2 - m^2 \phi^2 - \frac{\lambda}{2}\phi^4 \right)} \rightarrow \int \mathcal{D}\phi_\Lambda e^{-\int d^d x_E \mathcal{L}_0[\phi]},$$

$$\mathcal{L}_0 = \frac{1}{2}(\partial_\mu \phi)^2 + \frac{1}{2}m^2 \phi^2 + \frac{\lambda}{4}\phi^4. \tag{7.47}$$

Here we Wick-rotated to the Euclidean space-time with the coordinates x_E (we will suppress $_E$ from now on in this section) and changed variables in the functional integral to the fields in the momentum representation via discrete Fourier transform. Suppose we have a natural cutoff Λ (e.g. Plank scale...) and our theory contains only Euclidean $|k| < \Lambda$, i.e.

$$\int \mathcal{D}\phi_\Lambda = \Pi_{|k|<\Lambda} \int d\phi_k. \tag{7.48}$$

We can separate the fields into fast $b\Lambda < |k| < \Lambda$ and slow $b\Lambda > |k|$ with an arbitrary $0 < b < 1$ and take the functional integral over only fast fields.

$$\phi = \varphi + \hat{\phi}, \quad \varphi = \phi\theta(b\Lambda > |k|), \quad \hat{\phi} = \phi\theta(b\Lambda < |k| < \Lambda), \tag{7.49}$$

$$\int \mathcal{D}\phi_\Lambda e^{-\int d^d x \mathcal{L}_0[\phi]} = \int \mathcal{D}\varphi_{b\Lambda} e^{-\int d^d x \mathcal{L}_0[\varphi]}$$

$$\times \int \mathcal{D}\hat{\phi} e^{-\int d^d x \left(\frac{1}{2}(\partial_\mu \hat{\phi})^2 + \mathcal{L}_1[\varphi, \hat{\phi}] \right)}, \tag{7.50}$$

$$\mathcal{L}_1[\varphi, \hat{\phi}] = \frac{1}{2}m^2 \hat{\phi}^2 + \frac{\lambda}{4}(\hat{\phi}^4 + 4\hat{\phi}^3 \varphi + 6\hat{\phi}^2 \varphi^2 + 4\hat{\phi}\varphi^3). \tag{7.51}$$

Here we neglected

$$\int d^d x \left((\partial_\mu \hat{\phi})(\partial^\mu \varphi) + m^2 \varphi \hat{\phi} \right)$$

$$= \int d^d x \int \frac{d^d k_1}{(2\pi)^{2d}} e^{ik_1 x} \int d^d k_2 e^{ik_2 x} [m^2 - k_1 k_2] \hat{\phi}(k_1)\varphi(k_2)$$

$$= \int d^d k_2 \int \frac{d^d k_1}{(2\pi)^d} \delta^d(k_1 + k_2)[m^2 - k_1 k_2]\hat{\phi}(k_1)\varphi(k_2) = 0 \tag{7.52}$$

since $|k_1| \neq |k_2|$. Then,

$$\int \mathcal{D}\hat{\phi} e^{-\int d^d x \left(\frac{1}{2}(\partial_\mu \hat{\phi})^2 + \mathcal{L}_1[\varphi, \hat{\phi}]\right)}$$

$$= \int \mathcal{D}\hat{\phi} e^{-\frac{1}{2}\int d^d x (\partial_\mu \hat{\phi})^2} \left(1 - \int d^d x \mathcal{L}_1 + \frac{1}{2}\int d^d x \mathcal{L}_1 \int d^d x_1 \mathcal{L}_1 + ...\right).$$

(7.53)

All the integrals in this series are finite, e. g. (see problem 7.2)

$$\frac{1}{\hat{Z}}\int \mathcal{D}\hat{\phi} e^{-\frac{1}{2}\int d^d x (\partial_\mu \hat{\phi})^2} \int d^d x (-\lambda \frac{6}{4})(\hat{\phi}^2 \varphi^2) = const \int d^d x \varphi_x^2. \quad (7.54)$$

Therefore, one can calculate them and combine them back into the exponential thus getting the effective Lagrangian for the field φ

$$\int d^d x \mathcal{L}_{eff} = \int d^d x \left(\frac{1 + \Delta Z}{2}(\partial_\mu \varphi)^2 + \frac{m^2 + \Delta m^2}{2}\varphi^2\right.$$

$$\left. + \frac{\lambda + \Delta \lambda}{4}\varphi^4 + ... + \Delta C_{M,N}\partial_\mu^M \varphi^N + ...\right) \quad (7.55)$$

This effective Lagrangian depends only on the slow fields and describes the same theory as the initial one for slow external momenta. The effect of the fast modes is represented via nonlinear interactions of slow fields. One can rescale the momenta so that for $|k| < b\Lambda$

$$k = bk', \quad |k'| < \Lambda \quad \Longrightarrow \quad x = \frac{x'}{b}, \quad since \quad \int dk dx = \int dk' dx'.$$

(7.56)

In variable x'

$$\int d^d x \mathcal{L}_{eff} = \int d^d x' b^{-d} \mathcal{L}_{eff} = \int d^d x' b^{-d} \left(\frac{1}{2}(1 + \Delta Z)b^2(\partial_\mu \varphi)^2\right.$$

$$\left. + \frac{1}{2}(m^2 + \Delta m^2)\varphi^2 + \frac{\lambda + \Delta \lambda}{4}\varphi^4 + ... + \Delta C_{M,N}b^M \partial_\mu^M \varphi^N + ...\right)$$

$$= \int d^d x' \left(\frac{1}{2}(\partial_\mu \varphi')^2 + \frac{1}{2}m'^2 \varphi'^2 + \frac{\lambda'}{4}\varphi'^4 + ... + C'_{M,N}\partial_\mu^M \varphi'^N + ...\right)$$

(7.57)

we have the Lagrangian with the same kinetic term as the initial one with

$$\varphi' = \varphi(1 + \Delta Z)^{\frac{1}{2}} b^{1-\frac{d}{2}} \quad (7.58)$$

and

$$m'^2 = \frac{(m^2 + \Delta m^2)}{(1 + \Delta Z)} b^{-2} \simeq m^2 b^{-2}, \quad \lambda' = \frac{\lambda + \Delta\lambda}{(1 + \Delta Z)^2} b^{d-4} \simeq \lambda b^{d-4},$$

(7.59)

$$C'_{M,N} = \frac{(C_{M,N} + \Delta C_{M,N})}{(1 + \Delta Z)^{\frac{N}{2}}} b^{-(1-\frac{d}{2})N+M-d} \simeq C_{M,N} b^{d_{M,N}-d}, \quad (7.60)$$

$$d_{M,N} = (\frac{d}{2} - 1)N + M, \quad (7.61)$$

where $d_{M,N}$ is the dimension of operator $\partial_\mu^M \varphi^N$ in mass units. Here the second equalities give the approximate transformation law in the vicinity of a **fixed point** $\lambda = m = C_{M,N} = 0$. One can see that since $b < 1$, in the effective Lagrangian there are operators whose coefficients scale as a negative power of b, i.e. grow. They are called **relevant**, e.g. φ'^2, $\varphi'^4 (d < 4)$, $\partial_\mu^M \varphi^N (d_{M,N} < d)$. The ones whose coefficients scale as a positive power of b, i.e. decrease are called **irrelevant**, $\varphi'^4 (d < 4)$, $\partial_\mu^M \varphi^N (d_{M,N} > d)$. The ones whose coefficients scale as b^0 are referred to as **marginal**, e.g. $\varphi'^4 (d = 4)$. Their behavior has to be found from higher corrections. Therefore, a generic Lagrangian at the cut-off scale Λ with order of 1 coefficients will be equivalent to a Lagrangian with only relevant and marginal operators at low energy scale. Indeed, after numerous integrations of the fast fields all irrelevant operators will be suppressed.

EXERCISES

7.1 Calculate the propagator for the field $\hat{\phi}$.

SOLUTION. We have

$$\int d^d x (\partial_\mu \hat{\phi})^2 = \int d^d x \int \frac{d^d k}{(2\pi)^d} (-ik)^\mu e^{-ikx} \hat{\phi}_k \int \frac{d^d l}{(2\pi)^d} (-il)^\mu e^{-ilx} \hat{\phi}_l$$

$$= \int \frac{d^d k}{(2\pi)^d} \int \frac{d^d l}{(2\pi)^d} \hat{\phi}_l (2\pi)^2 \delta^d (k+l) k^2 \hat{\phi}_k$$

$$= \int \frac{d^d k}{(2\pi)^d} \hat{\phi}_{-k} k^2 \theta(b\Lambda < |k| < \Lambda) \hat{\phi}_k. \quad (\text{P7.1})$$

$$\frac{\int \mathcal{D}\hat{\phi} e^{-\frac{1}{2}\int d^d x (\partial_\mu \hat{\phi})^2} \hat{\phi}_k \hat{\phi}_l}{\int \mathcal{D}\hat{\phi} e^{-\frac{1}{2}\int d^d x (\partial_\mu \hat{\phi})^2}} = (2\pi)^d \delta(k+l) \frac{\theta(b\Lambda < |k| < \Lambda)}{k^2}. \quad (\text{P7.2})$$

Figure 7.1 Left: One-loop correction to the quark propagator. Right: Born and one-loop contributions to the local operator $\bar{\psi}_x \Gamma \psi_x$. The operator is a grey circle and k is the momentum conjugate to x.

7.2 Calculate integral (7.54).

SOLUTION. Using the propagator from the previous problem,

$$\frac{1}{\hat{Z}} \int \mathcal{D}\hat{\phi} e^{-\frac{1}{2}\int d^d x (\partial_\mu \hat{\phi})^2} \int d^d x (-\lambda \frac{6}{4})(\hat{\phi}^2 \varphi^2)$$

$$= (-\lambda \frac{3}{2}) \int d^d x \varphi_x^2 \int \frac{d^d k}{(2\pi)^d} e^{-ikx} \int \frac{d^d l}{(2\pi)^d} e^{-ilx} \underbrace{\hat{\phi}_k \hat{\phi}_l}$$

$$= (-\lambda \frac{3}{2}) \int d^d x \varphi_x^2 \int \frac{d^d k}{(2\pi)^d} e^{-ikx} \int \frac{d^d l}{(2\pi)^d} e^{-ilx} (2\pi)^d$$

$$\times \delta(k+l) \frac{\theta(b\Lambda < |k| < \Lambda)}{k^2}$$

$$= (-\lambda \frac{3}{2}) \int d^d x \varphi_x^2 \int \frac{d^d k}{(2\pi)^d} \frac{\theta(b\Lambda < |k| < \Lambda)}{k^2} = -\frac{\mu}{2} \int d^d x \varphi_x^2,$$

$$\mu = 3\lambda \int \frac{d^d k}{(2\pi)^d} \frac{\theta(b\Lambda < |k| < \Lambda)}{k^2} = 3\lambda\Omega_d \frac{\Lambda^{d-2}}{d-2}(1 - b^{d-2}). \qquad \text{(P7.3)}$$

7.3 Calculate the anomalous dimension of the quark field.

SOLUTION. Quark propagator with one-loop accuracy is shown in Figure 7.1. It reads ($d = 4 - 2\varepsilon$)

$$D(p) = \frac{i}{\hat{p}} + \frac{i}{\hat{p}} \int \frac{\mu^{4-d} d^d k}{(2\pi)^d k^2} \gamma^\mu \frac{-ii(ig)^2 C_F}{(\hat{p} - \hat{k})} \gamma_\mu \frac{i}{\hat{p}}. \qquad \text{(P7.4)}$$

The contribution to the renormalization constant comes from large loop momenta (referred to as ultraviolet or UV):

$$\int \frac{\mu^{4-d}d^d k}{(2\pi)^d k^2} \gamma^\mu \frac{-ii(ig)^2 C_F}{(\hat{p}-\hat{k})} \gamma_\mu = (d-2)g^2 C_F \int \frac{\mu^{4-d}d^d k}{(2\pi)^d k^2} \frac{(\hat{p}-\hat{k})}{k^2(1-\frac{2(pk)-p^2}{k^2})}$$

$$\overset{UV}{=} (d-2)g^2 C_F \int_{|k_E|>A} \frac{\mu^{4-d}d^d k}{(2\pi)^d k^4}(\hat{p}-\hat{k})(1+\frac{2(pk)-p^2}{k^2})$$

$$= (d-2)g^2 C_F \int_{|k_E|>A} \frac{\mu^{4-d}d^d k}{(2\pi)^d k^4}(\hat{p}-\hat{k}\frac{2(pk)}{k^2})$$

$$= \left| \int d^d r r^\alpha r^\beta f(r^2) = Bg^{\alpha\beta}, B = \frac{1}{d}\int d^d r r^2 f(r^2) \right|$$

$$= (d-2)(1-\frac{2}{d})g^2 C_F \hat{p} \int_{|k_E|>A} \frac{\mu^{4-d}d^d k}{(2\pi)^d k^4}$$

$$= (d-2)(1-\frac{2}{d})g^2 C_F i\hat{p}\Omega_d \int_{|k_E|>A} \frac{\mu^{4-d}k_E^{d-5}dk_E}{(2\pi)^d}$$

$$= (d-2)(1-\frac{2}{d})g^2 C_F i\hat{p} \frac{(\frac{A}{\mu})^{-2\varepsilon}}{\varepsilon(4\pi)^{\frac{d}{2}}\Gamma(\frac{d}{2})} \simeq g^2 C_F i\hat{p}\frac{(\frac{A}{\mu})^{-2\varepsilon}}{\varepsilon(4\pi)^2}. \qquad (P7.5)$$

$$D(p) \overset{UV}{=} \frac{i}{\hat{p}}\left[1+g^2 C_F \frac{(\frac{A}{\mu})^{-2\varepsilon}}{-\varepsilon(4\pi)^2}\right] \implies$$

$$\gamma_\psi = \frac{1}{2}\frac{\partial \ln Z}{\partial \ln M} = -\frac{1}{2}\frac{\partial}{\partial \ln \mu} \ln\left[1+g^2 C_F\frac{(\frac{A}{\mu})^{-2\varepsilon}}{-\varepsilon(4\pi)^2}\right] = \frac{g^2 C_F}{(4\pi)^2}. \qquad (P7.6)$$

7.4 Calculate the anomalous dimensions for $\bar{\psi}\psi$ and $\bar{\psi}\gamma^\rho\psi$.

SOLUTION. For the operator $\bar{\psi}_x \Gamma \psi_x$ with arbitrary Γ we build the Born matrix element schematically depicted in Figure 7.1

$$\langle \bar{\psi}_z \psi_y \bar{\psi}_x \Gamma \psi_x \rangle = -D(y-x)\Gamma D(x-z). \qquad (P7.7)$$

In the momentum space

$$\langle \bar{\psi}_{p_1} \psi_{p_2} \bar{\psi}_k \Gamma \psi_k \rangle = -\int dx dy dz e^{ip_1 z + ip_2 y + ikx} D(y-x)\Gamma D(x-z)$$

$$= -\int dx dy dz e^{ip_1(z+x)+ip_2(y+x)+ikx} D(y)\Gamma D(-z)$$

$$= -(2\pi)^d \delta(p_1+p_2+k)D(p_2)\Gamma D(-p_1)$$

$$= (2\pi)^d \delta(p_1+p_2+k)\frac{i}{\hat{p}_2}\Gamma\frac{i}{\hat{p}_1}. \qquad (P7.8)$$

The one-loop correction to this matrix element reads

$$\frac{(ig)^2}{2}\langle\bar{\psi}_z\psi_y\bar{\psi}_x\Gamma\psi_x\int du\bar{\psi}_u\hat{A}_u\psi_u\int dv\bar{\psi}_v\hat{A}_v\psi_v\rangle = I_1 + I_2 + I_3. \quad (P7.9)$$

$$I_1 = -(ig)^2 \int dxdydzdudv D^{\mu\nu ab}(u-v)$$
$$\times\, D(y-x)\Gamma D(x-u)\gamma_\mu t^a D(u-v)\gamma_\nu t^b D(v-z), \quad (P7.10)$$

$$I_2 = -(ig)^2 \int dxdydzdudv D^{\mu\nu ab}(u-v)$$
$$\times\, D(y-u)\gamma_\mu t^a D(u-v)\gamma_\nu t^b D(v-x)\Gamma D(x-z), \quad (P7.11)$$

$$I_3 = -(ig)^2 \int dxdydzdudv D^{\mu\nu ab}(u-v)$$
$$\times\, D(y-u)\gamma_\mu t^a D(u-x)\Gamma D(x-v)\gamma_\nu t^b D(v-z). \quad (P7.12)$$

In the momentum space

$$I_3 = -(ig)^2 \int dxdydzdudve^{ip_1 z+ip_2 y+ikx} D^{\mu\nu ab}(u-v)$$
$$\times\, D(y-u)\gamma_\mu t^a D(u-x)\Gamma D(x-v)\gamma_\nu t^b D(v-z)$$
$$= -(ig)^2 \int dxdydzdudv \int \frac{d^d l d^d q d^d r}{(2\pi)^{3d}} e^{ip_1(z+v)+ip_2(y+u)+ikx}$$
$$\times\, e^{-i(u-x)l-i(x-v)q-i(u-v)r} D(y)\gamma_\mu t^a D(l)\Gamma D(q)\gamma_\nu t^b D(-z)D^{\mu\nu ab}(r)$$
$$= (2\pi)^d \delta(p_1 + p_2 + k)$$
$$\times\, \frac{i}{\hat{p}_2}\left[(ig)^2 \int \frac{d^d r}{(2\pi)^d}\gamma_\mu t^a D(p_2 - r)\Gamma D(-p_1 - r)\gamma_\nu t^b D^{\mu\nu ab}(r)\right]\frac{i}{\hat{p}_1}.$$
$$(P7.13)$$

$$\left[\dots\right] = (ig)^2 C_F \int \frac{\mu^{4-d}d^d r}{(2\pi)^d}\gamma_\mu\frac{i}{\hat{p}_2 - \hat{r}}\Gamma\frac{-i}{\hat{p}_1 + \hat{r}}\gamma^\mu\frac{-i}{r^2}$$
$$\stackrel{UV}{=} -ig^2 C_F \int_{|r_E|>A}\frac{\mu^{4-d}d^d r}{(2\pi)^d r^2}\gamma_\mu\frac{1}{\hat{r}}\Gamma\frac{1}{\hat{r}}\gamma^\mu$$
$$= -\frac{ig^2 C_F}{d}\int_{|r_E|>A}\frac{\mu^{4-d}d^d r}{(2\pi)^d r^4}\gamma_\mu\gamma_\nu\Gamma\gamma^\nu\gamma^\mu$$
$$= -\frac{ig^2 C_F}{d}i\mu^{4-d}\Omega_d\int_{|r_E|>A}\frac{r^{d-5}dr}{(2\pi)^d}\gamma_\mu\gamma_\nu\Gamma\gamma^\nu\gamma^\mu$$
$$= \frac{ig^2 C_F}{d}\frac{i}{\Gamma(\frac{d}{2})(4\pi)^{\frac{d}{2}}}\frac{(\frac{A}{\mu})^{-2\varepsilon}}{-\varepsilon}\gamma_\mu\gamma_\nu\Gamma\gamma^\nu\gamma^\mu$$
$$\simeq -\frac{g^2 C_F}{(4\pi)^2}\frac{(\frac{A}{\mu})^{-2\varepsilon}}{-\varepsilon}\Gamma \times \begin{Bmatrix} 4, \Gamma = 1 \\ 1, \Gamma = \gamma^\rho \end{Bmatrix}. \quad (P7.14)$$

Therefore

$$-\frac{\partial}{\partial \ln \mu}\langle \bar{\psi}_{p_1}\psi_{p_2}\bar{\psi}_k\Gamma\psi_k\rangle_3 = \langle \bar{\psi}_{p_1}\psi_{p_2}\bar{\psi}_k\Gamma\psi_k\rangle_{Born}\frac{-g^2 C_F}{(4\pi)^2} \times \left\{ \begin{array}{l} 8, \Gamma = 1 \\ 2, \Gamma = \gamma^\rho \end{array} \right\}.$$
(P7.15)

The contribution of $I_1 + I_2$ equals $4\gamma_\psi$ (P7.6). As a result, the anomalous dimension of $\bar{\psi}\Gamma\psi$ reads

$$\gamma_\Gamma = -\frac{g^2 C_F}{(4\pi)^2}\left\{ \begin{array}{l} 8, \Gamma = 1 \\ 2, \Gamma = \gamma^\rho \end{array} \right\} + 4\gamma_\psi - 2\gamma_\psi = -\frac{g^2 C_F}{(4\pi)^2}\left\{ \begin{array}{l} 8, \Gamma = 1 \\ 2, \Gamma = \gamma^\rho \end{array} \right\} + 2\frac{g^2 C_F}{(4\pi)^2}$$

$$= -\frac{g^2 C_F}{(4\pi)^2}\left\{ \begin{array}{l} 6, \Gamma = 1 \\ 0, \Gamma = \gamma^\rho \end{array} \right\}.$$
(P7.16)

7.5 Explain why the vector current anomalous dimension is 0.

SOLUTION. The vector current is conserved

$$\partial_\mu \bar{\psi}\gamma^\mu \psi = 0 \quad \Longrightarrow \quad \int d^3 x \bar{\psi}\gamma^0 \psi = const \quad \Longrightarrow \quad Z = 1, \quad \gamma = 0.$$
(P7.17)

7.6 Show that the renormalization constants obey the following identities

$$\frac{Z_{1(\bar{\psi}\hat{A}\psi)}}{Z_3^{\frac{1}{2}} Z_2} = \frac{Z_{1(A^3)}}{Z_3^{\frac{3}{2}}} = \frac{Z_{1(A^4)}^{\frac{1}{2}}}{Z_3} = \frac{Z_{1(\bar{c}Ac)}}{Z_3^{\frac{1}{2}} Z_2^c} = Z_g = \frac{g_0}{g}.$$
(P7.18)

SOLUTION. The bare lagrangian is BRST invariant (see problem 4.21 in Chapter 4) only with the same g_0 in each vertex. After renormalization according to the procedure described in paragraph 1 of the lecture, the charge in each vertex in the lagrangian gets its own renormalization constant:

$$\mathcal{L} = -\frac{1}{4}(\partial_\mu A_{0\nu}^a - \partial_\nu A_{0\mu}^a)^2 - \frac{1}{2}g\frac{Z_{1(A^3)}}{Z_3^{\frac{3}{2}}}f^{abc}A_0^{\mu b}A_0^{\nu c}(\partial_\mu A_{0\nu}^a - \partial_\nu A_{0\mu}^a)$$

$$-\frac{1}{4}g^2\frac{Z_{1(A^4)}}{Z_3^2}f^{ab'c'}f^{abc}A_{0\mu}^b A_{0\nu}^c A_0^{\mu b'}A_0^{\nu c'} - \frac{1}{2\varsigma}(\partial_\mu A_0^{a\mu})^2$$

$$+ \bar{\psi}_0(i\hat{\partial} + \frac{Z_{1(\bar{\psi}\hat{A}\psi)}}{Z_3^{\frac{1}{2}} Z_2}g\hat{A}_0 - m_0)\psi_0$$

$$+ (\partial_\mu \bar{c}_0^a)(\delta^{ac}\partial^\mu + g\frac{Z_{1(\bar{c}\hat{A}c)}}{Z_3^{\frac{1}{2}} Z_2^c}f^{abc}A_0^{\mu b})c_0^c.$$
(P7.19)

Therefore, BRST invariance of this lagrangian demands (P7.18), i.e. one universal charge renormalization constant Z_g, which was used in the lecture.

7.7 Find dependence on the renormalization scale of the total cross-section of $e^+e^- \to hadrons$.

SOLUTION. Since the cross-section is an observable it does not have anomalous dimension. The Callan-Symanzik equation reads

$$\left(\frac{\partial}{\partial \ln M} + \beta(g)\frac{\partial}{\partial g}\right)\sigma(s, M, \alpha_s) = 0 \qquad \Longrightarrow \qquad \sigma(s, M) = f(\alpha_s(s)),$$

(P7.20)

where the arbitrary function f is to be found from the initial conditions at $M^2 = s$. At this scale one can calculate only several first terms of the perturbation theory

$$\sigma(s, s) = f(\alpha_s(s)) = \sigma^0_{ee\to\mu\mu}(s)N_c\sum_f Q_f^2(1 + \frac{\alpha_s(s)}{\pi}). \qquad (P7.21)$$

Then

$$\sigma(s, M) = \sigma^0_{ee\to\mu\mu}(s)N_c\sum_f Q_f^2(1 + \frac{\alpha_s(s)}{\pi}),$$

$$\alpha_s(s) = \frac{\alpha_s(M^2)}{1 + \frac{\alpha_s(M^2)}{4\pi}\beta_0 \ln \frac{s}{M^2}}. \qquad (P7.22)$$

7.8 Find dependence on the renormalization scale of the e^+e^- potential in QED.

SOLUTION. Since the potential is an observable it does not have anomalous dimension. The Callan-Symanzik equation reads

$$\left(\frac{\partial}{\partial \ln M} + \beta(e)\frac{\partial}{\partial e}\right)V(p^2, M, e) = 0 \qquad \Longrightarrow \qquad V(p^2, M) = f(e(p^2)),$$

(P7.23)

where the arbitrary function f is to be found from the initial conditions at $M^2 = p^2$. At this scale one can calculate only several first terms of the perturbation theory

$$V(p^2, p^2) = f(e(p^2)) = \frac{e^2(p^2)}{p^2}. \qquad (P7.24)$$

Then

$$V(p^2, M) = \frac{e^2(p^2)}{p^2}, \quad where \quad e^2(p^2) = \frac{e^2(M^2)}{1 + \frac{e^2(M^2)}{(4\pi)^2}\beta_0 \ln \frac{p^2}{M^2}}. \qquad (P7.25)$$

204 ■ Introduction to Strong Interactions: Theory and Applications

7.9 Find dependence on the renormalization scale of the scalar propagator.

SOLUTION. Since the propagator is not observable it does have an anomalous dimension. The Callan-Symanzik equation reads

$$\left(\frac{\partial}{\partial \ln M} + \beta(g)\frac{\partial}{\partial g} + 2\gamma(g)\right) G(p, M, g) = 0 \quad \Longrightarrow$$

$$G(p, M, g) = e^{-\int_{g(p)}^{g(M)} dg' \frac{2\gamma(g')}{\beta(g')}} f(g(p)), \tag{P7.26}$$

where the arbitrary function f is to be found from the initial conditions at $M^2 = p^2$. At this scale one can use the Born propagator

$$G(p, M, g) = f(g(p)) = \frac{i}{p^2} \quad \Longrightarrow \quad G(p, M, g) = \frac{i}{p^2} e^{-\int_{g(p)}^{g(M)} dg' \frac{2\gamma(g')}{\beta(g')}}. \tag{P7.27}$$

In one loop

$$G(p, M, g) = \frac{i}{p^2}\left(\frac{g(M)}{g(p)}\right)^{\frac{2\gamma^{(2)}}{\beta_0}} = \frac{i}{p^2}\left(1 + 2\frac{g^2(M)}{(4\pi)^2}\beta_0 \ln\frac{p}{M}\right)^{\frac{\gamma^{(2)}}{\beta_0}}$$

$$\simeq \frac{i}{p^2}\left(1 + 2\gamma^{(2)}\frac{g^2(M)}{(4\pi)^2}\ln\frac{p}{M}\right). \tag{P7.28}$$

Recalling that

$$\gamma = \frac{g^2}{(4\pi)^2}\gamma^{(2)} = \frac{1}{2}\frac{\partial \ln Z}{\partial \ln M}, \quad Z = 1 - \frac{g^2}{(4\pi)^2}\gamma^{(2)}\left(-\frac{1}{\varepsilon} - \gamma - \ln\frac{M^2}{4\pi\mu^2}\right), \tag{P7.29}$$

one can see that the solution of the Callan-Symanzik equation gives the logarithmically enhanced part of the renormalized propagator in one loop.

7.10 In $\overline{MS}$ scheme the 2-loop mass anomalous dimension reads

$$\gamma_{\bar{q}q} = \frac{\alpha_s}{4\pi}(\gamma_{\bar{q}q}^{(2)} + \alpha_s\gamma_{\bar{q}q}^{(4)}), \quad \gamma_{\bar{q}q}^{(2)} = -6C_F, \quad \gamma_{\bar{q}q}^{(4)} = -\frac{303 - 10n_f}{9\pi}. \tag{P7.30}$$

Find the current quark mass scale dependence at 2 loops.

SOLUTION. From the characteristic system (7.42) for Callan-Symanzik equation (7.41) and the 2-loop Gell-Mann - Low equation (P6.63) one gets the evolution equation for the quark current mass

$$\frac{d\ln m}{d\ln M} = \gamma_{\bar{q}q}(g) \quad \Longleftrightarrow \quad \frac{d\ln m}{dg} = \frac{\gamma_{\bar{q}q}(g)}{\beta(g)}, \tag{P7.31}$$

$$\frac{d\ln m}{d\alpha_s} = \frac{4\pi\gamma_{\bar{q}q}(g)}{2g\beta(g)} = \frac{\alpha_s(\gamma_{\bar{q}q}^{(2)} + \gamma_{\bar{q}q}^{(4)}\alpha_s)}{-2\alpha_s^2(\beta_0 + \alpha_s\beta_1)} = -\frac{\gamma_{\bar{q}q}^{(2)} + \gamma_{\bar{q}q}^{(4)}\alpha_s}{2\alpha_s(\beta_0 + \alpha_s\beta_1)} \quad \Longrightarrow \tag{P7.32}$$

$$\ln\frac{m(p^2)}{m(q^2)} = -\int_{\alpha_s(q^2)}^{\alpha_s(p^2)} \frac{d\alpha}{2} \frac{\gamma_{\bar{q}q}^{(2)} + \gamma_{\bar{q}q}^{(4)}\alpha}{\alpha(\beta_0 + \alpha\beta_1)}$$

$$= -\int_{\alpha_s(q^2)}^{\alpha_s(p^2)} \frac{d\alpha}{2\beta_0} \left(\frac{\gamma_{\bar{q}q}^{(2)}}{\alpha} + \frac{\gamma_{\bar{q}q}^{(4)}\beta_0 - \gamma_{\bar{q}q}^{(2)}\beta_1}{\beta_0 + \alpha\beta_1} \right) \quad \Longrightarrow \tag{P7.33}$$

$$\frac{m(p^2)}{m(q^2)} = \left(\frac{\alpha_s(p^2)}{\alpha_s(q^2)} \right)^{-\frac{\gamma_{\bar{q}q}^{(2)}}{2\beta_0}} \left(\frac{\beta_0 + \alpha_s(p^2)\beta_1}{\beta_0 + \alpha_s(q^2)\beta_1} \right)^{-\frac{\gamma_{\bar{q}q}^{(4)}\beta_0 - \gamma_{\bar{q}q}^{(2)}\beta_1}{2\beta_0}}. \tag{P7.34}$$

7.11 Show how Green functions change with the scaling of the external momenta at the fixed renormalization point M.

SOLUTION.

a. Using solution of the Kallan-Symanzic equation (7.31), for the n-point Green function with the external momenta p_i

$$G(p_1, \ldots, p_n, M, g(M), m(M)) = G(p_1, \ldots, p_n, P, g(P), m(P))$$
$$\times e^{\int_{g(P)}^{g(M)} \frac{dg'}{\beta(g')}(-n\gamma(g'))}, \tag{P7.35}$$

one takes the initial condition at $p_i \sim P$. Introducing λ :

$$P = \lambda M, \quad p_i = \lambda k_i, \quad k_i \sim M \tag{P7.36}$$

and the mass dimension d_G of the Green function as the sum of the mass dimensions of the fields (e.g. dim $A = 1$, dim $\psi = \frac{3}{2}$, etc.) and the necessary Fourier transforms:

$$G(lp_1, \ldots, lp_n, lM, g(M), lm(M))$$
$$= l^{d_G} G(p_1, \ldots, p_n, M, g(M), m(M)), \tag{P7.37}$$

we get

$$G(\lambda k_1, \ldots, \lambda k_n, M, g(M), m(M))$$

$$= G(\lambda k_1, \ldots, \lambda k_n, \lambda M, g(\lambda M), m(\lambda M)) e^{\int_{g(\lambda M)}^{g(M)} \frac{dg'}{\beta(g')}(-n\gamma(g'))}$$

$$= \lambda^{d_G} G(k_1, \ldots, k_n, M, g(\lambda M), \frac{m(\lambda M)}{\lambda}) e^{\int_{g(\lambda M)}^{g(M)} \frac{dg'}{\beta(g')}(-n\gamma(g'))}$$

$$= \lambda^{d_G} G(k_1, \ldots, k_n, M, g(\lambda M), \frac{m(\lambda M)}{\lambda}) \left(\frac{g(M)}{g(\lambda M)}\right)^{\frac{n\gamma^{(2)}}{\beta_0}}.$$

$$(P7.38)$$

Here the last line gives the Green function in one loop approximation.

b. For the renormalized Green function of the n fields in different points in the momentum space one has the Callan-Symanzic equation

$$\left(\frac{\partial}{\partial \ln M} + \beta(g)\frac{\partial}{\partial g} + \gamma_{\bar{q}q}\frac{\partial}{\partial \ln m} + n\gamma\right) G(p_1, \ldots, p_n, M, g, m) = 0.$$

$$(P7.39)$$

For the mass dimension d_G of the Green function one has

$$G(p_1, \ldots, p_n, M, g, m) = M^{d_G} G(\frac{p_1}{M}, \ldots, \frac{p_n}{M}, 1, g, \frac{m}{M}). \quad (P7.40)$$

Therefore

$$\frac{\partial}{\partial \ln M} G(\lambda p_1, \ldots, \lambda p_n, M, g, m) = \left(d_G - \frac{\partial}{\partial \ln \lambda} - \frac{\partial}{\partial \ln m}\right) G.$$

$$(P7.41)$$

Substituting it into the Callan-Symanzic equation, one gets

$$\left(-\frac{\partial}{\partial \ln \lambda} + \beta(g)\frac{\partial}{\partial g} + (\gamma_{\bar{q}q} - 1)\frac{\partial}{\partial \ln m} + d_G + n\gamma\right)$$

$$\times G(\lambda p_1, \ldots, \lambda p_n, M, g, m) = 0. \quad (P7.42)$$

Here γ and d_G enter this equation in the sum. Hence, the name of γ is the anomalous dimension, as it corrects the mass dimension of the Green function via renormalization. For local operators one gets the same result with the substitution $n\gamma + \gamma_O$. One often writes the Kallan-Symanzic equation in this form. Note that the matrix element of a local operator without inclusion of external fields is shift-invariant and $\sim \delta(p)$ in the momentum space. Therefore, the Kallan-Symanzic equation in this form for it has the form

$$\left(\beta(g)\frac{\partial}{\partial g} + (\gamma_{\bar{q}q} - 1)\frac{\partial}{\partial \ln m} + d_G\right) G(M, g, m) = 0. \quad (P7.43)$$

By a direct calculation show that equation (P7.42) has the solution given in the section a.

7.12 Show that $\langle m\bar{\psi}\psi \rangle$ does not depend on the renormalization scale.

SOLUTION. Using (7.43)

$$m(M) = m(P)e^{\int_{g(P)}^{g(M)} \frac{\gamma_{\bar{q}q}(g')}{\beta(g')} dg'} \qquad (P7.44)$$

and the solution of the Kallan-Symanzic equation for the $\langle \bar{\psi}\psi \rangle$ Green function

$$\langle \bar{\psi}\psi \rangle_M = \langle \bar{\psi}\psi \rangle_P e^{-\int_{g(P)}^{g(M)} \frac{\gamma_{\bar{q}q}(g')}{\beta(g')} dg'} \qquad (P7.45)$$

one can see that $\langle m\bar{\psi}\psi \rangle$ is renorm-invariant.

Renormalization in QCD is discussed in many textbooks on quantum field theory. The reader can find in the list below detailed discussion of renormalized perturbation theory and explicit calculation of renormalization constants.

FURTHER READING

[DKS03] G. Dissertori, I. G. Knowles, and M. Schmelling. *Quantum chromodynamics High energy experiments and theory.* Clarendon Press, Oxford, 2003.

[IZ80] C. Itzykson and J. B. Zuber. *Quantum field theory,* chapter Vol.2. International Series in Pure and Applied Physics. McGraw-Hill, New York, 1980.

[PS95] Michael E. Peskin and Daniel V. Schroeder. *An introduction to quantum field theory,* chapter 12.1–4, 17.2. Addison-Wesley, Reading, USA, 1995.

[Ram81] Pierre Ramond. *Field Theory. A Modern Primer,* volume 51 of *Frontiers in Physics.* 1981.

Scale Anomaly

I N this chapter we discuss scale transformation of the QCD lagrangian, introduce the dilatation current and calculate its divergence for the QCD effective action. Then we give the general idea of the trace anomaly and its relation to hadron masses. Problems deal with the Belinfante construction of the symmetric energy-momentum tensor and relation of its trace to the divergence of the dilatation current.

1. **Dilatation current**. Quantum corrections make the effective charge depend on the scale, i.e. they introduce dynamically generated scale (Λ_{QCD} or μ) into the theory. This phenomenon is refered to as **dimensional transmutation**. It breaks the scale invariance of the classical massless theory.

   The scale transformation of the fields reads

   $$A_\mu(x) \to A'_\mu(x') = e^{-a} A_\mu\left(e^{-a}x\right) \simeq A_\mu(x) - a(A_\mu(x) + x^\nu \partial_\nu A_\mu(x)),$$
   (8.1)

   $$\psi(x) \to \psi'(x') = e^{-\frac{3}{2}a} \psi\left(e^{-a}x\right) \simeq \psi(x) - a(\frac{3}{2}\psi(x) + x^\nu \partial_\nu \psi(x)).$$
   (8.2)

   One may equivalently transform the fields and the coordinate simultaneously

   $$x \to x' = e^a x, \quad \begin{cases} A(x) \to A'(x') = e^{-a} A(x), \\ \psi(x) \to \psi'(x') = e^{-\frac{3}{2}a} \psi(x), \end{cases}$$
   (8.3)

   If the theory is scale invariant the Lagrangian is transformed as

   $$\mathcal{L}(x) \to \mathcal{L}'(x') = e^{-4a} \mathcal{L}\left(e^{-a}x\right) \simeq \mathcal{L}(x) - a(\mathcal{L}(x) + x^\nu \partial_\nu \mathcal{L}(x))$$
   $$= \mathcal{L}(x) - a\partial_\nu(x^\nu \mathcal{L}(x)).$$
   (8.4)

   Then

   $$\int d^4x \mathcal{L}(x) \to \int d^4x e^{-4a} \mathcal{L}\left(e^{-a}x\right) = |x = e^a y| = \int d^4y \mathcal{L}(y), \quad (8.5)$$

DOI: 10.1201/9781003272403-8

and

$$-a\partial_\nu(x^\nu \mathcal{L}(x)) = \frac{\partial \mathcal{L}}{\partial(\partial_\mu A_\nu)}\delta\partial_\mu A_\nu + \frac{\partial \mathcal{L}}{\partial A_\nu}\delta A_\nu + \frac{\partial \mathcal{L}}{\partial(\partial_\mu\psi)}\delta\partial_\mu\psi + \frac{\partial \mathcal{L}}{\partial\psi}\delta\psi$$
$$+ \frac{\partial \mathcal{L}}{\partial(\partial_\mu\bar\psi)}\delta\partial_\mu\bar\psi + \frac{\partial \mathcal{L}}{\partial\bar\psi}\delta\bar\psi$$
$$\stackrel{\text{e.o.m.}}{=} \partial_\mu\left(\frac{\partial \mathcal{L}}{\partial(\partial_\mu A_\nu)}\delta A_\nu + \frac{\partial \mathcal{L}}{\partial(\partial_\mu\psi)}\delta\psi + \frac{\partial \mathcal{L}}{\partial(\partial_\mu\bar\psi)}\delta\bar\psi\right),$$
$$(8.6)$$

or the dilatation current

$$d^\mu = -\frac{\partial \mathcal{L}}{\partial(\partial_\mu A_\nu)}\delta A_\nu - \frac{\partial \mathcal{L}}{\partial(\partial_\mu\psi)}\delta\psi - \frac{\partial \mathcal{L}}{\partial(\partial_\mu\bar\psi)}\delta\bar\psi - g^{\mu\nu}x_\nu\mathcal{L} \qquad (8.7)$$

is conserved

$$\partial_\mu d^\mu = 0. \qquad (8.8)$$

2. Dilatation current in QCD. we follow the method of (DEM15). For QCD the effective action can be written as

$$\Gamma[b,\phi_R] = \Gamma_1 + \Gamma_2 + \Gamma_3 + \Gamma_4, \qquad (8.9)$$

$$\Gamma_1 = \int d^4x \left\{-\frac{1}{4g^2(-\mu^2)}F_{\mu\nu}^a(x)F^{a\mu\nu}(x) + \bar\phi_R(x)i\hat{D}\phi_R(x)\right\}, \qquad (8.10)$$

$$\Gamma_2 = -m(-\mu^2)\int d^4x\, \bar\phi_R(x)\phi_R(x), \qquad (8.11)$$

$$\Gamma_3 = +m(-\mu^2)\frac{\gamma_m}{2}\int \frac{d^4p}{(2\pi)^4}\tilde{\bar\phi}_R(-p)\tilde\phi_R(p)\ln\frac{-p^2}{\mu^2}, \qquad (8.12)$$

$$\gamma_m = g^2\left(-\mu^2\right)\frac{6C_F}{(4\pi)^2} = -\gamma_{\bar qq}, \qquad (8.13)$$

$$\Gamma_4 = -\frac{1}{4}\frac{\beta_0}{(4\pi)^2}\int\frac{d^4p}{(2\pi)^4}F_{\mu\nu}^a(-p)F^{a\mu\nu}(p)\ln\frac{-p^2}{\mu^2}. \qquad (8.14)$$

The first part Γ_1 is the massless classical one. Hence, the scale transformation reads:

$$\Gamma_1 \to \int d^4x \left\{-\frac{1}{4(-\mu^2)}\left[i\frac{\partial}{\partial x^\mu} + e^{-a}b_\mu(e^{-a}x), i\frac{\partial}{\partial x^\nu} + e^{-a}b_\nu(e^{-a}x)\right]^2\right.$$
$$\left. + e^{-3a}\bar\phi_R(e^{-a}x)\left[i\hat\partial + e^{-a}\hat b(e^{-a}x)\right]\phi_R(e^{-a}x)\right\} = |x = e^a y|$$
$$= \int d^4y\, e^{4a}\left\{-\frac{1}{4(-\mu^2)}\left[e^{-a}i\frac{\partial}{\partial y^\mu} + e^{-a}b_\mu(y), e^{-a}i\frac{\partial}{\partial y^\nu} + e^{-a}b_\nu(y)\right]^2\right.$$
$$\left. + e^{-3a}\bar\phi_R(y)\left[e^{-a}i\hat\partial + e^{-a}\hat b(y)\right]\phi_R(y)\right\} = \Gamma_1, \qquad (8.15)$$

i.e. Γ_1 is scale invariant, $\partial_\mu d_1^\mu = 0$.

$$\Gamma_2 = \int d^4x \mathcal{L}_2 \to -m(-\mu^2) \int d^4x \, e^{-3a}\bar{\phi}_R(e^{-a}x)\phi_R(e^{-a}x)$$

$$\simeq \Gamma_2 - am(-\mu^2) \int d^4x$$

$$\times \left\{ -3\bar{\phi}_R(x)\phi_R(x) - x^\mu(\partial_\mu\bar{\phi}_R(x))\phi_R(x) - \bar{\phi}_R(x)x^\mu\partial_\mu\phi_R(x) \right\}$$

$$= \Gamma_2 - am(-\mu^2) \int d^4x \, \left\{ \bar{\phi}_R(x)\phi_R(x) - \partial_\mu(x^\mu\bar{\phi}_R(x)\phi_R(x)) \right\}$$

$$= \Gamma_2 + \int d^4x \, \left\{ -am(-\mu^2)\bar{\phi}_R(x)\phi_R(x) - a\partial_\mu(x^\mu\mathcal{L}_2) \right\}, \quad (8.16)$$

i.e. Γ_2 is not scale invariant and adds $-m(-\mu^2)\bar{\phi}_R\phi_R$ to the divergence of the dilatation current

$$\partial_\mu d_2^\mu = m(-\mu^2)\bar{\phi}_R\phi_R. \quad (8.17)$$

This term breaks scale invariance at the classical level due to the quark mass. One can get this term also changing the variables:

$$\Gamma_2 \to -m(-\mu^2) \int d^4x \, e^{-3a}\bar{\phi}_R(e^{-a}x)\phi_R(e^{-a}x) = |x = e^a y|$$

$$= -m(-\mu^2)e^a \int d^4y \, \bar{\phi}_R(y)\phi_R(y) \simeq \Gamma_2 - am(-\mu^2) \int d^4y \, \bar{\phi}_R(y)\phi_R(y). \quad (8.18)$$

$$\Gamma_3 = m\frac{\gamma_m}{2} \int \frac{d^4p}{(2\pi)^4}\tilde{\bar{\phi}}_R(-p)\tilde{\phi}_R(p) \ln\frac{-p^2}{\mu^2}$$

$$= m\frac{\gamma_m}{2} \int d^4x \int d^4y \bar{\phi}_R(x) \int \frac{d^4p}{(2\pi)^4}e^{-ip(x-y)} \ln\frac{-p^2}{\mu^2}\phi_R(y)$$

$$= m\frac{\gamma_m}{2} \int d^4x \int d^4y \langle\bar{\phi}_R|x\rangle\langle x| \ln\frac{\partial^2}{\mu^2}|y\rangle\langle x|\phi_R\rangle$$

$$\to m\frac{\gamma_m}{2} \int d^4x \int d^4y \, e^{-3a}\bar{\phi}_R(e^{-a}x)\phi_R(e^{-a}y)$$

$$\times \int \frac{d^4p}{(2\pi)^4}e^{-ip(x-y)} \ln\frac{-p^2}{\mu^2} = |x = e^a x', y = e^a y', p = e^{-a}k|$$

$$= m\frac{\gamma_m}{2} \int d^4x' \int d^4y' e^a \bar{\phi}_R(x')\phi_R(y')$$

$$\times \int \frac{d^4k}{(2\pi)^4}e^{-ik(x'-y')}(\ln\frac{-k^2}{\mu^2} - 2a) = \Gamma_3 + a(\Gamma_3 + \gamma_m\Gamma_2), \quad (8.19)$$

i.e. Γ_3 is not scale invariant and adds to the divergence of the dilatation current

$$\partial_\mu d_3^\mu = \gamma_m m(-\mu^2)\bar{\phi}_R\phi_R - m(-\mu^2)\frac{\gamma_m}{2}\bar{\phi}_R(x)\int d^4y\langle x|\ln\frac{\partial^2}{\mu^2}|y\rangle\phi_R(y).$$

(8.20)

$$\Gamma_4 = -\frac{1}{4}\frac{\beta_0}{(4\pi)^2}\int\frac{d^4p}{(2\pi)^4}F_{\mu\nu}^a(-p)F^{a\mu\nu}(p)\ln\frac{-p^2}{\mu^2}$$

$$= -\frac{1}{4}\frac{\beta_0}{(4\pi)^2}\int d^4x\int d^4y F_{\mu\nu}^a(x)\int\frac{d^4p}{(2\pi)^4}e^{-ip(x-y)}\ln\frac{-p^2}{\mu^2}F^{a\mu\nu}(y)$$

$$= -\frac{1}{4}\frac{\beta_0}{(4\pi)^2}\int d^4x\int d^4y F_{\mu\nu}^a(x)\langle x|\ln\frac{\partial^2}{\mu^2}|y\rangle F^{a\mu\nu}(y)$$

$$\rightarrow -\frac{1}{4}\frac{\beta_0}{(4\pi)^2}$$

$$\times\int d^4x\int d^4y\left[i\frac{\partial}{\partial x^\mu}+e^{-a}A_\mu(e^{-a}x),i\frac{\partial}{\partial x^\nu}+e^{-a}A_\nu(e^{-a}x)\right]$$

$$\times\langle x|\ln\frac{\partial^2}{\mu^2}|y\rangle\left[i\frac{\partial}{\partial y_\mu}+e^{-a}A^\mu(e^{-a}y),i\frac{\partial}{\partial y_\nu}+e^{-a}A^\nu(e^{-a}y)\right]$$

$$= |x = e^a x', y = e^a y'|$$

$$= -\frac{1}{4}\frac{\beta_0}{(4\pi)^2}\int d^4x'\int d^4y' F_{\mu\nu}^a(x')\langle x'|\ln\frac{e^{-2a}\partial^2}{\mu^2}|y'\rangle F^{a\mu\nu}(y')$$

$$= \Gamma_4 + 2a\frac{1}{4}\frac{\beta_0}{(4\pi)^2}\int d^4x F_{\mu\nu}^a(x)F^{a\mu\nu}(x).$$

(8.21)

Therefore

$$\partial_\mu d_4^\mu = -\frac{1}{2}\frac{\beta_0}{(4\pi)^2}F_{\mu\nu}^a(x)F^{a\mu\nu}(x).$$

(8.22)

As a result, the divergence of the dilatation current reads

$$\partial_\mu d^\mu = -\frac{1}{2}\frac{\beta_0}{(4\pi)^2}F_{\mu\nu}^a(x)F^{a\mu\nu}(x)$$

$$+ m(-\mu^2)\bar{\phi}_R(x)\int d^4y\langle x|1-\frac{\gamma_m}{2}\ln\frac{\partial^2}{\mu^2}|y\rangle\phi_R(y)+\gamma_m m(-\mu^2)\bar{\phi}_R\phi_R$$

$$= -\frac{1}{2}\frac{\beta_0}{(4\pi)^2}F_{\mu\nu}^a(x)F^{a\mu\nu}(x)+\bar{\phi}_R(x)\int d^4y\langle x|m(\partial^2)|y\rangle\phi_R(y)(1+\gamma_m).$$

(8.23)

During this derivation we neglected the terms related to the gauge fixing and the terms vanishing via equations of motion since they will not give contribution to gauge invariant observables we will consider. Under this assumption the relation of the dilatation current divergence to the trace of the symmetric energy momentum tensor is the same as for the classical

Lagrangian:

$$\partial_\mu d^\mu = T^\mu_\mu = -\frac{1}{2}\frac{\beta_0}{(4\pi)^2}F^a_{\mu\nu}(x)F^{a\mu\nu}(x)$$

$$+ \bar{\phi}_R(x)\int d^4y\langle x|m(\partial^2)|y\rangle\phi_R(y)(1+\gamma_m). \qquad (8.24)$$

The first term in the r.h.s. of this equation is called the **trace anomaly**. The formula for the scale anomaly obtained here from the effective action is not convenient for our discussion below since it is nonlocal. Direct calculation of loop corrections to $\langle T^\mu_\mu \rangle$ between the on-shell gauge invariant states gives (ACD77; CDJ77; Nie77)

$$\langle T^\mu_\mu \rangle = \frac{1}{2}\frac{\beta(g)}{g}\langle F^a_{\mu\nu}F^{a\mu\nu}\rangle + m\langle\bar{\psi}\psi\rangle(1+\gamma_m(g)), \qquad (8.25)$$

$$\beta(g) = \frac{dg}{d\ln M} = -\frac{\beta_0 g^3}{(4\pi)^2} + ... \qquad (8.26)$$

Here all the operators, charge and mass are renormalized at the scale M, and charge is not taken into the field A.

3. **Hadron masses.** Consider a hadron H with the momentum p, spin s, and the state normalization

$$\langle H_{p',s'}||H_{p,s}\rangle = 2E_p\delta_{ss'}(2\pi)^3\delta^{(3)}(\vec{p}-\vec{p}'). \qquad (8.27)$$

The matrix element of the symmetric energy momentum tensor reads

$$\tau_{\mu\nu}(p,q,s) = \int d^4x e^{iqx}\langle H_{p',s}|T_{\mu\nu}(x)|H_{p,s}\rangle$$

$$= \int d^4x e^{iqx}\langle H_{p',s}|e^{i\hat{p}x}T_{\mu\nu}(0)e^{-i\hat{p}x}|H_{p,s}\rangle$$

$$= \int d^4x e^{i(q+p'-p)x}\langle H_{p',s}|T_{\mu\nu}(0)|H_{p,s}\rangle$$

$$= (2\pi)^4\delta^{(4)}(q+p'-p)\langle H_{p',s}|T_{\mu\nu}(0)|H_{p,s}\rangle. \qquad (8.28)$$

The latter matrix element can be decomposed into form-factors. Since $T_{\mu\nu}$ is a P-even conserved ($q^\mu T_{\mu\nu} = 0$) symmetric tensor, one has

$$\langle H_{p',s}|T_{\mu\nu}(0)|H_{p,s}\rangle$$
$$= A_0(q^2)P_\mu P_\nu + iA_1(q^2)(\varepsilon_{\mu\alpha\beta\sigma}P_\nu + \varepsilon_{\nu\alpha\beta\sigma}P_\mu)q^\alpha P^\beta s^\sigma + O(q^2), \qquad (8.29)$$

where $P_\mu = \frac{1}{2}(p_\mu + p'_\mu)$ and we neglect higher powers of q since we will be interested in the forward limit. Taking $q_\mu \to 0$,

$$\tau_{\mu\nu}(p,0,s) = (2\pi)^4\delta^{(4)}(0)A_0(0)p_\mu p_\nu. \qquad (8.30)$$

Recalling that the Hamiltonian

$$H = \int T^{00}(x) d^3 x, \tag{8.31}$$

we get

$$
\begin{aligned}
\tau_{00}(p, 0, s) &= \int d^4 x \langle H_{p,s} | T_{00}(x) | H_{p,s} \rangle = \int dt \langle H_{p,s} | H | H_{p,s} \rangle \\
&= E_p \int dt \langle H_{p,s} || H_{p,s} \rangle = 2E_p^2 (2\pi)^3 \delta^{(3)}(0) \int dt \\
&= 2E_p^2 (2\pi)^4 \delta^{(4)}(0) \quad \Longrightarrow \quad A_0(0) = 2.
\end{aligned} \tag{8.32}
$$

Hence

$$\langle H_{p,s} | T_{\mu\nu}(0) | H_{p,s} \rangle = 2 p_\mu p_\nu \quad \Longrightarrow \quad \langle H_{p,s} | T_\mu^\mu(0) | H_{p,s} \rangle = 2 p^2 = 2 m_H^2. \tag{8.33}$$

Neglecting the running of the mass in (8.24) one can write for the nucleon, e.g.

$$
\begin{aligned}
2 m_N^2 \simeq \langle N | &- \frac{1}{2} \frac{\beta_0 g^2}{(4\pi)^2} F_{\mu\nu}^a(0) F^{a\mu\nu}(0) \\
&+ m_u \bar{u}(0) u(0) + m_d \bar{d}(0) d(0) + m_s \bar{s}(0) s(0) | N \rangle.
\end{aligned} \tag{8.34}
$$

From different model calculations one can estimate the contributions of the parts of this equation (DGH14)

$$
\begin{aligned}
\frac{\langle N | - \frac{1}{2} \frac{\beta_0 g^2}{(4\pi)^2} F_{\mu\nu}^a F^{a\mu\nu} | N \rangle}{2 m_N} &\simeq 764 MeV, \\
\frac{\langle N | m_s \bar{s} s | N \rangle}{2 m_N} &\simeq 130 MeV, \\
\frac{\langle N | m_u \bar{u} u + m_d \bar{d} d | N \rangle}{2 m_N} &\simeq 44 MeV.
\end{aligned} \tag{8.35}
$$

We will get these results in Chapter 10.

EXERCISES

8.1 **Noether theorem.** Suppose we have an action invariant under some continuous infinitesimal transformation

$$\phi^A \to \phi'^A = \phi^A + \delta\phi^A, \quad \mathcal{L} \to \mathcal{L}' = \mathcal{L} + \delta\mathcal{L} = \mathcal{L} + \partial_\mu J^\mu. \tag{P8.1}$$

Find the corresponding conserved current.

SOLUTION. One can express the current J^μ via the fields

$$\mathcal{L}(\phi^A, \partial_\mu \phi^A) \to \mathcal{L}(\phi'^A, \partial_\mu \phi'^A)$$
$$= \mathcal{L}(\phi^A, \partial_\mu \phi^A) + \frac{\partial \mathcal{L}}{\partial \phi^A} \delta \phi^A + \frac{\partial \mathcal{L}}{\partial(\partial_\mu \phi^A)}(\partial_\mu \delta \phi^A) =$$
$$\mathcal{L}(\phi^A, \partial_\mu \phi^A) + \underbrace{\left[\frac{\partial \mathcal{L}}{\partial \phi^A} \delta \phi^A - \partial_\mu \frac{\partial \mathcal{L}}{\partial(\partial_\mu \phi^A)} \right]}_{=0 \text{ via } e.o.m.} \delta \phi^A + \partial_\mu \left[\frac{\partial \mathcal{L}}{\partial(\partial_\mu \phi^A)} \delta \phi^A \right]$$
$$= \partial_\mu \left[\Pi^\mu_A \delta \phi^A \right], \tag{P8.2}$$

where

$$\Pi^\mu_A = \frac{\partial \mathcal{L}}{\partial(\partial_\mu \phi^A)}. \tag{P8.3}$$

One can equate the 2 forms of the additional term in (P8.1) and (P8.2) and get the Noether theorem

$$\partial_\mu j^\mu = \partial_\mu(\Pi^\mu_A \delta \phi^A - J^\mu) = 0. \tag{P8.4}$$

8.2 Build the symmetric energy-momentum tensor for a general Lagrangian.

a. Build the canonical energy-momentum tensor $\theta^{\mu\nu}$.

b. Build the canonical angular momentum tensor $M^{\alpha\mu\nu}$.

c. Show that one can add to $\theta^{\mu\nu}$ total divergence $\partial_\alpha B^{\alpha\mu\nu}$: $B^{\alpha\mu\nu} = -B^{\mu\alpha\nu}$ without changing the conservation laws and the charges.

d. Show that if the theory is both Lorentz and shift invariant the anti-symmetric part of $\theta^{\nu\mu}$ must be total divergence.

e. Build the symmetric energy-momentum tensor $T^{\mu\nu}$ via the **Belinfante construction**

$$T^{\mu\nu} = \theta^{\mu\nu} + \partial_\alpha B^{\alpha\mu\nu}, \tag{P8.5}$$
$$B^{\alpha\mu\nu} = \frac{1}{2}(H^{\alpha\mu\nu} + H^{\mu\nu\alpha} - H^{\nu\alpha\mu}), \quad \theta^{\mu\nu} - \theta^{\nu\mu} = -\partial_\alpha H^{\alpha\mu\nu}. \tag{P8.6}$$

Find the **Belinfante tensor** $B^{\alpha\mu\nu}$ as a function of the fields.

f. Show that the conventional angular momentum tensor $J^{\alpha\mu\nu} = x^\mu T^{\alpha\nu} - x^\nu T^{\alpha\mu}$ has the same conservation laws and charges as $M^{\alpha\mu\nu}$.

SOLUTION. We follow the presentation of (Ban01).

a. Canonical energy-momentum tensor is a Noether current responsible for shift invarance

$$x^\mu \to x^\mu + a^\mu, \quad |a^\mu| \ll 1. \tag{P8.7}$$

The fields and the Lagrangian change as follows

$$\phi^A(x) \to \phi'^A(x-a) = \phi^A + a^\alpha \partial_\alpha \phi^A, \tag{P8.8}$$

$$\mathcal{L}(x) \to \mathcal{L}(x-a) = \mathcal{L} + a^\alpha \partial_\alpha \mathcal{L} = a^\alpha \partial_\mu \mathcal{L} g^{\mu\alpha}. \tag{P8.9}$$

Therefore, the Noether currents $\theta^{\mu\alpha}$ read

$$\partial_\mu \theta^{\mu\alpha} = \partial_\mu(\Pi^\mu_A \partial^\alpha \phi^A - \mathcal{L} g^{\mu\alpha}) = 0, \quad \Pi^\mu_A = \frac{\partial \mathcal{L}}{\partial(\partial_\mu \phi^A)} \quad \Longrightarrow$$
$$\tag{P8.10}$$

$$\theta^{\mu\alpha} = \Pi^\mu_A \partial^\alpha \phi^A - \mathcal{L} g^{\mu\alpha}, \quad \partial_\mu \theta^{\mu\alpha} = 0, \tag{P8.11}$$

$$\theta^{\mu\nu} - \theta^{\nu\mu} = \Pi^\mu_A \partial^\nu \phi^A - \Pi^\nu_A \partial^\mu \phi^A. \tag{P8.12}$$

Here $\theta^{\mu\alpha}$ is a canonical e.m. tensor. Note that it is not symmetric in general.

b. Canonical angular-momentum tensor is a Noether current responsible for rotational and boost invarance

$$x^\mu \to x^\mu + w^\mu{}_\nu x^\nu, \quad |w| \ll 1, \quad w_{\mu\nu} = -w_{\nu\mu}. \tag{P8.13}$$

The fields and the Lagrangian change as follows

$$\phi^A(x) \to \left(\delta^A_B + \frac{1}{2} w_{\mu\nu} (\Sigma^{\mu\nu})^A_B\right) \phi^B (x - w^\mu{}_\nu x^\nu)$$
$$= \phi^A + \frac{1}{2} w_{\mu\nu} \left\{(\Sigma^{\mu\nu})^A_B \phi^B - x^\nu \partial^\mu \phi^A + x^\mu \partial^\nu \phi^A\right\}, \quad \text{(P8.14)}$$

$$\mathcal{L}(x) \to \mathcal{L}(x - w^\mu{}_\nu x^\nu)$$
$$= \mathcal{L} - w^\mu{}_\nu x^\nu \partial_\mu \mathcal{L} = \mathcal{L} - \frac{1}{2} w_{\mu\nu}(x^\nu \partial^\mu \mathcal{L} - x^\mu \partial^\nu \mathcal{L})$$
$$= \mathcal{L} - \frac{1}{2} w_{\mu\nu}(\partial^\mu (x^\nu \mathcal{L}) - \partial^\nu (x^\mu \mathcal{L}))$$
$$= \mathcal{L} - \frac{1}{2} w_{\mu\nu} \partial_\alpha(g^{\alpha\mu} (x^\nu \mathcal{L}) - g^{\alpha\nu} (x^\mu \mathcal{L})). \tag{P8.15}$$

Here $(\Sigma^{\mu\nu})^A_B$ are generators of boosts in the representation of the field ϕ^A. The Noether currents $M^{\alpha\mu\nu}$ read

$$\partial_\alpha M^{\alpha\mu\nu} = \partial_\alpha(\Pi^\alpha_A \left\{(\Sigma^{\mu\nu})^A_B \phi^B - x^\nu \partial^\mu \phi^A + x^\mu \partial^\nu \phi^A\right\}$$
$$+ g^{\alpha\mu} x^\nu \mathcal{L} - g^{\alpha\nu} x^\mu \mathcal{L}) = 0 \tag{P8.16}$$

$$M^{\alpha\mu\nu} = \Pi^\alpha_A (\Sigma^{\mu\nu})^A_B \phi^B + x^\mu \theta^{\alpha\nu} - x^\nu \theta^{\alpha\mu} \quad \partial_\alpha M^{\alpha\mu\nu} = 0. \tag{P8.17}$$

Here $M^{\alpha\mu\nu}$ is a canonical angular momentum tensor.

c. One can add to $\theta^{\mu\nu}$ a total divergence

$$T^{\mu\nu} = \theta^{\mu\nu} + \partial_\alpha B^{\alpha\mu\nu}, \quad \partial_\alpha B^{\alpha\mu\nu} : B^{\alpha\mu\nu} = -B^{\mu\alpha\nu}, \quad \text{(P8.18)}$$

without changing the conservation laws (P8.11)

$$\partial_\mu T^{\mu\nu} = \partial_\mu \theta^{\mu\nu} + \partial_\mu \partial_\alpha B^{\alpha\mu\nu} = \partial_\mu \theta^{\mu\nu}. \quad \text{(P8.19)}$$

The charges

$$H = \int d^3x \theta^{00}, \quad P^\mu = \int d^3x \theta^{0\mu} \quad \text{(P8.20)}$$

do not change either for fields vanishing fast enough at infinity. Indeed,

$$\int d^3x T^{0\nu} = \int d^3x (\theta^{0\nu} + \partial_0 B^{00\nu} + \partial_i B^{\alpha i\nu})$$
$$= \int d^3x \theta^{0\nu} + \int_{S^i} d^2 B^{\alpha i\nu} = P^\nu. \quad \text{(P8.21)}$$

They are conserved for fields vanishing fast enough at infinity, e.g.

$$\partial_0 \int d^3x \theta^{00} = \int d^3x \left\{ \partial_0 \theta^{00} - \partial_j \theta^{j0} + \partial_j \theta^{j0} \right\}$$
$$= \int d^3x \partial_j \theta^{j0} = \int_{S^j} d^2 \theta^{j0} = 0. \quad \text{(P8.22)}$$

d. If the theory is both Lorentz and shift invariant there is a constraint on $\theta^{\nu\mu}$ from (P8.17)

$$\partial_\alpha M^{\alpha\mu\nu} = 0 \quad \Longrightarrow \quad \partial_\alpha(\Pi_A^\alpha (\Sigma^{\mu\nu})_B^A \phi^B) = \theta^{\nu\mu} - \theta^{\mu\nu}, \quad \text{(P8.23)}$$

i.e. its antisymmetric part must be a total divergence. This constraint can be rewritten via (P8.12) as

$$\theta^{\mu\nu} - \theta^{\nu\mu} = \Pi_A^\mu \partial^\nu \phi^A - \Pi_A^\nu \partial^\mu \phi^A = -\partial_\alpha(\Pi_A^\alpha (\Sigma^{\mu\nu})_B^A \phi^B). \quad \text{(P8.24)}$$

e. Let

$$\theta^{\mu\nu} - \theta^{\nu\mu} = -\partial_\alpha H^{\alpha\mu\nu}, \quad H^{\alpha\mu\nu} = -H^{\alpha\nu\mu} = \Pi_A^\alpha (\Sigma^{\mu\nu})_B^A \phi^B. \quad \text{(P8.25)}$$

One builds the **Belinfante tensor** $B^{\alpha\mu\nu}$ as

$$B^{\alpha\mu\nu} = \frac{1}{2}(H^{\alpha\mu\nu} + H^{\mu\nu\alpha} - H^{\nu\alpha\mu})$$
$$= -\frac{1}{2}(H^{\mu\alpha\nu} + H^{\alpha\nu\mu} - H^{\nu\mu\alpha}) = -B^{\mu\alpha\nu}, \quad \text{(P8.26)}$$
$$B^{\alpha\mu\nu} - B^{\alpha\nu\mu} = \frac{1}{2}\{H^{\alpha\mu\nu} + H^{\mu\nu\alpha} - H^{\nu\alpha\mu} - H^{\alpha\nu\mu} - H^{\nu\mu\alpha} + H^{\mu\alpha\nu}\}$$
$$= H^{\alpha\mu\nu}. \quad \text{(P8.27)}$$

Then the tensor $T^{\mu\nu} = \theta^{\mu\nu} + \partial_\alpha B^{\alpha\mu\nu}$ is symmetric

$$T^{\mu\nu} - T^{\nu\mu} = -\partial_\alpha H^{\alpha\mu\nu} + \partial_\alpha(B^{\alpha\mu\nu} - B^{\alpha\nu\mu})$$
$$= -\partial_\alpha H^{\alpha\mu\nu} + \partial_\alpha H^{\alpha\mu\nu} = 0. \qquad \text{(P8.28)}$$

In fields the Belinfante tensor reads

$$B^{\alpha\mu\nu} = \frac{1}{2}\left\{\Pi_A^\alpha(\Sigma^{\mu\nu})_B^A\phi^B + \Pi_A^\mu(\Sigma^{\nu\alpha})_B^A\phi^B - \Pi_A^\nu(\Sigma^{\alpha\mu})_B^A\phi^B\right\}. \qquad \text{(P8.29)}$$

f. Expressing the canonical energy-momentum tensor $\theta^{\alpha\nu}$ via the symmetric one $T^{\alpha\nu}$ in (P8.17),

$$
\begin{aligned}
M^{\alpha\mu\nu} &= \Pi_A^\alpha(\Sigma^{\mu\nu})_B^A\phi^B + x^\mu\theta^{\alpha\nu} - x^\nu\theta^{\alpha\mu} \\
&= \Pi_A^\alpha(\Sigma^{\mu\nu})_B^A\phi^B + x^\mu(T^{\alpha\nu} - \partial_\beta B^{\beta\alpha\nu}) - x^\nu(T^{\alpha\mu} - \partial_\beta B^{\beta\alpha\mu}) \\
&= x^\mu T^{\alpha\nu} - x^\nu T^{\alpha\mu} + H^{\alpha\mu\nu} \\
&\quad + \partial_\beta\left(x^\nu B^{\beta\alpha\mu} - x^\mu B^{\beta\alpha\nu}\right) - g_\beta^\nu B^{\beta\alpha\mu} + g_\beta^\mu B^{\beta\alpha\nu} \\
&= x^\mu T^{\alpha\nu} - x^\nu T^{\alpha\mu} + \partial_\beta\left(x^\nu B^{\beta\alpha\mu} - x^\mu B^{\beta\alpha\nu}\right) \\
&= J^{\alpha\mu\nu} + \partial_\beta\left(x^\nu B^{\beta\alpha\mu} - x^\mu B^{\beta\alpha\nu}\right), \qquad &\text{(P8.30)}
\end{aligned}
$$
$$J^{\alpha\mu\nu} = x^\mu T^{\alpha\nu} - x^\nu T^{\alpha\mu}. \qquad \text{(P8.31)}$$

The conservation law reads

$$\partial_\alpha J^{\alpha\mu\nu} = \partial_\alpha(x^\mu T^{\alpha\nu} - x^\nu T^{\alpha\mu}) = T^{\mu\nu} - T^{\nu\mu} = 0. \qquad \text{(P8.32)}$$

This equality holds identically without constraints. It is equivalent to $\partial_\alpha M^{\alpha\mu\nu} = 0$ thanks to the antisymmetry of $B^{\beta\alpha\nu}$ (P8.18). The conserved charges for fields vanishing fast enough at infinity are the same for $M^{\alpha\mu\nu}$ and $J^{\alpha\mu\nu}$. Indeed,

$$\int d^3x M^{0\mu\nu} = \int d^3x(J^{0\mu\nu} + \partial_\beta(x^\nu B^{\beta0\mu} - x^\mu B^{\beta0\nu})) = \int d^3x J^{0\mu\nu}$$

$$+ \int d^3x(\partial_0(x^\nu B^{00\mu} - x^\mu B^{00\nu}) + \partial_i(x^\nu B^{i0\mu} - x^\mu B^{i0\nu}))$$

$$= \int d^3x J^{0\mu\nu} + \int_{S^i} d^2x\left(x^\nu B^{i0\mu} - x^\mu B^{i0\nu}\right) = \int d^3x J^{0\mu\nu}. \qquad \text{(P8.33)}$$

8.3 Build the Belinfante tensor for QCD.

HINT. For the vector and spinor fields the generators of the Lorentz group read $(\Sigma^{\mu\nu})_A^B A^B = (g^{\mu A}g_B^\nu - g^{\nu A}g_B^\mu)A^B$, $\Sigma^{\mu\nu}\psi = \frac{1}{4}[\gamma^\mu\gamma^\nu]\psi = \frac{-i}{2}\sigma^{\mu\nu}\psi$, $\bar\psi\Sigma^{\mu\nu} = \psi\frac{i}{2}\sigma^{\mu\nu}$.

ANSWER. $B^{\alpha\mu\nu} = \frac{1}{g^2}F^{\mu\alpha}A^\nu + \frac{1}{8}\bar\psi\{\gamma^\alpha, \sigma^{\mu\nu}\}\psi$.

8.4 For the symmetric classical QCD action

$$S = \int d^4x \left(-\frac{1}{4g^2} F^{\mu\nu a} F^a_{\mu\nu} + \bar{\psi}(i\overleftrightarrow{D} - m)\psi \right),$$ (P8.34)

$$\overleftrightarrow{D} = \frac{(\vec{\partial} - iA) + (-\overleftarrow{\partial} - iA)}{2} = \overleftrightarrow{\partial} - iA = \frac{\vec{\partial} - \overleftarrow{\partial}}{2} - iA,$$ (P8.35)

a. Derive the **symmetric energy-momentum tensor** $T^{\mu\nu}$;

b. Calculate the dilatation current. Relate it to T^μ_μ.

SOLUTION.

a. Translation invariance gives the canonical energy momentum tensor from Noether theorem

$$\theta^{\mu\alpha} = \frac{\partial \mathcal{L}}{\partial(\partial_\mu A_\nu)} \partial^\alpha A_\nu + \frac{\partial \mathcal{L}}{\partial(\partial_\mu \psi)} \partial^\alpha \psi + \frac{\partial \mathcal{L}}{\partial(\partial_\mu \bar{\psi})} \partial^\alpha \bar{\psi} - g^{\mu\alpha} \mathcal{L}$$

$$= -\frac{1}{g^2} F^{\mu\nu} \partial^\alpha A_\nu + \bar{\psi}\gamma^\mu i \overleftrightarrow{\partial}^\alpha \psi - g^{\mu\alpha} \mathcal{L}.$$ (P8.36)

One builds the symmetric energy-momentum tensor $T^{\mu\nu}$ through the canonical one and the **Belinfante tensor** $B^{\alpha\mu\nu} = -B^{\mu\alpha\nu}$.

$$T^{\mu\nu} = \theta^{\mu\nu} + \partial_\alpha B^{\alpha\mu\nu}, \quad B^{\alpha\mu\nu} = \frac{1}{g^2} F^{\mu\alpha} A^\nu + \frac{1}{8} \bar{\psi} \{\gamma^\alpha, \sigma^{\mu\nu}\} \psi.$$ (P8.37)

This modification does not change the conservation law

$$\partial_\mu T^{\mu\nu} = \partial_\mu \theta^{\mu\nu}.$$ (P8.38)

It does not change the conserved charges

$$\int dx^3 T^{0\nu} = P^\nu = (H, P^i) = \int dx^3 \left(\theta^{0\nu} + \partial_\alpha B^{\alpha 0\nu} \right)$$

$$= \int dx^3 \left(\theta^{0\nu} + \partial_0 B^{00\nu} + \partial_i B^{i0\nu} \right)$$

$$= \int dx^3 \theta^{0\nu} + \int_{S^i} dx^2 B^{i0\nu} = \int dx^3 \theta^{0\nu}$$ (P8.39)

for the fields vanishing fast enough at infinity. Using e.o.m.

$$i\hat{\partial}\psi = (m - \hat{A})\psi, \quad \bar{\psi}i\overleftarrow{\hat{\partial}} = -\bar{\psi}(m - \hat{A}),$$ (P8.40)

$$\partial^\mu F^a_{\mu\nu} = iA^{c\mu} T^c_{ab} F^b_{\mu\nu} - g^2 \bar{\psi} t^a \gamma_\nu \psi,$$ (P8.41)

one gets

$$
\begin{aligned}
\partial_\alpha B^{\alpha\mu\nu} = &-\frac{1}{g^2}(iA^{c\alpha}T^c_{ab}F^b_{\alpha}{}^\mu - g^2\bar{\psi}t^a\gamma^\mu\psi)A^{a\nu} + \frac{1}{g^2}F^{\mu\alpha}\partial_\alpha A^\nu \\
&+\frac{-1}{4i}\bar{\psi}(m-\hat{A})\sigma^{\mu\nu}\psi + \frac{1}{8}\frac{i}{2}4(\partial_\alpha\bar{\psi})(g^{\alpha\nu}\gamma^\mu - g^{\alpha\mu}\gamma^\nu)\psi \\
&+\frac{1}{4i}\bar{\psi}\sigma^{\mu\nu}(m-\hat{A})\psi + \frac{1}{8}\frac{i}{2}4\bar{\psi}(g^{\alpha\mu}\gamma^\nu - g^{\alpha\nu}\gamma^\mu)\partial_\alpha\psi,
\end{aligned}
$$

$$\text{(P8.42)}$$

$$
\begin{aligned}
\partial_\alpha B^{\alpha\mu\nu} = &\ \bar{\psi}t^a\gamma^\mu\psi A^{a\nu} + \frac{1}{g^2}F^{a\mu\alpha}(\partial_\alpha A^{a\nu} - f^{abc}A^c_\alpha A^{b\nu}) \\
&+\frac{1}{i}\frac{i}{2}A^a_\alpha\bar{\psi}t^a(g^{\alpha\mu}\gamma^\nu - g^{\alpha\nu}\gamma^\mu)\psi \\
&+\frac{1}{2}\frac{i}{2}(\partial_\alpha\bar{\psi})(g^{\alpha\nu}\gamma^\mu - g^{\alpha\mu}\gamma^\nu)\psi \\
&+\frac{1}{2}\frac{i}{2}\bar{\psi}(g^{\alpha\mu}\gamma^\nu - g^{\alpha\nu}\gamma^\mu)\partial_\alpha\psi,
\end{aligned}
$$

$$\text{(P8.43)}$$

$$
\begin{aligned}
\partial_\alpha B^{\alpha\mu\nu} = &\ \frac{1}{g^2}F^{a\mu\alpha}(\partial_\alpha A^{a\nu} - f^{abc}A^c_\alpha A^{b\nu}) \\
&+\frac{1}{2}\bar{\psi}i\overset{\leftrightarrow}{D}{}^\mu\gamma^\nu\psi - \frac{1}{2}\bar{\psi}(i\overset{\leftrightarrow}{D}{}^\nu - 2A^\nu)\gamma^\mu\psi.
\end{aligned}
$$

$$\text{(P8.44)}$$

Then the **symmetric energy-momentm tensor** $T^{\mu\nu}$ reads:

$$
\begin{aligned}
T^{\mu\nu} &= \theta^{\mu\nu} + \partial_\alpha B^{\alpha\mu\nu} \\
&= \frac{1}{g^2}F^{a\mu\alpha}F^a_\alpha{}^\nu + \frac{1}{2}\bar{\psi}(i\overset{\leftrightarrow}{D}{}^\mu\gamma^\nu + i\overset{\leftrightarrow}{D}{}^\nu\gamma^\mu)\psi - g^{\mu\nu}\mathcal{L}.
\end{aligned}
$$

$$\text{(P8.45)}$$

Whence,

$$T^\mu_\mu = m\bar{\psi}\psi, \tag{P8.46}$$

on e.o.m.

b. The Lagrangian has dimension -4, therefore

$$
\begin{aligned}
\mathcal{L}(x) \to e^{-4a}\mathcal{L}(e^{-a}x) &\simeq \mathcal{L}(x) - a(4 + x^\nu\partial_\nu)\mathcal{L}(x) \\
&= \mathcal{L}(x) - a\partial_\nu(x^\nu\mathcal{L}(x)).
\end{aligned}
$$

$$\text{(P8.47)}$$

The corresponding Noether current

$$
\begin{aligned}
d^\mu &= -\frac{\partial\mathcal{L}}{\partial(\partial_\mu A_\nu)}\delta A_\nu - \frac{\partial\mathcal{L}}{\partial(\partial_\mu\psi)}\delta\psi - \frac{\partial\mathcal{L}}{\partial(\partial_\mu\bar{\psi})}\delta\bar{\psi} - g^{\mu\nu}x_\nu\mathcal{L} \\
&= \theta^{\mu\alpha}x_\alpha + \frac{\partial\mathcal{L}}{\partial(\partial_\mu A_\nu)}A_\nu + \frac{3}{2}\frac{\partial\mathcal{L}}{\partial(\partial_\mu\bar{\psi})}\bar{\psi} + \frac{3}{2}\frac{\partial\mathcal{L}}{\partial(\partial_\mu\psi)}\psi.
\end{aligned}
$$

$$\text{(P8.48)}$$

Here

$$\frac{\partial \mathcal{L}}{\partial(\partial_\mu A_\nu)} A_\nu + \frac{3}{2}\frac{\partial \mathcal{L}}{\partial(\partial_\mu \bar{\psi})}\bar{\psi} + \frac{3}{2}\frac{\partial \mathcal{L}}{\partial(\partial_\mu \psi)}\psi$$

$$= -\frac{1}{g^2}F^{\mu\nu}A_\nu + \frac{3}{2}\left(-\frac{i}{2}\bar{\psi}\gamma^\mu\psi + \frac{i}{2}\bar{\psi}\gamma^\mu\psi\right) = -\frac{1}{g^2}F^{\mu\nu}A_\nu. \quad \text{(P8.49)}$$

Rewriting canonical energy-momentum tensor $\theta^{\mu\alpha}$ through the symmetric one $T^{\mu\alpha}$ and the Belinfante tensor $B^{\alpha\beta\gamma}$ via (P8.37) one gets

$$d^\mu = T^{\mu\nu}x_\nu - x_\nu\partial_\alpha B^{\alpha\mu\nu} - \frac{1}{g^2}F^{\mu\nu}A_\nu$$

$$= T^{\mu\nu}x_\nu - \partial_\alpha(x_\nu B^{\alpha\mu\nu}) + B^{\alpha\mu}{}_\alpha - \frac{1}{g^2}F^{\mu\nu}A_\nu$$

$$= T^{\mu\nu}x_\nu - \partial_\alpha(x_\nu B^{\alpha\mu\nu}). \quad \text{(P8.50)}$$

The total derivative does not change the conservation law

$$\partial_\mu d^\mu = T^\mu_\mu = m\bar{\psi}\psi \underset{m=0}{=} 0. \quad \text{(P8.51)}$$

It does not change the conserved charge either

$$\int d^0 d^3x = \int (T^{0\nu}x_\nu - \partial_\alpha(x_\nu B^{\alpha 0\nu}))d^3x$$

$$= \int (T^{0\nu}x_\nu - \partial_0(x_\nu B^{00\nu}) - \partial_i(x_\nu B^{i0\nu}))d^3x$$

$$= \int T^{0\nu}x_\nu d^3x - \int_{S^i} x_\nu B^{i0\nu}d^2x = \int T^{0\nu}x_\nu d^3x, \quad \text{(P8.52)}$$

if we consider the fields vanishing fast enough at infinity. Therefore, one can use

$$d^\mu = T^{\mu\nu}x_\nu \quad \Longrightarrow \quad \partial_\mu d^\mu = \partial_\mu T^{\mu\nu}x_\nu = T^{\mu\nu}\partial_\mu x_\nu = T^\mu_\mu. \quad \text{(P8.53)}$$

8.5 The **Feynman-Hellmann** theorem. Show that

a. for a continuously changing parameter λ

$$\frac{\partial E_n}{\partial \lambda} = \frac{\langle \psi_n | \frac{\partial H}{\partial \lambda} | \psi_n \rangle}{\langle \psi_n | \psi_n \rangle}, \quad H|\psi_n\rangle = E_n|\psi_n\rangle; \quad \text{(P8.54)}$$

b. for the mass m_λ one-particle state with the relativistic normalization $|p, \lambda\rangle$

$$\frac{\partial m_\lambda^2}{\partial \lambda} = \langle p, \lambda | \frac{\partial \mathcal{H}(0)}{\partial \lambda} | p, \lambda \rangle, \quad H = \int d\vec{x}\,\mathcal{H}(\vec{x}), \quad \text{(P8.55)}$$

$$\langle p, \lambda | p', \lambda \rangle = 2p_\lambda^0 (2\pi)^3 \delta^{(3)}(p - p'); \quad \text{(P8.56)}$$

c. for the renormalized quark mass m, the renormalized mass operator of this quark $\bar{\psi}(0)\psi(0)$, the mass and the relativistically normalized state of hadron H m_H^2, $|H\rangle$, and the QCD contribution to the lambda term Λ (not the QCD scale!)

$$\frac{\partial m_H^2}{\partial \ln m} = \langle H|m\bar{\psi}(0)\psi(0)|H\rangle, \quad \frac{\partial \Lambda}{\partial \ln m} = \langle 0|m\bar{\psi}(0)\psi(0)|0\rangle. \quad \text{(P8.57)}$$

SOLUTION.

a.

$$0 = (H - E_n)|\psi_n\rangle \implies$$

$$0 = (\frac{\partial H}{\partial \lambda} - \frac{\partial E_n}{\partial \lambda})|\psi_n\rangle + (H - E_n)\frac{\partial}{\partial \lambda}|\psi_n\rangle \implies$$

$$0 = \langle\psi_n|(\frac{\partial H}{\partial \lambda} - \frac{\partial E_n}{\partial \lambda})|\psi_n\rangle + \langle\psi_n|(H - E_n)\frac{\partial}{\partial \lambda}|\psi_n\rangle. \quad \text{(P8.58)}$$

Next, one uses the hermiticity of the hamiltonian to get

$$\langle\psi_n|\psi_n\rangle\frac{\partial E_n}{\partial \lambda} = \langle\psi_n|\frac{\partial H}{\partial \lambda}|\psi_n\rangle. \quad \text{(P8.59)}$$

b. The relativistic normalization gives the total volume

$$(2\pi)^3\delta^{(3)}(0) = \int d\vec{x}e^{i0\vec{x}} = V. \quad \text{(P8.60)}$$

Therefore, using the result a. one has

$$2p_\lambda^0 V\frac{\partial p_\lambda^0}{\partial \lambda} = \langle p, \lambda|\frac{\partial \int d\vec{x}\mathcal{H}(\vec{x})}{\partial \lambda}|p, \lambda\rangle$$

$$= \langle p, \lambda|\frac{\partial \int d\vec{x}e^{-i\vec{P}\vec{x}}\mathcal{H}(0)e^{i\vec{P}\vec{x}}}{\partial \lambda}|p, \lambda\rangle$$

$$= \langle p, \lambda|\int d\vec{x}\frac{\partial \mathcal{H}(0)}{\partial \lambda}|p, \lambda\rangle = V\langle p, \lambda|\frac{\partial \mathcal{H}(0)}{\partial \lambda}|p, \lambda\rangle. \quad \text{(P8.61)}$$

Here we did not differentiate the exponentials since they contain the space components of the momentum operator, which are independent of λ. Next,

$$2p_\lambda^0\frac{\partial p_\lambda^0}{\partial \lambda} = \frac{\partial (p_\lambda^0)^2}{\partial \lambda} = \frac{\partial (m_\lambda^2 + \vec{p}^{\,2})}{\partial \lambda} = \frac{\partial m_\lambda^2}{\partial \lambda}. \quad \text{(P8.62)}$$

c. Using the result of b. one has

$$\frac{\partial m_H^2}{\partial \ln m} = \langle H|\frac{\partial \mathcal{H}_{QCD}(0)}{\partial \ln m}|H\rangle = \langle H|\frac{-\partial \mathcal{L}_{QCD}(0)}{\partial \ln m}|H\rangle$$

$$= \langle H|\frac{\partial m\bar{\psi}(0)\psi(0)}{\partial \ln m}|H\rangle = \langle H|m\bar{\psi}(0)\psi(0)|H\rangle. \quad \text{(P8.63)}$$

Here one assumes the mass-independent renormalization scheme so that renormalization constants Z do not depend on m. For Λ defined as

$$\langle 0|T_{\mu\nu}(0)|0\rangle = \Lambda g_{\mu\nu}, \tag{P8.64}$$

one gets likewise

$$\frac{\partial \Lambda}{\partial \ln m} = \langle 0|m\bar{\psi}(0)\psi(0)|0\rangle. \tag{P8.65}$$

8.6 Show that

$$\frac{\partial m_H^2}{\partial \ln g} = -\frac{1}{2}\langle F_{\mu\nu}^a(0)F^{\mu\nu a}(0)\rangle_H, \qquad \frac{\partial \Lambda}{\partial \ln g} = -\frac{1}{2}\langle F_{\mu\nu}^a(0)F^{\mu\nu a}(0)\rangle_0. \tag{P8.66}$$

Here M_H is the mass of hadron H, $\langle\dots\rangle_H$ is the connected matrix element between the relativistically normalized (P8.56) one particle states with this hadron, Λ is the QCD contribution to the Lambda term (P8.64), and $\langle\dots\rangle_0 = \langle 0|\dots|0\rangle$, $\langle 0|0\rangle = 1$.

SOLUTION. Here we follow (DDZ14). For the matrix elements of the energy momentum tensor (8.33) one has

$$\langle T_{\mu\nu}\rangle_H = 2p_\mu p_\nu \quad\Longrightarrow\quad \langle T_\mu^\mu\rangle_H = 2p^2 = 2m_H^2 \tag{P8.67}$$

$$\langle T_{\mu\nu}\rangle_0 = \Lambda g_{\mu\nu} \quad\Longrightarrow\quad \langle T_\mu^\mu\rangle_0 = 4\Lambda. \tag{P8.68}$$

Here we dropped the argument $x = 0$ of the operator in $\langle T_{\mu\nu}\rangle = \langle T_{\mu\nu}(0)\rangle$. On the other hand, the trace anomaly gives

$$\langle T_\mu^\mu\rangle_H = \frac{1}{2}\frac{\beta(g)}{g}\langle F_{\mu\nu}^a F^{a\mu\nu}\rangle_H + m\langle\bar{\psi}\psi\rangle_H(1 + \gamma_m), \tag{P8.69}$$

$$\langle T_\mu^\mu\rangle_0 = \frac{1}{2}\frac{\beta(g)}{g}\langle F_{\mu\nu}^a F^{a\mu\nu}\rangle_0 + m\langle\bar{\psi}\psi\rangle_0(1 + \gamma_m). \tag{P8.70}$$

One also has Callan-Symanzik equations (P7.43) for m_H^2 and Λ_{QCD}

$$\left(\beta(g)\frac{\partial}{\partial g} - (\gamma_m + 1)\frac{\partial}{\partial \ln m} + 2\right)m_H^2 = 0, \tag{P8.71}$$

$$\left(\beta(g)\frac{\partial}{\partial g} - (\gamma_m + 1)\frac{\partial}{\partial \ln m} + 4\right)\Lambda = 0. \tag{P8.72}$$

Here both m_H^2 and Λ_{QCD} are observable. Therefore, they do not have anomalous dimensions. Using (P8.57), one cancels the dependence on mass operator and gets

$$\beta(g)\frac{\partial m_H^2}{\partial g} = -\frac{1}{2}\frac{\beta(g)}{g}\langle F_{\mu\nu}^a F^{a\mu\nu}\rangle_H, \tag{P8.73}$$

$$\beta(g)\frac{\partial \Lambda}{\partial g} = -\frac{1}{2}\frac{\beta(g)}{g}\langle F_{\mu\nu}^a F^{a\mu\nu}\rangle_0. \tag{P8.74}$$

Cancelling the β-function, one has

$$g\frac{\partial m_H^2}{\partial g} = -\frac{1}{2}\langle F_{\mu\nu}^a F^{a\mu\nu}\rangle_H, \quad g\frac{\partial \Lambda}{\partial g} = -\frac{1}{2}\langle F_{\mu\nu}^a F^{a\mu\nu}\rangle_0. \qquad (\text{P8.75})$$

8.7 Show that $T^{\mu\nu}$ has zero anomalous dimension.

Direct calculation of the trace anomaly is given in (ACD77; CDJ77; Nie77). Derivation of the symmetric energy-momentum tensor via techniques of general relativity is introduced in (Sha16). Analysis and estimates of different parts of the hadronic matrix elements of the energy-momentum tensor is given in (Ji95). Angular momentum tensor for QCD is discussed in (JM90).

FURTHER READING

[ACD77] Stephen L. Adler, John C. Collins, and Anthony Duncan. Energy-momentum-tensor trace anomaly in spin 1/2 quantum electrodynamics. *Phys. Rev. D*, 15:1712, 1977.

[Ban01] Akash Bandyopadhyay. *Improvement of stress-energy tensor using space-time symmetries*. PhD thesis, University of Illinois at Urbana-Champaign, 2001. http://research.physics.illinois.edu/Publications/theses/copies/Bandyopadhyay/.

[Bel39] F. J. Belinfante. On the spin angular momentum of mesons. *Physica*, 6(7):887–898, 1939.

[CDJ77] John C. Collins, Anthony Duncan, and Satish D. Joglekar. Trace and dilatation anomalies in gauge theories. *Phys. Rev. D*, 16:438–449, 1977.

[DDZ14] Luigi Del Debbio and Roman Zwicky. Renormalisation group, trace anomaly and Feynman–Hellmann theorem. *Phys. Lett. B*, 734:107–110, 2014.

[DEM15] John F. Donoghue and Basem Kamal El-Menoufi. QED trace anomaly, non-local Lagrangians and quantum Equivalence Principle violations. *JHEP*, 05:118, 2015.

[DGH14] J. F. Donoghue, E. Golowich, and Barry R. Holstein. *Dynamics of the standard model*, volume 2, chapter III-4, XII-3. CUP, 2014.

[GZ80] J. Gasser and A. Zepeda. Approaching the chiral limit in QCD. *Nucl. Phys. B*, 174:445, 1980.

[Ji95] Xiang-Dong Ji. A QCD analysis of the mass structure of the nucleon. *Phys. Rev. Lett.*, 74:1071–1074, 1995.

[JM90] R. L. Jaffe and Aneesh Manohar. The G(1) Problem: fact and fantasy on the spin of the proton. *Nucl. Phys. B*, 337:509–546, 1990.

[Nie77] N. K. Nielsen. The energy momentum tensor in a nonabelian quark gluon theory. *Nucl. Phys. B*, 120:212–220, 1977.

[Sha16] Ilya L. Shapiro. Covariant derivative of fermions and all that. 11, 2016. gr-qc:1611.02263.

CHAPTER 9

Adler-Bell-Jakiw Anomaly

I N this chapter we calculate the Adler-Bell-Jakiw anomaly via the method proposed by K. Fujikawa. Then we the discuss the pion's decay to two photons and the strong CP problem.

1. Fermion generating functional reads

$$Z = \int \mathcal{D}\psi' \mathcal{D}\bar{\psi}' e^{i \int dx \bar{\psi}'(i\hat{D}-m)\psi'}. \tag{9.1}$$

Consider the **chiral transformation**

$$\psi'_x = e^{i\alpha_x \gamma^5} \psi_x \simeq (1 + i\alpha_x \gamma^5)\psi_x, \quad \bar{\psi}'_x = \bar{\psi}_x e^{i\alpha_x \gamma^5} \simeq \bar{\psi}_x(1 + i\alpha_x \gamma^5). \tag{9.2}$$

Then,

$$\begin{aligned}
\bar{\psi}'(i\hat{D} - m)\psi' &= \bar{\psi}e^{i\alpha_x\gamma^5}(i\hat{\partial} + \hat{A} - m)e^{i\alpha_x\gamma^5}\psi \\
&= \bar{\psi}(-e^{i\alpha_x\gamma^5}(\hat{\partial}\alpha)\gamma^5 e^{i\alpha_x\gamma^5} + e^{i\alpha_x\gamma^5}\gamma_\mu e^{i\alpha_x\gamma^5}(i\partial^\mu + A^\mu) \\
&\quad - me^{2i\alpha_x\gamma^5})\psi \\
&= \bar{\psi}(-e^{i\alpha_x\gamma^5}(\hat{\partial}\alpha)\gamma^5 e^{i\alpha_x\gamma^5} + i\hat{\partial} + \hat{A} - me^{2i\alpha_x\gamma^5})\psi \tag{9.3} \\
&\simeq \bar{\psi}(i\hat{\partial} + \hat{A} - m)\psi - 2i\alpha m\bar{\psi}\gamma^5\psi - (\partial^\mu\alpha)\bar{\psi}\gamma_\mu\gamma^5\psi. \tag{9.4}
\end{aligned}$$

Therefore, for **constant α ($\equiv$ globally) this transformation is a symmetry of the classical massless action.**

2. However, the **integration measure changes**

$$Z = \int \mathcal{D}\psi \mathcal{D}\bar{\psi} e^{i \int dx(\bar{\psi}(i\hat{\partial}+\hat{A}-m)\psi - 2i\alpha m\bar{\psi}\gamma^5\psi - (\partial^\mu\alpha)\bar{\psi}\gamma_\mu\gamma^5\psi)} \det\frac{\delta\psi}{\delta\psi'} \det\frac{\delta\bar{\psi}}{\delta\bar{\psi}'}. \tag{9.5}$$

DOI: 10.1201/9781003272403-9

Indeed, expand the fields into the complete set of $\hat{D}$ eigenfunctions ϕ_n and $\bar{\phi}_n$

$$\hat{D}\phi_n = \lambda_n, \quad \bar{\phi}_n\overleftarrow{\hat{D}} = \lambda_n\bar{\phi}_n, \quad \int dx\bar{\phi}_n^i(x)\phi_m^j(x) = \delta^{ij}\delta_{mn}, \qquad (9.6)$$

$$\sum_n^{\infty}\bar{\phi}_n^i(x)\phi_n^j(y) = \delta^{ij}\delta(x-y). \qquad (9.7)$$

$$\psi_x^i = \sum_n c_n^i\phi_n^i, \quad \psi_x'^i = \sum_n c_n'^i\phi_n^i = \sum_{n,j}(1+i\alpha_x\gamma^5)^{ij}c_n^j\phi_n^j, \qquad (9.8)$$

$$\bar{\psi}_x^i = \sum_n \bar{c}_n^i\bar{\phi}_n^i, \quad \bar{\psi}_x'^i = \sum_n \bar{c}_n'^i\bar{\phi}_n^i = \sum_{n,j}\bar{c}_n^j\bar{\phi}_n^j(1+i\alpha_x\gamma^5)^{ji}, \qquad (9.9)$$

with Grassmann $c_n^{i=1,2,3,4}$ and $\bar{c}_n^i$. Multiplying the expansions for ψ' and $\bar{\psi}'$ with $\bar{\phi}_m$ and ϕ_m correspondingly and integrating via orthogonality relation (9.7), one gets

$$\sum_n c_n'^i\int dx\bar{\phi}_m^i\phi_n^i = \int dx\bar{\phi}_m^i\sum_{n,j}(1+i\alpha_x\gamma^5)^{ij}c_n^j\phi_n^j \qquad (9.10)$$

$$c_m'^i = \int dx\bar{\phi}_m^i\sum_{n,j}(1+i\alpha_x\gamma^5)^{ij}c_n^j\phi_n^j = \sum_{n,j}(1+A)_{mn}^{ij}c_n^j, \qquad (9.11)$$

$$A_{mn}^{ij} = i\int dx\alpha_x\bar{\phi}_m^i(\gamma^5)^{ij}\phi_n^j, \qquad (9.12)$$

$$\sum_n \bar{c}_n'^i\int dx\bar{\phi}_n^i\phi_m^i = \sum_n \bar{c}_n^j\int dx\bar{\phi}_n^j(1+i\alpha_x\gamma^5)^{ji}\phi_m^i \qquad (9.13)$$

$$\bar{c}_m'^i = \sum_n \bar{c}_n^j\int dx\bar{\phi}_n^j(1+i\alpha_x\gamma^5)^{ji}\phi_m^i = \sum_n \bar{c}_n^j(1+A)_{nm}^{ji}. \qquad (9.14)$$

Then,

$$\det\frac{\delta\psi'}{\delta\psi} = \det\frac{\delta\psi'}{\delta c'}\det\frac{\delta c'}{\delta c}\det\frac{\delta c}{\delta\psi} = \det(\phi_n^i)\det\frac{\delta c'}{\delta c}\det(\phi_n^i)^{-1}$$

$$= \det\frac{\delta c'}{\delta c} = \det(1+A)_{mn}^{ij} = \det\frac{\delta\bar{c}'}{\delta\bar{c}} = \det\frac{\delta\bar{\psi}'}{\delta\bar{\psi}}$$

$$= e^{tr\ln(1+A)} \simeq e^{tr(A)}. \qquad (9.15)$$

This expression needs regularization since

$$e^{tr(A)} = e^{tr(i\int dx\alpha_x\bar{\phi}_m^i(\gamma^5)^{ij}\phi_n^j)} = e^{i(\gamma^5)^{ii}\int dx\alpha_x\bar{\phi}_n^i\phi_n^i}$$

$$= e^{i(\gamma^5)^{ii}\delta(0)\int dx\alpha_x} = e^{0\infty} =? \qquad (9.16)$$

3. One can regularize this expression by δ-**function smearing**, i.e. summing over incomplete set of eigenfunctions in the completeness relation

$$\sum_n^N \bar{\phi}_n^i\,(x)\,\phi_n^j\,(y) = f((\frac{i\hat{D}}{M})^2)^{ij}\delta(x-y),$$

$$f(0) = 1, \quad f(\infty) = f'(0) = f''(0) = 0. \tag{9.17}$$

Taking the whole sum, i.e. as $N \to \infty$ or $M \to \infty$, one recovers the correct completeness relation. Such a regularization keeps gauge invariance in traces. As a result,

$$tr(i\int dx\alpha_x\bar{\phi}_m^i(\gamma^5)^{ij}\phi_n^j)$$

$$= \lim_{M\to\infty} tr(i\int dy\delta(x-y)\int dx\alpha_x\bar{\phi}_m^i(x)(\gamma^5)^{ij}\phi_n^j(y)$$

$$= \lim_{M\to\infty} tr(i\int dy\delta(x-y)\int dx\alpha_x(\gamma^5)^{ij}f((\frac{i\hat{D}}{M})^2)^{ji})\delta(x-y)$$

$$= \lim_{M\to\infty} i\int dy\delta(x-y)\int dx\alpha_x(\gamma^5)^{ij}f((\frac{i\hat{\partial}+\hat{A}}{M})^2)^{ji}\int \frac{d^4p}{(2\pi)^4}e^{-ip(x-y)}$$

$$= \lim_{M\to\infty} i\int \frac{d^4p}{(2\pi)^4}\int dx\alpha_x(\gamma^5)^{ij}f((\frac{\hat{p}+i\hat{\partial}+\hat{A}}{M})^2)^{ji}), \tag{9.18}$$

Here f is understood as the series expansion where the derivative acts on A or reduces to 0. Finally,

$$tr(i\int dx\alpha_x\bar{\phi}_m^i(\gamma^5)^{ij}\phi_n^j) = |p \to Mp|$$

$$= \lim_{M\to\infty} iM^4\int \frac{d^4p}{(2\pi)^4}\int dx(\gamma^5)^{ij}f((\hat{p}+\frac{i\hat{D}}{M})^2)^{ji})\alpha_x. \tag{9.19}$$

Using

$$(\hat{p}+\frac{i\hat{D}}{M})^2 = p^2 + \frac{\hat{p}i\hat{D}+i\hat{D}\hat{p}}{M} + (\frac{i\hat{D}}{M})^2 = p^2 + \frac{2(piD)}{M} + \frac{(iD)^2 + \frac{1}{2}\sigma F}{M^2}, \tag{9.20}$$

$$tr\left[\gamma^5 f((\hat{p} + \frac{i\hat{D}}{M})^2)\right] = tr[\gamma^5] f(p^2)$$

$$+ f'(p^2) tr[\gamma^5 \left(\frac{2(piD)}{M} + \frac{(iD)^2 + \frac{1}{2}\sigma F}{M^2}\right)]$$

$$+ \frac{1}{2} f''(p^2) tr[\gamma^5 \left(\frac{2(piD)}{M} + \frac{(iD)^2 + \frac{1}{2}\sigma F}{M^2}\right)^2]$$

$$+ \frac{1}{6} f'''(p^2) tr[\gamma^5 \left(\frac{2(piD)}{M}\right)^3]$$

$$+ \frac{1}{24} f^{(4)}(p^2) tr[\gamma^5 \left(\frac{2(piD)}{M}\right)^4] + o\left(\frac{1}{M^4}\right) \qquad (9.21)$$

$$= \frac{F^{\mu\nu} F^{\alpha\beta}}{8M^4} f''(p^2) tr[\gamma^5 \sigma_{\mu\nu} \sigma_{\alpha\beta}] + o\left(\frac{1}{M^4}\right)$$

$$= i\varepsilon_{\mu\nu\alpha\beta} \frac{F^{\mu\nu} F^{\alpha\beta}}{2M^4} f''(p^2) + o\left(\frac{1}{M^4}\right), \qquad (9.22)$$

we have

$$tr(i \int dx \alpha_x \bar{\phi}_m^i (\gamma^5)^{ij} \phi_n^j) = -\varepsilon_{\mu\nu\alpha\beta} \int dx F^{\mu\nu} F^{\alpha\beta} \alpha_x \int \frac{d^4 p}{2(2\pi)^4} f''(p^2)$$

$$= -\frac{\pi^2 i}{2(2\pi)^4} \varepsilon_{\mu\nu\alpha\beta} \int dx F^{\mu\nu} F^{\alpha\beta} \alpha_x. \qquad (9.23)$$

Finally,

$$\det \frac{\delta\psi'}{\delta\psi} = e^{\frac{-\pi^2 i}{2(2\pi)^4} \varepsilon_{\mu\nu\alpha\beta} \int dx F^{\mu\nu} F^{\alpha\beta} \alpha_x}, \qquad (9.24)$$

and we have

$$Z = \int \mathcal{D}\psi \mathcal{D}\bar{\psi}$$

$$\times e^{i\int dx (\bar{\psi}(i\hat{\partial} + \hat{A} - m)\psi + \alpha\{(\partial^\mu \bar{\psi}\gamma_\mu\gamma^5\psi) - 2im\bar{\psi}\gamma^5\psi + \frac{1}{16\pi^2}\varepsilon_{\mu\nu\alpha\beta}F^{\mu\nu}F^{\alpha\beta}\})}. \qquad (9.25)$$

Since α is arbitrary and we only changed the integration variables, we get

$$\partial_\mu J_A^\mu = 2im\bar{\psi}\gamma^5\psi - \frac{e^2}{16\pi^2}\varepsilon_{\mu\nu\alpha\beta}F^{\mu\nu}F^{\alpha\beta}, \quad J_A^\mu = \bar{\psi}\gamma^\mu\gamma^5\psi, \qquad (9.26)$$

where we returned to the usual field normalization $F \to eF$. In other words **even in the massless case the axial current is not** conserved but has an anomaly known as axial, chiral or Adler-Bell-Jakiv (ABJ) anomaly. In QCD one has an additional factor

$$F^{\mu\nu}F^{\alpha\beta} \to tr[t^a t^b]F^{a\mu\nu}F^{b\alpha\beta} = \frac{1}{2}F^{a\mu\nu}F^{a\alpha\beta}. \qquad (9.27)$$

$$\partial_\mu J_A^\mu = 2im\bar{\psi}\gamma^5\psi - \frac{g^2}{32\pi^2}\varepsilon_{\mu\nu\alpha\beta}F^{a\mu\nu}F^{a\alpha\beta}. \qquad (9.28)$$

EXERCISES

9.1 Show that

$$e^{i\alpha\gamma^5}\gamma_\mu e^{i\alpha\gamma^5} = \gamma_\mu. \qquad (P9.1)$$

SOLUTION. Indeed,

$$f(\alpha) = e^{i\alpha\gamma^5}\gamma_\mu e^{i\alpha\gamma^5} \implies$$
$$f(0) = \gamma_\mu, \quad f'(\alpha) = i\alpha e^{i\alpha\gamma^5}\{\gamma^5\gamma_\mu\}e^{i\alpha\gamma^5} = 0 \implies$$
$$f(\alpha) = f(0). \qquad (P9.2)$$

9.2 Show that

$$tr[\gamma^5\sigma^{\mu\nu}\sigma^{\alpha\beta}] = 4i\varepsilon^{\mu\nu\alpha\beta}, \quad \sigma^{\mu\nu} = \frac{i}{2}[\gamma^\mu, \gamma^\nu]. \qquad (P9.3)$$

9.3 Calculate the integral $\int \frac{d^4p}{(2\pi)^4}f''(p^2)$ allowing for $f(0) = 1, \quad f'(0) = f''(0) = f(\infty) = 0$.

SOLUTION.

$$\int \frac{d^4p}{(2\pi)^4}f''(p^2) = 2\pi^2 i \int \frac{p_E^3 dp_E}{(2\pi)^4}f''(-p_E^2) = \pi^2 i \int \frac{p_E^2 dp_E^2}{(2\pi)^4}f''(-p_E^2)$$
$$= -\pi^2 i \int \frac{p_E^2 df'(-p_E^2)}{(2\pi)^4} = -\pi^2 i \frac{p_E^2 df'(-p_E^2)}{(2\pi)^4}|_0^\infty$$
$$+ \pi^2 i \int f'(-p_E^2)\frac{dp_E^2}{(2\pi)^4} = -\pi^2 i \int \frac{df(-p_E^2)}{(2\pi)^4}$$
$$= -\pi^2 i \frac{f(-p_E^2)}{(2\pi)^4}|_0^\infty = \frac{i}{(4\pi)^2}. \qquad (P9.4)$$

9.4 Calculate the scale anomaly in massless QED by the Fujikawa method.

9.5 Derive the axial anomaly for the third component of the nonsinglet axial current for the u and d quarks. Show that other components of the nonsinglet axial current do not have the anomaly.

SOLUTION. Consider the EM current and the third component of the nonsinglet axial current for the u and d quarks:

$$J_{EM}^{\mu}(x) = \frac{2}{3}\bar{u}\gamma^{\mu}u - \frac{1}{3}\bar{d}\gamma^{\mu}d = \bar{Q}\gamma^{\mu}\hat{Q}_{EM}Q, \tag{P9.5}$$

$$J_A^{\mu 3}(x) = \frac{1}{2}(\bar{u}\gamma^{\mu}\gamma^5 u - \bar{d}\gamma^{\mu}\gamma^5 d) = \bar{Q}\gamma^{\mu}\gamma^5\frac{\tau^3}{2}Q, \tag{P9.6}$$

where

$$Q = \begin{pmatrix} u \\ d \end{pmatrix}, \quad \hat{Q}_{EM} = \frac{1}{3}\begin{pmatrix} 2 & \\ & -1 \end{pmatrix}, \quad \tau^3 = \begin{pmatrix} 1 & \\ & -1 \end{pmatrix}. \tag{P9.7}$$

The EM axial anomaly for $J_A^{\mu 3}$ is

$$\partial_{\mu}J_A^{\mu 3} = -\frac{e^2}{16\pi^2}\varepsilon_{\mu\nu\alpha\beta}F^{\mu\nu}F^{\alpha\beta} \times \frac{N_c}{2}\left(\left(\frac{2}{3}\right)^2 - \left(\frac{1}{3}\right)^2\right)$$

$$= -\frac{e^2}{16\pi^2}\varepsilon_{\mu\nu\alpha\beta}F^{\mu\nu}F^{\alpha\beta} \times \frac{N_c}{2}\mathrm{tr}(\hat{Q}_{EM}^2\tau^3)$$

$$= -\frac{e^2 N_c}{96\pi^2}\varepsilon_{\mu\nu\alpha\beta}F^{\mu\nu}F^{\alpha\beta}. \tag{P9.8}$$

9.6 Calculate the $\pi^0 \to \gamma\gamma$ decay width.

SOLUTION. First, this is an electromagnetic (EM) decay. Therefore, the effective Lagrangian for this decay should be CP-even and gauge invariant for both photons. Recall that π^0 has quantum numbers 0^{-+}. The only option is

$$\mathcal{L} = A\pi^0 \varepsilon_{\mu\nu\alpha\beta}F^{\mu\nu}F^{\alpha\beta}. \tag{P9.9}$$

Then

$$\langle\gamma_{k_1\lambda_1}\gamma_{k_2\lambda_2}|i\int dx A\pi_x^0\varepsilon_{\mu\nu\alpha\beta}F_x^{\mu\nu}F_x^{\alpha\beta}|\pi_p^0\rangle$$

$$= (2\pi)^4\delta^{(4)}(p - k_1 - k_2)\frac{-8iA\varepsilon_{\mu\nu\alpha\beta}k_1^{\mu}\epsilon_1^{\nu*}k_2^{\alpha}\epsilon_2^{\beta*}}{\sqrt{2E_p}\sqrt{2\omega_1}\sqrt{2\omega_2}}, \tag{P9.10}$$

and

$$\Gamma_{\pi^0\to\gamma\gamma} = \frac{|A|^2}{\pi}m_{\pi^0}^3. \tag{P9.11}$$

Next, the matrix element of the $J_A^{\mu 3}(q)$ to π_p^0 transition reads

$$
\begin{aligned}
\langle \pi_p^0 | J_A^{\mu 3}(q) | 0 \rangle &= \langle \pi_p^0 | \int dx e^{-iqx} J_A^{\mu 3}(x) | 0 \rangle \\
&= \langle \pi_p^0 | \int dx e^{-iqx} e^{i\hat{P}x} J_A^{\mu 3}(0) e^{-i\hat{P}x} | 0 \rangle \\
&= \langle \pi_p^0 | \int dx e^{i(p-q)x} J_A^{\mu 3}(0) | 0 \rangle \\
&= (2\pi)^4 \delta(p-q) \langle \pi_p^0 | J_A^{\mu 3}(0) | 0 \rangle \\
&= (2\pi)^4 \delta(p-q) i M_{J_A^\mu(q) \to \pi_p^0}.
\end{aligned}
\tag{P9.12}
$$

Here the latter matrix element is parametrized as

$$
\langle \pi_p^0 | J_A^{\mu 3}(0) | 0 \rangle = i F_{\pi^0} p^\mu, \qquad F_{\pi^0} = 93 MeV,
\tag{P9.13}
$$

where F_{π^0} is called the **pion decay constant** and can be measured in the weak pion decay. The state normalization of the pion is

$$
\langle \pi_p^0 | \pi_q^0 \rangle = 2 E_p (2\pi)^3 \delta^{(3)}(p-q),
\tag{P9.14}
$$

i.e. the factor $\dfrac{1}{\sqrt{2E_p}}$ is already taken out.

Then, consider the $J_A^{\mu 3}(q)$ to 2 photon matrix element:

$$
\begin{aligned}
\langle \gamma_{k_1 \lambda_1} \gamma_{k_2 \lambda_2} | J_A^{\mu 3}(q) | 0 \rangle &= \langle \gamma_{k_1 \lambda_1} \gamma_{k_2 \lambda_2} | \int dx e^{-iqx} J_A^{\mu 3}(x) | 0 \rangle \\
&= \langle \gamma_{k_1 \lambda_1} \gamma_{k_2 \lambda_2} | \int dx e^{-iqx} e^{i\hat{P}x} J_A^{\mu 3}(0) e^{-i\hat{P}x} | 0 \rangle \\
&= \langle \gamma_{k_1 \lambda_1} \gamma_{k_2 \lambda_2} | \int dx e^{i(k_1+k_2-q)x} J_A^{\mu 3}(0) | 0 \rangle \\
&= (2\pi)^4 \delta(k_1 + k_2 - q) \langle \gamma_{k_1 \lambda_1} \gamma_{k_2 \lambda_2} | J_A^{\mu 3}(0) | 0 \rangle \\
&= \frac{(2\pi)^4 \delta(k_1 + k_2 - q)}{\sqrt{2\omega_1}\sqrt{2\omega_2}} \epsilon_{1\nu}^* \epsilon_{2\rho}^* i M^{\mu\nu\rho}.
\end{aligned}
\tag{P9.15}
$$

One can find its imaginary part via the unitarity relation. At $q^2 = m_{\pi^0}^2$ we expect that the dominant contribution comes from the intermadiate on-shell pion:

$$2 \operatorname{Im} M_{J_A^\mu(q) \to \gamma\gamma} = \sum_n \int M^*_{n \to \gamma\gamma} M_{n \to J_A^\mu(q)} d\rho_n$$

$$\simeq \int (2\pi)^4 \delta^{(4)}(q-p) M^*_{\pi_p^0 \to \gamma\gamma} M_{\pi_p^0 \to J_A^\mu(q)} \frac{d^3 p}{(2\pi)^3 2E_p}$$

$$= 2\pi \delta^{(4)}(p^2 - m_{\pi^0}^2) M^*_{\pi_p^0 \to \gamma\gamma} M_{\pi_p^0 \to J_A^\mu(q)}$$

$$= 2\pi \delta^{(4)}(p^2 - m_{\pi^0}^2) \left(-8A\varepsilon_{\mu\nu\alpha\beta} k_1^\mu \epsilon_1^{\nu*} k_2^\alpha \epsilon_2^{\beta*}\right) F_{\pi^0} p^\mu$$

$$= \operatorname{Im} \frac{1}{p^2 - m_{\pi^0}^2 + i0} 2 \left(8A\varepsilon_{\mu\nu\alpha\beta} k_1^\mu \epsilon_1^{\nu*} k_2^\alpha \epsilon_2^{\beta*}\right) F_{\pi^0} p^\mu.$$

$$(P9.16)$$

Therefore, in the massless pion approximation

$$M_{J_A^\mu(q) \to \gamma\gamma} = \frac{1}{p^2 + i0} \left(8A\varepsilon_{\mu\nu\alpha\beta} k_1^\mu \epsilon_1^{\nu*} k_2^\alpha \epsilon_2^{\beta*}\right) F_{\pi^0} p^\mu \quad \Longrightarrow \tag{P9.17}$$

$$q_\mu M_{J_A^\mu(q) \to \gamma\gamma} = \left(8A\varepsilon_{\mu\nu\alpha\beta} k_1^\mu \epsilon_1^{\nu*} k_2^\alpha \epsilon_2^{\beta*}\right) F_{\pi^0}. \tag{P9.18}$$

On the other hand anomaly relation (P9.8) gives for the latter matrix element

$$\langle \gamma_{k_1 \lambda_1} \gamma_{k_2 \lambda_2} | q_\mu J_A^{\mu 3}(q) | 0 \rangle$$

$$= -i \langle \gamma_{k_1 \lambda_1} \gamma_{k_2 \lambda_2} | - \frac{e^2 N_c}{96\pi^2} \varepsilon_{\mu\nu\alpha\beta} \int dx F_x^{\mu\nu} F_x^{\alpha\beta} e^{-iqx} | 0 \rangle. \tag{P9.19}$$

Hence

$$q_\mu M^{\mu\nu\rho} = \frac{e^2 N_c}{96\pi^2} \times 8\varepsilon^{\nu\rho\alpha\beta} k_{1\alpha} k_{2\beta} \quad \Longrightarrow \quad A = \frac{-e^2 N_c}{96\pi^2 F_{\pi^0}}. \tag{P9.20}$$

Finally,

$$\Gamma_{\pi_0 \to \gamma\gamma} = \frac{|A|^2}{\pi} m_{\pi^0}^3 = \frac{\alpha^2 N_c^2}{9(4\pi)^3 F_{\pi^0}^2} m_{\pi^0}^3 |_{N_c=3} \simeq 7.63 eV. \tag{P9.21}$$

This result gives the dominant contribution to the experimental width $\Gamma_{\pi_0 \to \gamma\gamma} \simeq 7.82 eV$. Since the anomaly coefficient in (P9.8) is proportional to the number of colors N_c, this width is proportional to N_c^2. Hence, observation of this decay allows us to measure N_c directly.

9.7 What orbital momentum, C and P parity do the 2 photons have in the $\pi^0 \to \gamma\gamma$ decay?

SOLUTION. Photon's quantum numbers are 1^{--}. $C_{\gamma\gamma} = (-1)^2 = +1 = C_{\pi^0}$. $P_{\pi^0} = -1 = P_{\gamma\gamma} = (-1)^2(-1)^L \implies L = 2k + 1$. However, spins of 2 photons can form the following states: $1 \otimes 1 = 0 \oplus 1 \oplus 2$. Then the total momentum is $(0 \oplus 1 \oplus 2) \otimes L$. This product contains $0 = S_{\pi^0}$ only if $L = 1$.

9.8 Find the $SU(3)_F$ relation between the decay constants for $\pi^0 \to \gamma\gamma$ and $\eta \to \gamma\gamma$, the ratio of the corresponding widths, and the ratio of the cross-sections $\frac{\sigma_{e^+e^- \to \gamma\pi^0}}{\sigma_{e^+e^- \to \gamma\eta^0}}$.

ANSWER. $g_{\pi^0 \to \gamma\gamma} = \sqrt{3}g_{\eta^0 \to \gamma\gamma}$, $\frac{\Gamma_{\pi^0 \to \gamma\gamma}}{\Gamma_{\eta^0 \to \gamma\gamma}} = 3\frac{m_{\pi^0}^3}{m_{\eta^0}^3}$, $\frac{\sigma_{e^+e^- \to \gamma\pi^0}}{\sigma_{e^+e^- \to \gamma\eta^0}} =$
$3\frac{s+m_\eta^2}{\left(s-m_\eta^2\right)^3}\frac{\left(s-m_{\pi^0}^2\right)^3}{s+m_{\pi^0}^2}$.

9.9 Derive the Swinger-Dyson equation for the $\langle \partial_{\mu y} J_A^\mu(y)(\bar\psi_x \gamma^5 \psi_x)\rangle$.

SOLUTION. Repeating the derivation from the lecture for the average of the operator $\bar\psi_x \gamma^5 \psi_x$, one gets

$$\langle \bar\psi_x \gamma^5 \psi_x \rangle = \frac{1}{Z} \int \mathcal{D}\psi \mathcal{D}\bar\psi \bar\psi_x (\gamma^5 + 2i\alpha\gamma^5\gamma^5)\psi_x$$
$$\times e^{i\int dx \left(\bar\psi(i\hat\partial + \hat A - m)\psi + \alpha\{(\partial^\mu\bar\psi\gamma_\mu\gamma^5\psi) - 2im\bar\psi\gamma^5\psi + \frac{1}{16\pi^2}\varepsilon_{\mu\nu\alpha\beta}F^{\mu\nu}F^{\alpha\beta}\}\right)}$$
$$= \frac{1}{Z} \int \mathcal{D}\psi \mathcal{D}\bar\psi e^{iS} \left(2i\alpha_x\bar\psi_x\psi_x + \bar\psi_x\gamma^5\psi_x \left(1\right.\right.$$
$$\left.\left. + i\int\alpha_y\left\{\partial_y^\mu J_{\mu y}^A - 2im\bar\psi_y\gamma^5\psi_y + \frac{\varepsilon_{\mu\nu\alpha\beta}}{16\pi^2}F_y^{\mu\nu}F_y^{\alpha\beta}\right\}\right)\right).$$
$$\tag{P9.22}$$

Therefore, we have

$$\langle \bar\psi_x \gamma^5 \psi_x \partial_y^\mu J_{\mu y}^A \rangle = \langle \left\{2im\bar\psi_y\gamma^5\psi_y - \frac{\varepsilon_{\mu\nu\alpha\beta}}{16\pi^2}F_y^{\mu\nu}F_y^{\alpha\beta}\right\}\bar\psi_x\gamma^5\psi_x\rangle$$
$$- 2\langle\bar\psi_x\psi_x\rangle\delta(x - y). \tag{P9.23}$$

9.10 **Theta vacuum.** Construct a vacuum state invariant under both small and large gauge transformations up to an overall phase.

SOLUTION. Working in the temporal gauge $A^0 = 0$ in problem 2.10 from Chapter 2, we constructed the small gauge transformation operator as $e^{-\frac{i}{g}G_\alpha}$. Here G_α (P2.47) is the integral of the Gauss law (P2.38) with the function $\alpha(x)$ generating the gauge transformation $U = e^{i\alpha}$: $\alpha(\vec{x}) \to 0$ as $\vec{x} \to \infty$. We showed that the physical states vanishing under the action of the Gauss law do not change under the small gauge transformations

$$e^{-\frac{i}{g}G_\alpha}|n\rangle_{phys} = |n\rangle_{phys}. \tag{P9.24}$$

By n here we denote the vacuum state of the fields with winding number n. One may construct the large gauge transformation operator as $e^{-\frac{i}{g}\tilde{G}_\alpha}$ with

$$\tilde{G}_\alpha = \int d^3x \; \left[-(D_i^{ac}\alpha^c(\vec{x}))E^{ia}(x) + g\psi(x)^\dagger t^c \alpha^c(\vec{x})\psi(x)\right]. \quad (P9.25)$$

For small gauge transformations this definition reduces to the Gauss law after integration by parts. For the large ones it does not since the total divergence does not vanish. Problem 2.16 in Chapter 2 shows that the large gauge transformation Λ_1 with the winding number 1 changes the winding number of the state

$$e^{-\frac{i}{g}\tilde{G}_{\ln\Lambda_1}}|n\rangle_{phys} = |n+1\rangle_{phys}. \quad (P9.26)$$

Therefore, the state

$$|\theta\rangle = \sum_{n=-\infty}^{+\infty} e^{-in\theta}|n\rangle_{phys} \quad (P9.27)$$

does not change under the small gauge transformations and changes under the large gauge transformations by on overall phase

$$e^{-\frac{i}{g}\tilde{G}_{\ln\Lambda_1}}|\theta\rangle = \sum_n e^{-in\theta}|n+1\rangle_{phys} = e^{i\theta}\sum_n e^{-i(n+1)\theta}|n+1\rangle_{phys} = e^{i\theta}|\theta\rangle,$$

$$(P9.28)$$

which is not important in matrix elements. Such a vacuum is gauge invariant up to an overall phase. It is called **the theta vacuum.**

9.11 Show that matrix elements between the vacuum states with $\theta \neq 0$ can be reduced to the same matrix elements between the vacuum states with $\theta = 0$ with the QCD action S_{QCD}^θ modified by **the θ term**

$$S_{QCD}^\theta = S_{QCD}^{\theta=0} + \theta \frac{g^2}{64\pi^2} \int \varepsilon_{\mu\nu\alpha\beta} F^{a\mu\nu} F^{a\alpha\beta} d^4x. \quad (P9.29)$$

SOLUTION. An arbitrary transition amplitude in the $\theta = 0$ vacuum described by the operator X is proportional to

$$\langle\theta = 0|X|\theta = 0\rangle = \int \mathcal{D}A\mathcal{D}\psi\mathcal{D}\bar{\psi}Xe^{iS_{QCD}} = \sum_{n,m}\langle m|X|n\rangle. \quad (P9.30)$$

If $\theta \neq 0$ this expression changes to

$$\langle\theta|X|\theta\rangle = \int \mathcal{D}A\mathcal{D}\psi\mathcal{D}\bar{\psi}Xe^{iS_{QCD}} = \sum_{n,m}\langle m|X|n\rangle e^{i\theta(m-n)}$$

$$= \int \mathcal{D}A\mathcal{D}\psi\mathcal{D}\bar{\psi}Xe^{iS_{QCD}+i\theta\frac{g^2}{64\pi^2}\int \varepsilon_{\mu\nu\alpha\beta}F^{a\mu\nu}F^{a\alpha\beta}d^4x}. \quad (P9.31)$$

Here we used the result of problem 2.14 from Chapter 2 to express the difference of the winding numbers in terms of the expectation value of the topological charge operator (P2.68). Hence, the calculation in the presence of the θ vacuum with $\theta \neq 0$ differs from the calculation in the $\theta = 0$ vacuum by adding to the QCD action **the θ term.** Note that **this term is P-odd but C-even** (see problem 2.1 in Chapter 2) **and leads to CP violation**.

9.12 Show that if there is a massless quark in the spectrum, all the vacua with different winding numbers can be transformed into one another by chiral rotations.

SOLUTION. Indeed, one can do the axial transformation of the massless quark field with a constant phase α. Such a transformation only adds to the Lagrangian a contribution proportional to the axial anomaly via the change in the integration measure. Then one can choose $\alpha = -\frac{\theta}{2}$ to cancel the existing θ-term.

9.13 Show that in the theory with n_f massless quarks the modified axial current $\tilde{J}_A^\mu = J_A^\mu + \frac{n_f g^2}{32\pi^2} K^\mu$, where J_A^μ is defined in (9.26) and K^μ in (P2.13), has zero divergence and its charge transforms under large gauge transformation Λ_1 (P2.54) in the following way

$$\tilde{Q}_A \to \tilde{Q}_A' = \tilde{Q}_A + 2n_f. \tag{P9.32}$$

SOLUTION. With n_f massless quarks the $U(1)$ anomaly reads

$$\partial_\mu J_A^\mu = -\frac{n_f g^2}{32\pi^2} \varepsilon_{\mu\nu\alpha\beta} F^{a\mu\nu} F^{a\alpha\beta} = -\frac{n_f g^2}{32\pi^2} \partial_\mu K^\mu \implies \tag{P9.33}$$

$$\partial_\mu \tilde{J}_A^\mu = \partial_\mu (J_A^\mu + \frac{n_f g^2}{32\pi^2} K^\mu) = 0. \tag{P9.34}$$

Thus one gets a new conserved current, which however is not gauge invariant. Indeed, the corresponding conserved charge

$$\tilde{Q}_A = \int d^3x \tilde{J}_A^0 \tag{P9.35}$$

transforms with Λ_1 as (see problem 2.7 in Chapter 2)

$$\tilde{Q}_A \to \tilde{Q}_A' = \int d^3x J_A^0 + \frac{n_f g^2}{32\pi^2} \int d^3x K^0$$

$$+ \frac{n_f g^2}{32\pi^2} \varepsilon^{0\nu\alpha\beta} \int d^3x \left(\frac{8}{3g^2} tr[\Lambda_1 \{\partial_\nu \Lambda_1^\dagger\} \Lambda_1 \{\partial_\alpha \Lambda_1^\dagger\} \Lambda_1 \{\partial_\beta \Lambda_1^\dagger\}] \right.$$

$$\left. - \frac{8}{ig} \partial_\alpha tr[\{\partial_\nu \Lambda_1^\dagger\} \Lambda_1 A_\beta] \right)$$

$$= \tilde{Q}_A + 2n_f - \frac{n_f g^2}{32\pi^2} \frac{8}{ig} \varepsilon^{0\nu\alpha\beta} \int_{S_\infty^\alpha} d^2x tr[\{\partial_\nu \Lambda_1^\dagger\} \Lambda_1 A_\beta]$$

$$= \tilde{Q}_A + 2n_f, \tag{P9.36}$$

since $\Lambda_1 \sim -e^{i\pi \hat{n}\vec{\tau}}$ at spacial infinity (P2.56, P2.78).

9.14 Show that in the limit of n_f massless quarks the state $e^{i\tilde{Q}_A}|\theta\rangle$ transforms under gauge transformation Λ_1 in the same way as the vacuum state $|\theta + 2n_f\rangle$.

SOLUTION.

$$e^{-\frac{i}{g}\tilde{G}_{\ln}\Lambda_1}e^{i\tilde{Q}_A}|\theta\rangle = e^{i(\tilde{Q}_A+2n_f)}e^{-\frac{i}{g}\tilde{G}_{\ln}\Lambda_1}|\theta\rangle = e^{i(\theta+2n_f)}e^{i\tilde{Q}_A}|\theta\rangle. \quad (P9.37)$$

9.15 Show how the theta term depends on the Higgs Yukawa couplings to fermions.

SOLUTION. The quark masses come from the Higgs Yukawa couplings to fermions

$$\begin{aligned}
-\mathcal{L}_m &= \bar{\psi}_L'^i M^{ij}\psi_R'^j + h.c. \\
&= \bar{\psi}_L' S_L S_L^\dagger M S_R S_R^\dagger \psi_R' + \bar{\psi}_R' S_R S_R^\dagger M^\dagger S_L S_L^\dagger \psi_L' \\
&= \bar{\psi}_L m \psi_R + \bar{\psi}_R m^\dagger \psi_L,
\end{aligned} \quad (P9.38)$$

where one rotates the fields to the mass eigenstates

$$\psi_R = S_R^\dagger \psi_R', \quad \psi_L = S_L^\dagger \psi_L', \quad m = S_L^\dagger M S_R. \quad (P9.39)$$

In general,

$$S_{L,R} = e^{i\phi_{L,R}}\bar{S}_{L,R}, \quad \bar{S}_{L,R} \in SU(n_f). \quad (P9.40)$$

Since the masses are real,

$$\begin{aligned}
0 = \arg\det m &= \arg\det S_L^\dagger M S_R = \arg\left[e^{in_f(\phi_R-\phi_L)}\det M\right] \\
&= \arg\left[e^{in_f(\phi_R-\phi_L)}\det M\right] = n_f(\phi_R - \phi_L) + \arg\det M. \quad (P9.41)
\end{aligned}$$

Therefore, if in the initial theory one had the initial mass matrix M : $\arg\det M \neq 0$, then one had to perform the left and right axial transformations with different phases to get the quarks to the mass eigenstates. Such a transformation will change the initial θ vacuum or the parity violating θ-term in the lagrangian so that

$$\theta \to \bar{\theta} = \theta + \arg\det M. \quad (P9.42)$$

Indeed, this transformation can be written as

$$\begin{aligned}
\psi' &= (e^{i\phi_L}P_L + e^{i\phi_L}P_R)\psi = (e^{i\phi_L}P_L + e^{i\phi_R}P_R)\psi \\
&= \left(\frac{e^{i\phi_L}+e^{i\phi_R}}{2} + \gamma^5\frac{e^{i\phi_R}-e^{i\phi_L}}{2}\right)\psi \\
&= e^{i\frac{\phi_L+\phi_R}{2}}\left(\cos(\frac{\phi_L-\phi_R}{2}) + i\gamma^5\sin(\frac{\phi_R-\phi_L}{2})\right)\psi \\
&= e^{i\frac{\phi_L+\phi_R}{2}}e^{i\gamma^5\frac{\phi_R-\phi_L}{2}}\psi. \quad (P9.43)
\end{aligned}$$

The vector transformation $e^{i\frac{\phi_L+\phi_R}{2}}$ does not change the vacuum, while the axial rotation leads to the appearance of

$$-n_f\frac{\phi_R-\phi_L}{2}\frac{g^2}{32\pi^2}\varepsilon_{\mu\nu\alpha\beta}F^{a\mu\nu}F^{a\alpha\beta} \qquad (P9.44)$$

in the Lagrangian or equivalently to the change in θ

$$\theta \to \bar\theta = \theta - n_f(\phi_R - \phi_L) = \theta + \arg\det M. \qquad (P9.45)$$

Experimentally $\bar\theta < 10^{-10}$. **The strong CP problem is the question why $\bar\theta$ is so small** that we do not see the CP-violation in QCD.

A self-contained comprehensive introduction to anomalies one can find in (Bil08). Derivation of the anomaly via Feynman graphs is also given (PS95). Strong CP-problem is discussed in (Cre78; DGH14).

FURTHER READING

[Bil08] Adel Bilal. Lectures on anomalies. 2, 2008. hep-th:0802.0634.

[Cre78] R. J. Crewther. Effects of topological charge in gauge theories. *Acta Phys. Austriaca Suppl.*, 19:47–153, 1978.

[DGH14] J. F. Donoghue, E. Golowich, and Barry R. Holstein. *Dynamics of the standard model*, volume 2, chapter III,IV,VI, IX-4,5. CUP, 2014.

[Fuj79] Kazuo Fujikawa. Path integral measure for gauge invariant fermion theories. *Phys. Rev. Lett.*, 42:1195–1198, 1979.

[Fuj80] Kazuo Fujikawa. Path integral for gauge theories with fermions. *Phys. Rev. D*, 21:2848, 1980. [Erratum: *Phys. Rev. D 22*, 1499 (1980)].

[PS95] Michael E. Peskin and Daniel V. Schroeder. *An introduction to quantum field theory*, chapter 19.1–3. Addison-Wesley, Reading, USA, 1995.

CHAPTER 10

Continuous Non-Color Symmetries

I N this chapter we discuss the symmetries of QCD, which are not related to color. We start with the Goldstone theorem, explore the conservation laws for vector and axial symmetries of the QCD lagrangian, and build the chiral lagrangian.

1. **Linear σ-model.** Consider a real scalar n-component field $\vec{\phi} = (\phi_1, ...\phi_n)$ with the action

$$S[\vec{\phi}] = \int dx \left(\frac{1}{2}(\partial\vec{\phi})^2 - V(\vec{\phi}) \right), \quad V(\vec{\phi}) = \frac{1}{2}m^2\vec{\phi}^{\,2} + \frac{\lambda}{4}\vec{\phi}^{\,4}, \quad (10.1)$$

invariant under the $O(n)$ group. This theory known as the **linear σ-model**. For $\forall R \in O(3)$ the action is invariant under global transformations

$$\phi_i \to \phi_i' = R_{ij}\phi_j. \qquad (10.2)$$

Suppose

$$m^2 = -\mu^2, \quad \mu^2 > 0. \qquad (10.3)$$

Then the vacuum, i.e. the state with the minimal energy corresponds to the nonzero $\vec{\phi}_0$:

$$\frac{\partial V}{\partial \vec{\phi}^{\,2}} = 0 \quad \Longrightarrow \quad -\frac{1}{2}\mu^2 + \frac{\lambda}{2}\vec{\phi}^{\,2} = 0 \quad \Longrightarrow \quad \vec{\phi}_0^{\,2} = \frac{\mu^2}{\lambda} = v^2,$$

$$(10.4)$$

$$V(\vec{\phi}) = -\frac{1}{2}\mu^2\vec{\phi}^{\,2} + \frac{\lambda}{4}\vec{\phi}^{\,4} = \frac{\lambda}{4}\vec{\phi}^{\,2}\left(\vec{\phi}^{\,2} - 2v^2\right). \qquad (10.5)$$

This is a continuum of states (sphere) and the system chooses one of them as the real vacuum spontaneously. **Such a phenomenon when**

DOI: 10.1201/9781003272403-10

the theory with a symmetric Lagrangian gets a nonsymmetric vacuum state is called spontaneous symmetry breaking (or hidden symmetry). One can expand the fields around this genuine vacuum, which we chose on the n-th axis

$$\vec{\phi} = (\pi_1, ..., \pi_{n-1}, v + \sigma). \tag{10.6}$$

In terms of the new fields $\vec{\pi}$ and σ

$$
\begin{aligned}
V &= \frac{\lambda}{4}(\vec{\pi}^{\,2} + \sigma^2 + 2\sigma v + v^2)\left(\vec{\pi}^{\,2} + \sigma^2 + 2\sigma v - v^2\right) \\
&= \frac{\lambda}{4}((\vec{\pi}^{\,2} + \sigma^2 + 2\sigma v)^2 - v^4) \\
&= \sigma^2 \lambda v^2 + \lambda v\sigma(\vec{\pi}^{\,2} + \sigma^2) + \frac{\lambda}{4}(\vec{\pi}^{\,2} + \sigma^2)^2 - \lambda\frac{v^4}{4}.
\end{aligned}
\tag{10.7}
$$

$$
\mathcal{L} = \frac{1}{2}(\partial\vec{\pi})^2 + \frac{1}{2}(\partial\sigma)^2 - \frac{m_\sigma^2}{2}\sigma^2 - \lambda v\sigma(\vec{\pi}^{\,2} + \sigma^2) - \frac{\lambda}{4}(\vec{\pi}^{\,2} + \sigma^2)^2, \quad m_\sigma = v\sqrt{2\lambda}.
\tag{10.8}
$$

Since the potential on this sphere is the same as the vacuum energy, the $\vec{\pi}$ fields responsible for changing the system configuration on this sphere are massless.

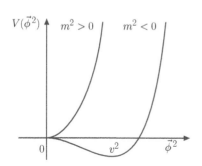
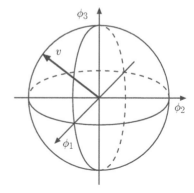

Figure 10.1 Left: Potential for the σ-model for apparent ($m^2 > 0$) and hidden ($m^2 < 0$) symmetry. Right: Configuration space for σ-model with n=3. The sphere of radius v is the degenerate vacuum in the spontaneously broken case. In the unbroken case, the only vacuum state is $\vec{\phi} = 0$.

2. **Goldstone theorem.** For every spontaneously broken continuous symmetry there appears a massless particle in the theory called the **Goldstone boson.**

PROOF. Here we follow (PS95). Suppose the theory has several fields ϕ^i and the Lagrangian

$$\mathcal{L} = K - V(\phi^i), \tag{10.9}$$

where K stands for the terms with derivatives. Let ϕ_0^i be the constant field minimizing V.

$$\frac{\partial V}{\partial \phi^i}\big|_{\phi^i=\phi_0^i} = 0 \implies$$

$$V(\phi) = V(\phi^0) + \frac{1}{2}(\phi^i - \phi_0^i)(\phi^j - \phi_0^j)\underbrace{\frac{\partial^2 V}{\partial \phi^i \partial \phi^j}\big|_{\phi^i=\phi_0^i}}_{m_{ij}^2} + \ldots \tag{10.10}$$

Continuous symmetry implies for constant fields

$$\phi^i \to \phi^i + \alpha \Delta^i(\phi)$$

$$\implies \quad V(\phi^i) = V(\phi^i + \alpha\Delta^i(\phi)) \quad \implies \quad \Delta^i(\phi)\frac{\partial V}{\partial \phi^i} = 0 \tag{10.11}$$

$$\implies \quad \frac{\partial \Delta^i(\phi)}{\partial \phi^j}\frac{\partial V}{\partial \phi^i} + \Delta^i(\phi)\frac{\partial^2 V}{\partial \phi^i \partial \phi^j} = 0 \tag{10.12}$$

$$\implies \quad \frac{\partial \Delta^i(\phi)}{\partial \phi^j}\frac{\partial V}{\partial \phi^i}\big|_{\phi^i=\phi_0^i} + \Delta^i(\phi)\frac{\partial^2 V}{\partial \phi^i \partial \phi^j}\big|_{\phi^i=\phi_0^i} = 0$$

$$\implies \quad \Delta^i(\phi_0)\frac{\partial^2 V}{\partial \phi^i \partial \phi^j}\big|_{\phi^i=\phi_0^i} = 0. \tag{10.13}$$

Hence, there are 2 options. First, $\Delta^i(\phi_0) = 0 \implies$ the vacuum is symmetric, i.e. turnes into itself under the symmetry transformation. Or, in other words the symmetry is not broken. Second, $\Delta^i(\phi_0) \neq 0 \implies$ the vacuum is not symmetric, i.e. it changes under the symmetry transformation. Or, in other words the symmetry is broken spontaneously. As a result, $\Delta^i(\phi_0)$ is an eigenvector of m_{ij}^2 with 0 eigenvalue, i.e. the **theory has a massless particle**.

However, quantum corrections may spoil this proof. Therefore, one has to consider the effective action to take them into account. In the absence of the external sources,

$$\frac{\delta \Gamma[\phi^{cl}]}{\delta \phi_i^{cl}(x)} = -J^i(x) = 0. \tag{10.14}$$

Translationally invariant solution to this equation is the vacuum field, which does not depend on x. Therefore,

$$\Gamma[\phi^{cl}] = -(VT) \times V_{eff}(\phi^{cl}) \quad \implies \quad \frac{\partial V_{eff}(\phi^{cl})}{\partial \phi_i^{cl}} = 0. \tag{10.15}$$

Here $V_{eff}(\phi^{cl})$ is called the **effective potential** and the classical field defined by this equation gives the ground state of the system as in the classical case. Repeating the classical arguments for V_{eff} one gets that the spontaneously broken symmetry demands that $\frac{\partial^2 V_{eff}}{\partial \phi_i^{cl} \partial \phi_j^{cl}}$ have a zero eigenvalue. For one field

$$\frac{\delta^2 \Gamma}{\delta \phi_z^{cl} \delta \phi_x^{cl}} = i D^{-1}(z,x), \quad \overset{Fourier}{\Longrightarrow} \quad i \tilde{D}^{-1}(p) = -i(p^2 - m^2(p)), \quad (10.16)$$

and the mass can be found from the equation

$$i \tilde{D}^{-1}(p)|_{p^2=m^2} = 0, \quad (10.17)$$

i.e. the mass is the pole of the propagator. For a set of fields ϕ_i

$$\int dx e^{ip(x-z)} \frac{\delta^2 \Gamma}{\delta \phi_{i\,z}^{cl} \delta \phi_{j\,x}^{cl}} \quad \rightarrow \quad -i \begin{pmatrix} p^2 - m_1^2(p) & & 0 \\ & \cdots & \\ 0 & & p^2 - m_n^2(p) \end{pmatrix}. \quad (10.18)$$

after diagonalization. Taking $p = 0$ leads to vanishing of all terms with derivatives in the effective action, where only the potential terms remain. They do not depend on the coordinates. Therefore,

$$\int dx \frac{\delta^2 \Gamma}{\delta \phi_{i\,z}^{cl} \delta \phi_{j\,x}^{cl}} = - \int dx \frac{\delta^2 V}{\delta \phi_{i\,z}^{cl} \delta \phi_{j\,x}^{cl}} = -(VT) \frac{\partial^2 V_{eff}}{\partial \phi_i^{cl} \partial \phi_j^{cl}} \quad (10.19)$$

$$\underset{\substack{after \\ diagonalization}}{\Longrightarrow} \quad -i \begin{pmatrix} m_1^2(0) & & 0 \\ & \cdots & \\ 0 & & m_n^2(0) \end{pmatrix}. \quad (10.20)$$

Hence, if $\frac{\partial^2 V_{eff}}{\partial \phi_i^{cl} \partial \phi_j^{cl}}$ has a zero eigenvalue then the theory has a massless particle. Indeed, it means that for sme i $i\tilde{D}_i^{-1}(p)|_{p^2=0} = 0$.

One often says that if a symmetry group G is spontaneously broken to a subgroup H then the Goldstone bosons live in the coset G/H and the number of the Goldstone fields is $\dim G/H$, or the number of broken generators. This means that the Goldstone bosons are proportional to the generators of G that break the vacuum, i.e. all generators which do not form H. These generators give the infinitesimal transformations $\Delta^i(\phi_0) \neq 0$ above, while the generators forming H give $\Delta^i(\phi_0) = 0$.

3. **Symmetries in QCD.** Consider the classical QCD Lagrangian with quarks of 2 lightest flavors u and d. Its fermion part reads

$$\mathcal{L} = \bar{u} i \hat{D} u + \bar{v} i D v - m_u \bar{u} u - m_d \bar{d} d = \bar{Q}(i\hat{D} - M)Q, \quad (10.21)$$

$$Q = \begin{pmatrix} u \\ d \end{pmatrix}, \quad M = \begin{pmatrix} m_u & 0 \\ 0 & m_d \end{pmatrix}. \quad (10.22)$$

At the energy scale much larger than the quark masses one can neglect them. Then the left and right fields decouple.

$$\mathcal{L} \to \bar{Q}i\hat{D}Q = \bar{Q}_L i\hat{D}Q_L + \bar{Q}_R i\hat{D}Q_R, \quad Q_L = P_L Q, \qquad (10.23)$$

$$Q_R = P_R Q, \quad P_R = \frac{1+\gamma^5}{2}, \quad P_L = \frac{1-\gamma^5}{2}. \qquad (10.24)$$

Since $\hat{D}$ is flavor-blind the **Lagrangian has the global symmetry**

$$Q \to Q' = U_V Q \simeq (1 + i\alpha + i\alpha^a \frac{1}{2}\tau^a)Q, \qquad (10.25)$$

$$U_V = e^{i\alpha + i\alpha^a \frac{1}{2}\tau^a} \in U(2) = SU(2) \times U(1), \qquad (10.26)$$

where τ are the σ-matrices acting in the space of flavor indices. The $SU(2)$ **part of this symmetry is called the isospin symmetry**. In fact, since left and right fields decouple one can transform them separately. Therefore, the symmetry group is $SU(2)_L \times U(1)_L \times SU(2)_R \times U(1)_R$. (Often called also $SU(2)_V \times U(1)_V \times SU(2)_A \times U(1)_A$, although $SU(2)_A \times U(1)_A$ is not actually a group.) According to Noether theorem these symmetries must lead to conserved currents. They are

$$j^\mu = \frac{\partial \mathcal{L}}{\partial(\partial_\mu Q)}(i\alpha + i\alpha^a \frac{1}{2}\tau^a)Q \quad \Longrightarrow$$

$$j_L^\mu = \bar{Q}_L \gamma^\mu Q_L, \quad j_L^{a\mu} = \bar{Q}_L \frac{1}{2}\tau^a \gamma^\mu Q_L, \qquad (10.27)$$

$$j_R^\mu = \bar{Q}_R \gamma^\mu Q_R, \quad j_R^{a\mu} = \bar{Q}_R \frac{1}{2}\tau^a \gamma^\mu Q_R, \qquad (10.28)$$

$$\partial_\mu j_{R,L}^{(a)\mu} = 0 \text{ at the classical level.} \qquad (10.29)$$

Alternatively, one can say that the Lagrangian is symmetric under the vector

$$Q \to U_V Q \quad \Longrightarrow \quad \bar{Q} \to \bar{Q} U_V^\dagger, \qquad (10.30)$$

and axial rotations

$$Q \to U_A Q, \quad \Longrightarrow \quad \bar{Q} \to Q^\dagger U_A^\dagger \gamma^0 = \bar{Q} U_A \quad U_A = e^{i\beta\gamma^5 + i\frac{\beta^a}{2}\tau^a \gamma^5}. \qquad (10.31)$$

Then, changing variables in the generating functional one gets for the **vector rotations**

$$Z_0 = \int DQD\bar{Q} e^{i\int dx \bar{Q}(i\hat{D}-M)Q}$$

$$= \int DQD\bar{Q} \det U_V \det U_V^\dagger e^{i\int dx(\bar{Q}i\hat{D}Q - \bar{Q}U_V^\dagger MU_V Q + \bar{Q}U_V^\dagger i(\hat{\partial}U_V)Q)}$$

$$= \int DQD\bar{Q} e^{i\int dx(\bar{Q}(i\hat{D}-M)Q - i\alpha^a \bar{Q}[M,\frac{1}{2}\tau^a]Q + \bar{Q}(i(\partial_\mu \alpha) + i(\partial_\mu \alpha^a)\frac{1}{2}\tau^a)\gamma^\mu Q)}$$

$$= \int DQD\bar{Q}$$

$$\times e^{i\int dx(\bar{Q}(i\hat{D}-M)Q - i\alpha^a \{\bar{Q}[M,\frac{1}{2}\tau^a]Q + i\partial_\mu(\bar{Q}\frac{1}{2}\tau^a\gamma^\mu Q)\} - i\alpha\partial_\mu(\bar{Q}\gamma^\mu Q))}.$$

$$(10.32)$$

Since α are arbitrary one gets equations for 2 currents

$$\partial_\mu(\bar{Q}\gamma^\mu Q) = 0, \quad \partial_\mu(\bar{Q}\frac{1}{2}\tau^a\gamma^\mu Q) = i\bar{Q}[M,\frac{1}{2}\tau^a]Q. \quad (10.33)$$

- The first one is the **baryon number current** and the corresponding conserved charge is the **baryon number**

$$j_B^\mu = \bar{Q}\gamma^\mu Q = \bar{Q}_L \gamma^\mu Q_L + \bar{Q}_R \gamma^\mu Q_R, \quad \partial_\mu j_B^\mu = 0, \quad (10.34)$$

$$B = \int d^3x \bar{Q}\gamma^0 Q, \quad \frac{dB}{dt} = 0. \quad (10.35)$$

Baryon number current is conserved in QCD exactly for any quark masses. (But is violated by the Electroweak interactions.)

$$B = \int d^3x \bar{Q}\gamma^0 Q = \int d^3x (u^\dagger u + d^\dagger d)$$

$$= \int d^3x \int \frac{d^3p}{(2\pi)^3\sqrt{2E_p}} \sum_s (a_{sp}u_{up}^s e^{-ipx} + b_{sp}^\dagger v_{up}^s e^{ipx})^\dagger$$

$$\times \int \frac{d^3k}{(2\pi)^3\sqrt{2E_k}} \sum_\lambda (a_{\lambda k}u_{uk}^\lambda e^{-ikx} + b_{\lambda k}^\dagger v_{uk}^\lambda e^{ikx}) + (u \leftrightarrow d)$$

$$= \left| \begin{array}{c} u_p^{s\dagger}u_p^\lambda = v_p^{s\dagger}v_p^\lambda = 2E_p\delta^{s\lambda} \\ u_{\vec{p}}^{s\dagger}v_{-\vec{p}}^\lambda = v_{\vec{p}}^{s\dagger}u_{-\vec{p}}^\lambda = 0 \end{array} \right|$$

$$= \sum_s \int \frac{d^3p}{(2\pi)^3}(a_{up}^{s\dagger}a_{up}^s - b_{up}^{s\dagger}b_{up}^s + \{b_{up}^{s\dagger}b_{up}^s\}) + (u \leftrightarrow d)),$$

$$(10.36)$$

$$\implies \langle B \rangle = \#quarks - \#antiquarks + const. \quad (10.37)$$

- The second is the **isospin vector current** and the corresponding conserved charges (neglecting the quark masses) are the **isospin components**

$$j_V^{a\mu} = \bar{Q}\frac{\tau^a}{2}\gamma^\mu Q = \bar{Q}_L\frac{\tau^a}{2}\gamma^\mu Q_L + \bar{Q}_R\frac{\tau^a}{2}\gamma^\mu Q_R, \quad \partial_\mu j_V^{a\mu} = 0,$$

$$I^a = \int d^3x\,\bar{Q}\frac{\tau^a}{2}\gamma^0 Q, \quad \frac{dI^a}{dt} = 0. \tag{10.38}$$

The isospin vector symmetry is violated by the quark mass difference, which at the Λ_{QCD} scale gives $\sim 1\%$ accuracy, and is also broken by EM and weak interactions. For the unbroken symmetry

$$\frac{dI^a}{dt} = i[H_{QCD}, I^a] = 0 \implies$$

$$for \quad \forall \quad bound \quad state \quad |h\rangle : H_{QCD}|h\rangle = E|h\rangle \tag{10.39}$$

$$H_{QCD}e^{iI^a\alpha^a}|h\rangle = e^{iI^a\alpha^a}H_{QCD}|h\rangle = Ee^{iI^a\alpha^a}|h\rangle, \tag{10.40}$$

i.e. the bound states related by vector isospin rotations have the same masses. This phenomenon is observed as close to degenerate mass spectrum of particles within one multiplet corresponding to a fixed I^2, e.g. for $\pi^{\pm,0}$ and p, n, etc. So one concludes that the vector isospin symmetry is apparent rather than hidden and the vacuum is symmetric under $SU(2)_V$ transformations

$$U|0\rangle = |0\rangle, \quad U \in SU(2)_V. \tag{10.41}$$

Next, changing variables in the generating functional for the **axial rotations** one gets

$$Z_0 = \int \mathcal{D}Q\mathcal{D}\bar{Q}e^{i\int dx\bar{Q}(i\hat{D}-M)Q}$$

$$= \int \mathcal{D}Q\mathcal{D}\bar{Q}(\det U_A)^{-2}e^{i\int dx(\bar{Q}i\hat{D}Q-\bar{Q}U_AMU_AQ+\bar{Q}U_Ai(\hat{\partial}U_A)Q)}$$

$$= \int \mathcal{D}Q\mathcal{D}\bar{Q}(\det U_A)^{-2}$$

$$\times e^{i\int dx(\bar{Q}(i\hat{D}-M)Q-2i\beta\bar{Q}\gamma^5MQ-i\beta^a\bar{Q}\gamma^5\{M,\frac{\tau^a}{2}\}Q+\bar{Q}(i(i\hat{\partial}\beta)+i(i\hat{\partial}\beta^a)\frac{\tau^a}{2})\gamma^5Q)}. \tag{10.42}$$

Here **the Jacobian gives the Adler-Bell-Jakiw anomaly**

$$(\det U_A)^{-2} = e^{-i\frac{g^2}{16\pi^2}\int dx\varepsilon^{\alpha\beta\mu\nu}F_{\alpha\beta}^b F_{\mu\nu}^c(\beta^a tr[\frac{1}{2}\tau^a t^b t^c]+\beta tr[t^b t^c])}$$

$$= e^{-i\frac{g^2}{32\pi^2}\int dx\varepsilon^{\alpha\beta\mu\nu}F_{\alpha\beta}^a F_{\mu\nu}^a\beta}. \tag{10.43}$$

Then,

$$Z_0 = \int DQ D\bar{Q} e^{i \int dx \bar{Q}(i\hat{D}-M)Q} e^{i \int dx (-i\beta^a \{\bar{Q}\gamma^5 \{M, \frac{\tau^a}{2}\}Q + i\partial_\mu (\bar{Q}\frac{\tau^a}{2}\gamma^\mu\gamma^5 Q)\})}$$

$$\times e^{i \int dx (-i\beta \{i\partial_\mu (\bar{Q}\gamma^\mu\gamma^5 Q) + 2\bar{Q}\gamma^5 MQ - \frac{2g^2}{32\pi^2} \varepsilon^{\alpha\beta\mu\nu} F^a_{\alpha\beta} F^a_{\mu\nu}\})}. \tag{10.44}$$

Since β are arbitrary one gets equations for 2 currents

$$\partial_\mu (\bar{Q}\gamma^\mu\gamma^5 Q) = 2i\bar{Q}\gamma^5 MQ - \frac{g^2}{16\pi^2} \varepsilon^{\alpha\beta\mu\nu} F^a_{\alpha\beta} F^a_{\mu\nu}, \tag{10.45}$$

$$\partial_\mu (\bar{Q}\frac{1}{2}\tau^a\gamma^\mu\gamma^5 Q) = i\bar{Q}\gamma^5 \{M, \frac{1}{2}\tau^a\}Q. \tag{10.46}$$

- The first one is the **axial current, which is not conserved due to the ABJ anomaly** even in the massless limit.

$$j^\mu_A = \bar{Q}\gamma^\mu\gamma^5 Q = \bar{Q}_R\gamma^\mu Q_R - \bar{Q}_L\gamma^\mu Q_L, \tag{10.47}$$

$$\partial_\mu j^\mu_A \xrightarrow[M\to 0]{} -\frac{g^2}{16\pi^2} \varepsilon^{\alpha\beta\mu\nu} F^a_{\alpha\beta} F^a_{\mu\nu}, \tag{10.48}$$

- The second one is the **isospin axial current** and the corresponding charge is the **axial isospin charge**

$$j^{a\mu}_A = \bar{Q}\frac{\tau^a}{2}\gamma^\mu\gamma^5 Q = \bar{Q}_R\frac{\tau^a}{2}\gamma^\mu Q_R - \bar{Q}_L\frac{\tau^a}{2}\gamma^\mu Q_L, \tag{10.49}$$

$$\partial_\mu j^{\mu a}_A \xrightarrow[M\to 0]{} 0, \quad I^a_A = \int d^3x \bar{Q}\frac{\tau^a}{2}\gamma^0\gamma^5 Q, \quad \frac{dI^a_A}{dt} = 0. \tag{10.50}$$

This current is conserved in the massless limit. Since γ^5 changes parity in the quark bilinears, the axial transformations like (10.40) mix bound states with different parity. As a result the QCD spectrum should be degenerate in parity in the massless quark limit, i.e. each state should have a partner with the opposite parity and almost the same mass. However, this phenomenon is not observed. Therefore, it is considered that this symmetry also called **chiral symmetry is broken spontaneously**.

4. **Spontaneous breaking of chiral symmetry** of $SU(2)_R \times SU(2)_L$ to $SU(2)_V$ implies then that there are 3 almost Goldstone bosons, i.e. almost massless particles in the theory with the parity of the corresponding broken currents. They are identified with the π meson triplet, whose masses are much smaller than the masses of other hadrons. In the massless quark limit, we expect therefore the pion masses to vanish. In the massless quark limit also the existence of the massless pseudoscalar π meson means that adding a very soft pion to any state will not change its energy = mass but will change its parity. Therefore, the spectrum of the theory is indeed P-degenerate.

Spontaneous symmetry breaking indicates that there is an **order parameter with a nonzero vacuum expectation value** like $\vec{\phi}_n$ in the σ-model above. We can build this order parameter from quark colorless bilinear operators, which are P-even Lorentz scalars and do not change under transformations $U \in SU(2)_V$. Indeed, a vacuum expectation value of such an operator O

$$\langle 0|O|0\rangle = \langle 0|U^\dagger U O U^\dagger U|0\rangle = \left| U|0\rangle = |0\rangle, \quad U O U^\dagger = O \right| = \langle 0|O|0\rangle \tag{10.51}$$

can be nonzero since it does not change under an $SU(2)_V$ rotation since the vacuum is invariant under them. For an $SU(2)_V$ noninvariant operator such a v.e.v. is to be 0 since U is arbitrary. The simplest of such $SU(2)_V$ invariant operators is

$$\bar{Q}Q = \bar{Q}_L Q_R + \bar{Q}_R Q_L. \tag{10.52}$$

Its vacuum expectation value is called the **quark condensate,**

$$\langle 0|\bar{Q}Q|0\rangle = \langle 0|\bar{u}u + \bar{d}d|0\rangle \underset{\substack{SU_V(2) \\ invariance}}{\simeq} 2\langle 0|\bar{u}u|0\rangle, \tag{10.53}$$

$$\langle 0|\bar{u}u|0\rangle|_{2GeV,\overline{MS}} \simeq -(250MeV)^3. \tag{10.54}$$

Operator $\bar{Q}Q$ is invariant under the vacuum symmetry group $SU(2)_V$ but transforms nontrivially under the full group $SU(2)_R \times SU(2)_L$. Vacuum expectation values of such operators are called the **order parameters**. Returning to the σ-model discussed above $SU(2)_V$ transformations are analogous to the rotations about the n-th axis of the field (10.6). They do not change the vacuum expectation value of ϕ_n . By contrast, $SU(2)_A$ transformations are analogous to all other rotations, which change ϕ_n. Then ϕ_n is the order parameter and $\phi_j, j \neq n$ have zero v.e.v.'s.

5. **Chiral effective Lagrangian.** At low energy the quark and gluon d. o. f. are not good since the charge is large and the theory is confined. However, we know that in the massless quark limit the Lagrangian has $SU(2)_R \times SU(2)_L$ invariance and the vacuum has only the $SU(2)_V$ invariance. Therefore, one can try to build an effective theory with fields living in the coset space $(SU(2)_R \times SU(2)_L)/SU(2)_V$ and demand that the effective action be invariant under the $SU(2)_R \times SU(2)_L$. Then we expect that correlation functions calculated via these effective fields are equal to the QCD ones. Such a theory is known as the **chiral effective**

theory. Indeed, for

$$g \in SU(2)_R \times SU(2)_L \implies$$
$$g = (V_R, V_L) = (V_R V_L^\dagger V_L, V_L) = (V_R V_L^\dagger, 1)(V_L, V_L) = h\gamma, \quad (10.55)$$
$$\gamma = (V_L, V_L) \in SU(2)_V,$$
$$h = (V_R V_L^\dagger, 1) = (U, 1) \in (SU(2)_R \times SU(2)_L)/SU(2)_V, \quad (10.56)$$

i.e. any transformation from the full group can be decomposed into a diagonal $SU(2)_V$ transformation and a transformation from the coset. The $SU(2)_V$ transformations keep the vacuum invariant, while transformations from the coset do not. It means that the latter transformations can be built from the spontaneously broken generators, which are proportional to the Goldstone bosons' fields. Action of any $j \in SU(2)_R \times SU(2)_L$ on the $h \in (SU(2)_R \times SU(2)_L)/SU(2)_V$ reads

$$jh = (V_R, V_L)(U, 1) = (V_R U, V_L) = (V_R U V_L^\dagger, 1)(V_L, V_L) = h'\gamma'$$
$$\implies \quad U \to V_R U V_L^\dagger. \quad (10.57)$$

We will use U in the exponential parametrization, i.e. consider the $SU(2)$ matrix

$$U = e^{\frac{i\vec{\tau}\vec{\pi}}{F}} = e^{\frac{i}{F}\begin{pmatrix} \pi^3 & \pi^1 - i\pi^2 \\ \pi^1 + i\pi^2 & -\pi^3 \end{pmatrix}} = e^{\frac{i}{F}\begin{pmatrix} \pi^0 & \sqrt{2}\pi^+ \\ \sqrt{2}\pi^- & -\pi^0 \end{pmatrix}} \in SU(2).$$
$$(10.58)$$

The Lagrangian, symmetric under such transformations

$$\mathcal{L}_{eff} = \mathcal{L}(U, \partial U, \partial^2 U...) = \underbrace{f_0(U)}_{\sim O(1)} + \underbrace{f_1(U) \times \partial^2 U}_{\sim O(p^2)} + \underbrace{f_2(U) \times \partial_\mu U \times \partial^\mu U}_{\sim O(p^2)} + ...$$
$$(10.59)$$

Here one assumes arbitrary functions $f_i(U)$ including $U^\dagger$, and $\times$ means that the matrix indices are somehow contracted. The first term may be dropped since it can be rotated to a constant taking $V_L = U$ and $V_R = 1$

$$f_0(U) = f_0(V_R U V_L^\dagger) = |V_L = U| = f_0(1). \quad (10.60)$$

The second term can be absorbed into the third one by partial integration

$$f_1(U) \times \partial^2 U = -\partial^\mu U \times \partial_\mu f_1(U) = -f_1'(U) \times \partial^\mu U \times \partial_\mu U. \quad (10.61)$$

The third term may be rewritten as

$$f_2(U) \times \partial_\mu U \times \partial^\mu U = \tilde{f}_2(U) \times \Delta_\mu \times \Delta^\mu, \quad \Delta_\mu = U^\dagger \partial_\mu U. \quad (10.62)$$

Then, Δ_μ is invariant under global $SU(2)_R$

$$\Delta_\mu \to U^\dagger V_R^\dagger \partial_\mu (V_R U) = U^\dagger V_R^\dagger V_R \partial_\mu (U) = \Delta_\mu, \quad (10.63)$$

and the invariance of the Lagrangian under these rotations demands that $\tilde{f}_2$ be independent of U, i.e. constant. Under global $SU(2)_L$

$$\Delta_\mu \to V_L U^\dagger \partial_\mu (U V_L^\dagger) = V_L \Delta_\mu V_L^\dagger, \qquad (10.64)$$

i.e. Δ_μ transforms under the adjoint ($\equiv$ vector $\equiv H^1$) representation of $SU(2)_L$. Therefore, the product $\Delta_\mu \times \Delta^\mu$ transforms as

$$H^1 \otimes H^1 = H^0 \oplus H^1 \oplus H^2. \qquad (10.65)$$

So there is only one scalar component in this product (trace), which one takes into the Lagrangian. Indeed,

$$tr[\Delta_\mu \Delta^\mu] = tr[(U^\dagger \partial_\mu U) U^\dagger \partial^\mu U] = -tr[(\partial_\mu U^\dagger) \partial^\mu U]. \qquad (10.66)$$

As a result, the **lowest order** $\sim O(p^2)$ **chiral effective Lagrangian**

$$\mathcal{L}_{eff} = \frac{F^2}{4} tr[(\partial_\mu U^\dagger) \partial^\mu U], \qquad (10.67)$$

where the coefficient is taken to get the standard normalization of the kinetic term. Indeed,

$$
\begin{aligned}
\mathcal{L}_{eff} &= \frac{F^2}{4} tr[(\partial_\mu U^\dagger) \partial^\mu U] \simeq \frac{F^2}{4} tr[(\partial_\mu (-\frac{i \vec{\tau}\vec{\pi}}{F} - \frac{\vec{\pi}^2}{2F^2} + \frac{i \vec{\tau}\vec{\pi}}{6F^3}\vec{\pi}^2 + ...) \\
&\times \partial^\mu (\frac{i \vec{\tau}\vec{\pi}}{F} - \frac{\vec{\pi}^2}{2F^2} - \frac{i \vec{\tau}\vec{\pi}}{6F^3}\vec{\pi}^2 + ...)] \\
&= \frac{F^2}{4} \left(2\frac{\partial_\mu \vec{\pi}}{F}\frac{\partial^\mu \vec{\pi}}{F} + 2\frac{\partial_\mu \vec{\pi}^2}{2F^2}\frac{\partial^\mu \vec{\pi}^2}{2F^2} - 4\frac{\partial_\mu \vec{\pi}}{F}\frac{\partial^\mu (\vec{\pi}\vec{\pi}^2)}{6F^3} + ... \right) \\
&= \frac{(\partial_\mu \vec{\pi})^2}{2} + \frac{(\partial_\mu \vec{\pi}^2)^2}{8F^2} - \frac{(\partial_\mu \vec{\pi})\partial^\mu (\vec{\pi}\vec{\pi}^2)}{6F^2} + ... \\
&= \frac{(\partial_\mu \vec{\pi})^2}{2} + \frac{(\vec{\pi}\partial_\mu \vec{\pi})(\vec{\pi}\partial^\mu \vec{\pi}) - \vec{\pi}^2(\partial_\mu \vec{\pi})(\partial^\mu \vec{\pi})}{6F^2} + ... \qquad (10.68)
\end{aligned}
$$

6. **Explicit chiral symmetry breaking by quark masses.** Mass matrix M explicitly brakes both $SU(2)_R \times SU(2)_L$ and $SU(2)_V$. One may try to introduce it into the effective Lagrangian as a perturbation. It can be done in the following way. Suppose that M were not a constant matrix but an external matrix-valued field transforming as

$$M \to V_R M V_L^\dagger. \qquad (10.69)$$

Then the mass term in the QCD Lagrangian would be invariant under $SU(2)_R \times SU(2)_L$ rotations

$$
\begin{aligned}
-\mathcal{L}_M = \bar{Q} M Q &= \bar{Q}_R M Q_L + \bar{Q}_L M^\dagger Q_R \\
&\to \bar{Q}_R V_R^\dagger V_R M V_L^\dagger V_L Q_L + \bar{Q}_L V_L^\dagger V_L M^\dagger V_R^\dagger V_R Q_R = -\mathcal{L}_M. \\
&\qquad (10.70)
\end{aligned}
$$

Here we took into account that $M = M^\dagger$ for (10.22). As a result, one would have an extra invariant structure to build the effective Lagrangian from

$$\mathcal{L}_{eff} = \mathcal{L}(U, \partial U, \partial^2 U..., M) = \underbrace{\mathcal{L}_{eff}^{M=0}}_{\sim O(1)} + \underbrace{f(U) \times M}_{\sim O(M)} + ... \quad (10.71)$$

Hence, we build the effective Lagrangian in double expansion in both momentum and M. Such expansion is called **the chiral expansion** of the effective Lagrangian. In the leading (0) power of the momentum there are only 2 terms with mass

$$tr[MU^\dagger] \quad \text{and} \quad tr[UM^\dagger]. \quad (10.72)$$

Since strong interactions respect parity, the effective Lagrangian is to be invariant under P transformation $\vec{\pi} \to -\vec{\pi}$, i.e. $U \to U^\dagger$. Therefore,

$$\mathcal{L}_{eff} = \frac{F^2}{4} tr[(\partial_\mu U^\dagger)\partial^\mu U] + \frac{F^2 B}{2} tr[M(U^\dagger + U)] \quad (10.73)$$

$$\simeq \frac{(\partial_\mu \vec{\pi})^2}{2} + \frac{F^2 B}{2} tr[M](2 - \frac{\vec{\pi}^2}{F^2}) + ...$$

$$= \frac{(\partial_\mu \vec{\pi})^2}{2} + (m_u + m_d)B(F^2 - \frac{\vec{\pi}^2}{2}) + ..., \quad (10.74)$$

where the effective Lagrangian now depends on 2 constants F and B. One can at once see that

$$m_\pi^2 = B(m_u + m_d). \quad (10.75)$$

Equating the vacuum expectation of the symmetry violating terms calculated in the QCD Lagrangian and in the effective one, we get to the leading order

$$\langle 0|\frac{\partial \mathcal{L}_{QCD}}{\partial m_u}|0\rangle = \langle 0|\frac{\partial \mathcal{L}_{eff}}{\partial m_u}|0\rangle \implies -\langle 0|\bar{u}u|0\rangle = BF^2. \quad (10.76)$$

As a result, we come to the **Gell-Mann - Oakes - Renner formula**

$$m_\pi^2 = -\frac{\langle 0|\bar{u}u|0\rangle}{F^2}(m_u + m_d), \quad (10.77)$$

where we see that the pion mass originates from both the explicit ($m_u + m_d$) and the spontaneous ($\langle 0|\bar{u}u|0\rangle$) chiral symmetry breaking. For the on-shell pion's momentum

$$p^2 = m_\pi^2 = B(m_u + m_d), \quad (10.78)$$

i.e. in the chiral expansion the mass term is to be counted as 2 powers of momentum.

7. **Source fields.** To calculate matrix elements of different composite operators one can add source fields of these operators to the Lagrangian and then differentiate w.r.t. them. Let us write the quark part of the QCD Lagrangian in the presence of the source fields as

$$\mathcal{L}_{ext} = \mathcal{L}_{quark} + \bar{Q}\gamma^{\mu}(v_{\mu} + a_{\mu}\gamma^5)Q - \bar{Q}(s - ip\gamma^5)Q \tag{10.79}$$
$$= \bar{Q}_L\gamma^{\mu}(i\partial_{\mu} + v_{\mu} - a_{\mu})Q_L + \bar{Q}_R\gamma^{\mu}(i\partial_{\mu} + v_{\mu} + a_{\mu})Q_R$$
$$- \bar{Q}_R(s + ip)Q_L - \bar{Q}_L(s - ip)Q_R, \tag{10.80}$$

$$v_{\mu} = \vec{v}_{\mu}\frac{\vec{\tau}}{2}, \quad a_{\mu} = \vec{a}_{\mu}\frac{\vec{\tau}}{2}, \quad s = s^0 + \vec{s}\vec{\tau}, \quad p = p^0 + \vec{p}\vec{\tau}, \tag{10.81}$$

$$s^{\dagger} = s, \quad p^{\dagger} = p, \tag{10.82}$$

$$s^0(x) = \frac{m_u + m_d}{2} + ..., \quad s^3(x) = \frac{m_u - m_d}{2} + ... \tag{10.83}$$

Demanding the P-invariance in the extended Lagrangian $\mathcal{L}_{ext}$ we see that under P transformation $p \to -p$, $s \to s$. Hence, under P

$$\chi = 2B(s^0 + i\vec{p}\vec{\tau}) \to \chi^{\dagger}, \quad \tilde{\chi} = 2B(p^0 - i\vec{s}\vec{\tau}) \to -\tilde{\chi}^{\dagger}. \tag{10.84}$$

Choosing the external fields such that under the chiral rotations

$$s + ip = \frac{1}{2B}[\chi + i\tilde{\chi}] \to V_R(s + ip)V_L^{\dagger}, \tag{10.85}$$

one can see that their contribution to the Lagrangian is invariant

$$\bar{Q}_R(s + ip)Q_L + h.c. \to \bar{Q}_R V_R^{\dagger} V_R(s + ip)V_L^{\dagger}V_L Q_L + h.c.$$
$$\to \bar{Q}_R(s + ip)Q_L + h.c. \tag{10.86}$$

and

$$\chi \to V_R\chi V_L^{\dagger}, \quad \tilde{\chi} \to V_R\tilde{\chi}V_L^{\dagger}. \tag{10.87}$$

Therefore, one can take only the following structures invariant both under P and chiral transformations into the chiral Lagrangian in the second order

$$tr[\chi U^{\dagger} + \chi^{\dagger}U], \quad \text{and} \quad tr[\tilde{\chi}U^{\dagger} - \tilde{\chi}^{\dagger}U] \equiv 0. \tag{10.88}$$

To incorporate the other source fields into the effective Lagrangian one assumes that the source fields transform under the local $SU(2)_L \times SU(2)_R$ so as

$$v_{\mu} - a_{\mu} \to V_L(v_{\mu} - a_{\mu})V_L^{\dagger} - i(\partial_{\mu}V_L)V_L^{\dagger}, \tag{10.89}$$
$$v_{\mu} + a_{\mu} \to V_R(v_{\mu} + a_{\mu})V_R^{\dagger} - i(\partial_{\mu}V_R)V_R^{\dagger}. \tag{10.90}$$

Then the extended Lagrangian would be invariant under $SU(2)_R \times SU(2)_L$ rotations

$$\bar{Q}_L V_L^\dagger \gamma^\mu (i\partial_\mu + V_L(v_\mu - a_\mu)V_L^\dagger - i\,(\partial_\mu V_L)\,V_L^\dagger)V_L Q_L$$

$$+ \bar{Q}_R V_R^\dagger \gamma^\mu (i\partial_\mu + V_R(v_\mu + a_\mu)V_R^\dagger - i\,(\partial_\mu V_R)\,V_R^\dagger)V_R Q_R$$

$$= \bar{Q}_L \gamma^\mu (i\partial_\mu + v_\mu - a_\mu)Q_L + \bar{Q}_R \gamma^\mu (i\partial_\mu + v_\mu + a_\mu)Q_R. \qquad (10.91)$$

In the effective Lagrangian, we had the transformation law for the global symmetry

$$U \to V_R U V_L^\dagger, \qquad (10.92)$$

which remains for the coordinate-dependent V_R and V_L. However, to keep this law for the derivative terms one has to promote the derivative to the covariant derivative

$$\partial_\mu U \to D_\mu U = \partial_\mu U + iU(v_\mu - a_\mu) - i(v_\mu + a_\mu)U. \qquad (10.93)$$

Then under the local rotation

$$D_\mu U \to \partial_\mu(V_R U V_L^\dagger) + iV_R U V_L^\dagger(V_L(v_\mu - a_\mu)V_L^\dagger - i\,(\partial_\mu V_L)\,V_L^\dagger)$$

$$- i(V_R(v_\mu + a_\mu)V_R^\dagger - i\,(\partial_\mu V_R)\,V_R^\dagger)V_R U V_L^\dagger$$

$$= V_R(\partial_\mu U)V_L^\dagger + iV_R U V_L^\dagger(V_L(v_\mu - a_\mu)V_L^\dagger) - i(V_R(v_\mu + a_\mu)V_R^\dagger)V_R U V_L^\dagger$$

$$= V_R(\partial_\mu U + iU(v_\mu - a_\mu) - i(v_\mu + a_\mu)U)V_L^\dagger = V_R(D_\mu U)V_L^\dagger. \quad (10.94)$$

Finally, the leading term in the chiral Lagrangian reads

$$\mathcal{L}_{eff}[U, a, v, s, p] = \frac{F^2}{4} tr[(D_\mu U^\dagger)D^\mu U + \chi U^\dagger + U\chi^\dagger]. \qquad (10.95)$$

This term contains 2 constants F and B. In the next order, there are ~10 constants, in the next to next about 100. The central claim of the theory is that at low energy when only pions are active d.o.f. the **effective theory gives alternative representation of the QCD generating functional**

$$Z[A, Q, a, v, s, p] = \int \mathcal{D}Q\mathcal{D}\bar{Q}\mathcal{D}A e^{i\int dx \mathcal{L}_{ext}[A,Q,a,v,s,p]}$$

$$= const \int \mathcal{D}\pi e^{i\int dx \mathcal{L}_{eff}[\pi,a,v,s,p]}. \qquad (10.96)$$

8. **Pion decay constant** F. Consider the axial vector correlator

$$\langle 0|\bar{Q}(x)\gamma^\mu\gamma^5\frac{T^a}{2}Q(x)|\pi^i(p)\rangle$$

$$= \langle 0|\frac{\delta e^{i\int dx \mathcal{L}_{ext}[A,Q,a,v,s,p]}}{i\delta a_\mu^a(x)}|\pi^i(p)\rangle|_{a=v=p=0,s=M}$$

$$= \langle 0|\frac{\delta e^{i\int dx \mathcal{L}_{eff}[\pi,a,v,s,p]}}{i\delta a_\mu^a(x)}|\pi^i(p)\rangle|_{a=v=p=0,s=M}$$

$$= \frac{F^2}{4}\langle 0|tr[2(-2i\frac{T^a}{2})(-\frac{i\partial_\mu\vec{\pi}\vec{\tau}}{F})]|\pi^i(p)\rangle = -F\langle 0|\partial_\mu\pi^a|\pi^i(p)\rangle$$

$$= -F\delta^{ai}\partial_\mu e^{-ipx} = F\delta^{ai}ip_\mu e^{-ipx}. \tag{10.97}$$

Finally,

$$\langle 0|\bar{Q}(0)\gamma^\mu\gamma^5\frac{T^a}{2}Q(0)|\pi^i(p)\rangle = F\delta^{ai}ip_\mu, \tag{10.98}$$

which gives a definition for the pion decay constant F. This matrix element can be directly measured in the weak pion decay. We met it already in the $\pi^0 \to \gamma\gamma$ decay in the previous section. Here the normalization is assumed

$$\langle\pi^i(p)|\pi^j(k)\rangle = [a_p^i a_k^{j\dagger}] = 2E_p\delta^{ij}(2\pi)^3\delta(\vec{p}-\vec{k}), \tag{10.99}$$

$$\pi^i(x) = \int\frac{dp}{2E_p(2\pi)^3}(a_p^i e^{-ipx} + a_p^{i\dagger}e^{ipx}). \tag{10.100}$$

EXERCISES

10.1 Show that

$$\bar{u}u = \bar{u}_L u_R + \bar{u}_R u_L, \tag{P10.1}$$

$$\bar{u}\gamma^5 u = \bar{u}_L u_R - \bar{u}_R u_L, \tag{P10.2}$$

$$\bar{u}\hat{D}u = \bar{u}_L\hat{D}u_L + \bar{u}_R\hat{D}u_R, \tag{P10.3}$$

$$(\bar{u}_R v_L)^\dagger = \bar{v}_L u_R, \tag{P10.4}$$

$$(\bar{u}_R\gamma^\mu v_R)^\dagger = \bar{v}_R\gamma^\mu u_R, \tag{P10.5}$$

$$\bar{Q}\gamma^\mu(v_\mu + a_\mu\gamma^5)Q = \bar{Q}_R\gamma^\mu(v_\mu + a_\mu)Q_R + \bar{Q}_L\gamma^\mu(v_\mu - a_\mu)Q_L, \tag{P10.6}$$

$$\bar{Q}(s - ip\gamma^5)Q = \bar{Q}_R(s + ip)Q_L + \bar{Q}_L(s - ip)Q_R, \tag{P10.7}$$

$$[\bar{Q}_R(s + ip)Q_L]^\dagger = \bar{Q}_L(s - ip)Q_R. \tag{P10.8}$$

SOLUTION. Indeed,

$$P_L^2 = P_L, \quad P_L + P_R = 1, \tag{P10.9}$$

$$P_R\gamma^\mu = \gamma^\mu P_L, \quad P_L^\dagger = P_L, \quad \gamma^0 P_L \gamma^0 = P_R, \tag{P10.10}$$

$$u_L = P_L u, \quad \bar{u}_L = u^\dagger P_L \gamma^0 = u^\dagger \gamma^0 P_R = \bar{u} P_R, \tag{P10.11}$$

$$P_R\gamma^5 = \gamma^5 P_R = P_R, \quad P_L\gamma^5 = \gamma^5 P_L = -P_L. \tag{P10.12}$$

Then,

$$\bar{u}u = \bar{u}(P_L + P_R)u = \bar{u}(P_L^2 + P_R^2)u = \bar{u}_L u_R + \bar{u}_R u_L, \tag{P10.13}$$

$$\bar{u}\hat{D}u = \bar{u}\hat{D}(P_L + P_R)u = \bar{u}\hat{D}(P_L^2 + P_R^2)u$$
$$= \bar{u}(P_R\hat{D}P_L + P_L\hat{D}P_R)u = \bar{u}_L\hat{D}u_L + \bar{u}_R\hat{D}u_R, \tag{P10.14}$$

$$(\bar{u}_R v_L)^\dagger = (\bar{u}P_L v)^\dagger = v^\dagger P_L^\dagger \gamma^{0\dagger} u$$
$$= v^\dagger \gamma^0 \gamma^0 P_L \gamma^0 u = \bar{v}\gamma^0 P_L \gamma^0 u = \bar{v}P_R u = \bar{v}_L u_R, \tag{P10.15}$$

$$(\bar{u}_L \gamma^\mu v_L)^\dagger = (\bar{u}\gamma^\mu P_L v)^\dagger = v^\dagger P_L \gamma^{\mu\dagger} \gamma^{0\dagger} u$$
$$= v^\dagger \gamma^0 \gamma^0 P_L \gamma^0 \gamma^0 \gamma^{\mu\dagger} \gamma^0 u = \bar{v}P_R \gamma^\mu u = \bar{v}_L \gamma^\mu u_L. \tag{P10.16}$$

$$\bar{Q}\gamma^\mu(v_\mu + a_\mu\gamma^5)Q = \bar{Q}(P_L + P_R)\gamma^\mu(v_\mu + a_\mu\gamma^5)Q$$
$$= \bar{Q}P_L\gamma^\mu(v_\mu + a_\mu\gamma^5)Q + \bar{Q}P_R\gamma^\mu(v_\mu + a_\mu\gamma^5)Q$$
$$= \bar{Q}_R\gamma^\mu(v_\mu + a_\mu)Q_R + \bar{Q}_L\gamma^\mu(v_\mu - a_\mu)Q_L. \tag{P10.17}$$

$$\bar{Q}(s - ip\gamma^5)Q = \bar{Q}(P_L + P_R)(s - ip\gamma^5)Q$$
$$= \bar{Q}P_L(s - ip\gamma^5)Q + \bar{Q}P_R(s - ip\gamma^5)Q$$
$$= \bar{Q}_R(s + ip)Q_L + \bar{Q}_L(s - ip)Q_R. \tag{P10.18}$$

$$\left[\bar{Q}_R(s + ip)Q_L\right]^\dagger = \left[\bar{Q}P_L(s + ip)Q\right]^\dagger = Q^\dagger \gamma^0 \gamma^0 (s - ip) P_L^\dagger \gamma^{0\dagger} Q$$
$$= \bar{Q}P_R(s - ip)Q = \bar{Q}_L(s - ip)Q_R. \tag{P10.19}$$

10.2 Find how many continuous symmetries were broken in the σ-model from the lecture.

SOLUTION.

$$R \in O(n) \implies R^T = R^{-1} \quad R = e^A \implies$$

$$A^T = -A \implies \dim O(n) = \frac{n^2 - n}{2}. \tag{P10.20}$$

The symmetry group was broken from $O(n)$ to $O(n-1)$. Therefore, the number of broken symmetries, or broken generators is

$$\dim O(n)/O(n-1) = \dim O(n) - \dim O(n-1)$$
$$= \frac{n(n-1)}{2} - \frac{(n-2)(n-1)}{2} = n-1, \quad \text{(P10.21)}$$

and there are $n-1$ Goldstone bosons.

10.3 Show that the Goldstone fields in the σ-model from the lecture live in the coset.

SOLUTION. Indeed, near the vacuum one can write

$$\{\pi_1, ..\pi_{n-1}, v\} = \{0, ..0, v\}R + O(\beta^2), \quad R = e^{iT^i\beta^i}, \quad \text{(P10.22)}$$

where T^i are broken generators. We have exactly $n-1$ ones, rotating in the $i-n$ planes. If one rotates first in the $i-j$ plane, $i \neq n \neq j$, the vacuum remains unchanged. Therefore,

$$\{\pi_1, ..\pi_{n-1}, v\} + O(\beta^2) = \{0, ..0, v\}R = \{0, ..0, 1\}rR,$$
$$r = e^{it^i\alpha^i} \in O(n-1), \quad \text{(P10.23)}$$

where t^i are unbroken generators. So the Goldstone fields are defined up to the action of the element of the unbroken subgroup from the left, i.e. they belong to the left coset $O(n)/O(n-1)$.

10.4 Find relation between V_R and V_L for vector and axial transformations.

SOLUTION.

$$Q = Q_L + Q_R \to V_L Q_L + V_R Q_R = \frac{1}{2}(V_L + V_R + (V_R - V_L)\gamma^5)Q \quad \text{(P10.24)}$$

vector (diagonal) subgroup $V_L = V_R = V \implies Q \to VQ$, (P10.25)

axial rotations $V_L^\dagger = V_R = e^{i\alpha} \implies$

$$Q \to (\cos\alpha + i\sin\alpha\gamma^5)Q = e^{i\alpha\gamma^5}Q. \quad \text{(P10.26)}$$

10.5 Show that the 3-rd component of the vector isospin current counts the number of u quarks – the number of d quarks:

$$I^3 = \frac{1}{2}(\#uquarks - \#\bar{u}antiquarks) - \frac{1}{2}(\#dquarks - \#\bar{d}antiquarks). \quad \text{(P10.27)}$$

10.6 Which operators can have a nonzero vacuum expectation value: $\bar{Q}\gamma^5 Q$, $(\bar{Q}\gamma^5 Q)^2$, $Q^\dagger Q$, $\bar{Q}\gamma^\mu Q$, $(\bar{Q}\gamma^\mu Q)^2$, $\bar{Q}\gamma^\mu\gamma^\nu F_{\mu\nu}Q$, $\bar{Q}\tau^a Q$, $(\bar{Q}\tau^a Q)^2$? Why?

10.7 Show that $tr[\tilde{\chi}U^\dagger - \tilde{\chi}^\dagger U] \equiv 0$.

SOLUTION. Indeed

$$tr[\tilde{\chi}U^\dagger - \tilde{\chi}^\dagger U] = 2Btr[(p^0 - i\vec{s}\vec{\tau})e^{-i\frac{\vec{\pi}\vec{\tau}}{F}} - e^{i\frac{\vec{\pi}\vec{\tau}}{F}}(p^0 + i\vec{s}\vec{\tau})]$$

$$= 2Btr[p^0(e^{-i\frac{\vec{\pi}\vec{\tau}}{F}} - e^{i\frac{\vec{\pi}\vec{\tau}}{F}}) - i\vec{s}\vec{\tau}(e^{-i\frac{\vec{\pi}\vec{\tau}}{F}} + e^{i\frac{\vec{\pi}\vec{\tau}}{F}})]$$

$$= 2Btr[-2ip^0\frac{\vec{\pi}\vec{\tau}}{|\vec{\pi}|}\sin(\frac{\sqrt{\vec{\pi}^2}}{F}) - 2i\vec{s}\vec{\tau}\cos(\frac{\sqrt{\vec{\pi}^2}}{F})] = 0.$$

$$\text{(P10.28)}$$

10.8 Calculate the EM current in terms of pion fields in the leading order.

SOLUTION. In the quark Lagrangian, the EM interaction reads

$$\mathcal{L}_{Q,EM} = e\bar{Q}\gamma^\mu q A_\mu Q, \quad q = \begin{pmatrix} \frac{2}{3} & \\ & -\frac{1}{3} \end{pmatrix} = \frac{1 + 3\tau^3}{6}, \quad \text{(P10.29)}$$

where A is the photon field and the e is the electron charge. One gets this contribution in the extended Lagrangian taking $v = eqA$

$$J_\mu^{EM} = \frac{\partial \mathcal{L}_{ext}[A^a, Q, a, v = eqA, s, p]}{\partial e A_\mu(x)}\Big|_{a=0,A=0,s+ip=M}$$

$$= \frac{\partial \mathcal{L}_{eff}[\pi, a, v = eqA, s, p]}{\partial e A_\mu(x)}\Big|_{a=0,A=0,s+ip=M}$$

$$= \frac{F^2}{4}tr[(-iqU^\dagger + iU^\dagger q)(\partial_\mu U) + (\partial_\mu U^\dagger)(iUq - iqU)] =$$

$$= \frac{F^2}{4}tr[(-iq\frac{-i\vec{\pi}\vec{\tau}}{F} + i\frac{-i\vec{\pi}\vec{\tau}}{F}q)(\partial_\mu\frac{i\vec{\pi}\vec{\tau}}{F})$$

$$+ (-\partial_\mu\frac{i\vec{\pi}\vec{\tau}}{F})(i\frac{i\vec{\pi}\vec{\tau}}{F}q - iq\frac{i\vec{\pi}\vec{\tau}}{F})]$$

$$= \frac{i}{4}tr[(-q\vec{\pi}\vec{\tau} + \vec{\pi}\vec{\tau}q)(\partial_\mu\vec{\pi}\vec{\tau}) + (-\partial_\mu\vec{\pi}\vec{\tau})(-\vec{\pi}\vec{\tau}q + q\vec{\pi}\vec{\tau})]$$

$$= \frac{i}{4}tr[(-\tau^3\vec{\pi}\vec{\tau} + \vec{\pi}\vec{\tau}\tau^3)(\partial_\mu\vec{\pi}\vec{\tau})] = -\frac{1}{2}\varepsilon^{i3j}((\partial_\mu\pi^j)\pi^i - (\partial_\mu\pi^i)\pi^j)$$

$$= (\partial_\mu\pi^2)\pi^1 - (\partial_\mu\pi^1)\pi^2$$

$$= (\partial_\mu i\frac{\pi^+ - \pi^-}{\sqrt{2}})\frac{\pi^+ + \pi^-}{\sqrt{2}} - (\partial_\mu\frac{\pi^+ + \pi^-}{\sqrt{2}})i\frac{\pi^+ - \pi^-}{\sqrt{2}}$$

$$= i(\partial_\mu\pi^+)\pi^- - i(\partial_\mu\pi^-)\pi^+. \quad \text{(P10.30)}$$

10.9 Consider the $SU(3)_R \times SU(3)_L$ chiral Lagrangian to find the mass relations for the octet of pseudo-Goldsone pseudoscalar bosons.

SOLUTION. The octet is parametrized as

$$U = e^{2i\frac{\phi^a t^a}{F}} = e^{i\frac{\phi^a \lambda^a}{F}}, \quad \frac{\phi^a \lambda^a}{\sqrt{2}} = \begin{pmatrix} \frac{\pi^0}{\sqrt{2}} + \frac{\eta_8}{\sqrt{6}} & \pi^+ & K^+ \\ \pi^- & -\frac{\pi^0}{\sqrt{2}} + \frac{\eta_8}{\sqrt{6}} & K^0 \\ K^- & \bar{K}^0 & -\frac{2}{\sqrt{6}}\eta_8 \end{pmatrix}$$

$$\text{(P10.31)}$$

and the corresponding mass matrix reads

$$M = \begin{pmatrix} m_u & & \\ & m_d & \\ & & m_s \end{pmatrix}. \qquad \text{(P10.32)}$$

Then the mass comes from the following term in the chiral Lagrangian

$$\mathcal{L}_{eff,M} = \frac{F^2 B}{2} tr[M(U^\dagger + U)] = -B\frac{(m_u + m_d)}{2}\pi_0^2 - B\frac{(m_u - m_d)}{\sqrt{3}}\pi_0\eta_8$$
$$- \frac{1}{3}B\frac{(m_u + m_d + 4m_s)}{2}\eta_8^2 - B(m_d + m_s)K^0\bar{K}^0$$
$$- B(m_u + m_s)K^-K^+ - B(m_u + m_s)\pi^-\pi^+ + \dots \qquad \text{(P10.33)}$$

Therefore, neglecting the isospin violating effect $m_u - m_d \simeq 0$,

$$m_\pi^2 = B(m_u + m_d), \quad m_{\eta_8}^2 = B\frac{m_u + m_d + 4m_s}{3}, \qquad \text{(P10.34)}$$
$$m_{K^0}^2 = m_{\bar{K}^0}^2 = B(m_d + m_s), \quad m_{K^+}^2 = m_{K^-}^2 = B(m_u + m_s). \qquad \text{(P10.35)}$$

Then, we have the **Gell-Mann – Okubo formula**

$$m_{\eta_8}^2 = \frac{1}{3}(4m_K^2 - m_\pi^2) \simeq (565 MeV)^2, \qquad \text{(P10.36)}$$

while we have $m_\eta = 549 MeV$.

10.10 Estimate the ratio

$$\frac{\hat{m}}{m_s}, \quad \hat{m} = \frac{m_u + m_d}{2} \qquad \text{(P10.37)}$$

from the chiral Lagrangian.

SOLUTION. One has

$$\frac{\hat{m}}{m_s} = \frac{m_\pi^2}{2m_K^2 - m_\pi^2} \simeq \frac{1}{24}. \qquad \text{(P10.38)}$$

10.11 Taking into account the electromagnetic interaction estimate the quark mass ratios

$$\frac{m_u}{m_d}, \quad \frac{m_s}{m_d}, \qquad \text{(P10.39)}$$

and assuming that the $SU(3)$ symmetry breaking is majorly due to the large s quark mass estimate the absolute values of the quark masses.

SOLUTION. We follow (Wei77). Electromagnetic interaction of d and s quarks is the same. Therefore, assuming that the mass correction from it is zero for neutral mesons and the same for the charged ones, one gets

$$m_{\pi^0}^2 = B(m_u + m_d), \tag{P10.40}$$

$$m_{\pi^+}^2 = B(m_u + m_d) + C_{em}, \tag{P10.41}$$

$$m_{K^0}^2 = B(m_d + m_s), \tag{P10.42}$$

$$m_{K^+}^2 = B(m_u + m_s) + C_{em}. \tag{P10.43}$$

Therefore

$$\frac{m_u}{m_d} = \frac{m_{K^+}^2 + 2m_{\pi^0}^2 - m_{K^0}^2 - m_{\pi^+}^2}{m_{K^0}^2 - m_{K^+}^2 + m_{\pi^+}^2} = 0.56, \tag{P10.44}$$

$$\frac{m_s}{m_d} = \frac{m_{K^+}^2 + m_{K^0}^2 - m_{\pi^+}^2}{m_{K^0}^2 - m_{K^+}^2 + m_{\pi^+}^2} = 20.2. \tag{P10.45}$$

Assuming that in one multiplet the mass splitting is due to the heavy s quark,

$$m_H = m_0 + m_s N_s, \tag{P10.46}$$

where N_s is the number of strange quarks in the hadron H, one estimates

$$m_s = \frac{m_\phi - m_\rho}{2} \simeq 120\,MeV. \tag{P10.47}$$

Using the ratios obtained above, one gets

$$m_d \simeq 6\,MeV, \quad m_u \simeq 3.4\,MeV. \tag{P10.48}$$

10.12 The chiral Lagrangian for the interaction of goldstone mesons with baryons (On the basis of (DGH14)).

a. Show that one can also define the coset space $(SU(2)_R \times SU(2)_L)/SU(2)_V$ so as

$$g \in SU(2)_R \times SU(2)_L \implies g = (V_R', V_L') = (\xi, \xi^\dagger)(\eta, \eta) = h\gamma, \tag{P10.49}$$

$$\gamma = (\eta, \eta) \in SU(2)_V, \quad h = (\xi, \xi^\dagger) \in (SU(2)_R \times SU(2)_L)/SU(2)_V, \tag{P10.50}$$

i.e. find ξ and η in terms of V_R' and V_L'.

b. Show that action of any $j = (V_R, V_L) \in SU(2)_R \times SU(2)_L$ on $h = (\xi, \xi^\dagger)$

$$jh = (\xi', \xi'^\dagger)(V, V) \tag{P10.51}$$

leads to the following transformation rule for ξ :

$$\xi \to \xi' = V\xi V_L^\dagger = V_R \xi V^\dagger. \tag{P10.52}$$

c. Show that

$$\xi = \sqrt{U} = e^{\frac{i\vec{\tau}\vec{\pi}}{2F}} \qquad \text{(P10.53)}$$

and find V in terms of π.

d. Show that the Lagrangian

$$\mathcal{L}_N^0 = \bar{N}(i\hat{D} - g_A\bar{\hat{A}}\gamma^5 - m_0 I)N, \quad N = \begin{pmatrix} p \\ n \end{pmatrix}, \qquad \text{(P10.54)}$$

$$D_\mu = \partial_\mu + i\bar{V}_\mu, \quad \bar{V}_\mu = -\frac{i}{2}(\xi^\dagger\partial_\mu\xi + \xi\partial_\mu\xi^\dagger), \qquad \text{(P10.55)}$$

$$\bar{A}_\mu = -\frac{i}{2}(\xi^\dagger\partial_\mu\xi - \xi\partial_\mu\xi^\dagger), \qquad \text{(P10.56)}$$

is invariant under the global transformations from $SU(2)_R \times SU(2)_L$, where nucleons transform under the **fundamental representation** of $SU(2)_V$, i.e.

$$N \to N' = VN, \quad V \in SU(2)_V. \qquad \text{(P10.57)}$$

e. Find the πNN coupling constant (**Goldberger-Treiman relation**):

$$g_{\pi NN} = \frac{g_A m_N}{F_\pi}. \qquad \text{(P10.58)}$$

f. Show that one can introduce the chiral symmetry breaking terms proportional to quark masses in the following way

$$\mathcal{L}_N^1 = -\frac{Z_0}{2}\bar{N}(\xi\hat{M}\xi + \xi^\dagger\hat{M}\xi^\dagger)N - \frac{Z_1}{2}\bar{N}N \; tr(U\hat{M} + \hat{M}U^\dagger),$$

$$\hat{M} = \begin{pmatrix} m_u & \\ & m_d \end{pmatrix} \simeq \hat{m}I \qquad \text{(P10.59)}$$

and show that the nucleon mass in the $SU(2)$ chiral limit m_0 gets the correction

$$m_N = m_0 + \sigma, \quad \sigma = \hat{m}(Z_0 + 2Z_1) = \hat{m}\frac{\langle N|\bar{u}u + \bar{d}d|N\rangle}{2m_N} \simeq 45 MeV.$$

$$\text{(P10.60)}$$

The experimental value $\sigma \simeq 45 MeV$ is discussed in (GLS91).

g. Generalize the chiral Lagrangian to the $SU(3)$ octet of baryons

$$\mathcal{B} = \frac{B^a\lambda^a}{\sqrt{2}} = \begin{pmatrix} \frac{\Sigma^0}{\sqrt{2}} + \frac{\Lambda}{\sqrt{6}} & \Sigma^+ & p \\ \Sigma^- & -\frac{\Sigma^0}{\sqrt{2}} + \frac{\Lambda}{\sqrt{6}} & n \\ \Xi^- & \Xi^0 & -\frac{2}{\sqrt{6}}\Lambda \end{pmatrix}, \qquad \text{(P10.61)}$$

which transform under the **adjoint representation** of $SU(3)_V$, i.e. show that the following Lagrangian is invariant under the $SU(3)_R \times SU(3)_L$ global transformations

$$\mathcal{L}_\mathcal{B}^0 = tr(\bar{\mathcal{B}}(i\hat{D} - \bar{m}_0 I)\mathcal{B} - D(\bar{\mathcal{B}}\gamma^\mu\gamma^5\{\bar{A}_\mu, \mathcal{B}\}) - F(\bar{\mathcal{B}}\gamma^\mu\gamma^5[\bar{A}_\mu, \mathcal{B}])),$$
$$D_\mu\mathcal{B} = \partial_\mu\mathcal{B} + i[\bar{V}_\mu\mathcal{B}], \quad \xi = \sqrt{U}, \tag{P10.62}$$

where U is defined in (P10.31) and $\bar{A}$ and $\bar{V}$ are defined as in (P10.56).

h. Show that one can introduce the chiral symmetry breaking terms proportional to quark masses in the following way

$$\mathcal{L}_\mathcal{B}^0 = -\frac{Z_0}{2}tr\left(d\left(\bar{\mathcal{B}}\{\xi m\xi + \xi^\dagger m\xi^\dagger, \mathcal{B}\}\right) + f\left(\bar{\mathcal{B}}[\xi m\xi + \xi^\dagger m\xi^\dagger, \mathcal{B}]\right)\right)$$
$$- \frac{Z_1}{2}tr(\bar{\mathcal{B}}\mathcal{B})tr\left(mU + U^\dagger m\right),$$
$$m = \begin{pmatrix} \hat{m} & & \\ & \hat{m} & \\ & & m_s \end{pmatrix} = \frac{I}{3}(2\hat{m} + m_s) + \frac{\hat{m} - m_s}{\sqrt{3}}\lambda_8. \tag{P10.63}$$

i. Show that consistency with the nucleon Lagrangian demands

$$F + D = g_A, \quad f + d = 1, \quad m_0 = \bar{m}_0 + m_s(Z_0(d-f) + Z_1). \tag{P10.64}$$

j. Show that this Lagrangian gives the Gell-Mann–Okubo relation

$$m_\Sigma - m_N = \frac{1}{2}(m_\Xi - m_N) + \frac{3}{4}(m_\Sigma - m_\Lambda), \tag{P10.65}$$

and check it with data.

k. Show that

$$Z_0(m_s - \hat{m}) = 132 MeV, \quad \frac{d}{f} = -0.31 \tag{P10.66}$$

give few percent fit for these mass differences $(m_\Sigma - m_N)$, $(m_\Xi - m_N)$, $(m_\Sigma - m_\Lambda)$.

l. Show that

$$(m_s - \hat{m})\frac{\langle p|\bar{u}u + \bar{d}d - 2\bar{s}s|p\rangle}{2m_N} = (m_s - \hat{m})Z_0(3f - d)$$
$$\simeq m_\Sigma + m_\Xi - 2m_N \simeq 634 MeV \tag{P10.67}$$
$$\implies \hat{m}\frac{\langle p|\bar{u}u + \bar{d}d - 2\bar{s}s|p\rangle}{2m_N} \simeq 26 MeV. \tag{P10.68}$$

m. Using this value, the value for the σ term given above (P10.60), and the decomposition of the nucleon mass via the trace anomaly (8.34) estimate the contribution of light quarks, s quark, and gluon to the nucleon mass (cf. (8.35)).

10.13 Estimate the theta term from the bounds on the electric dipole moment of the neutron.

SOLUTION. We follow (DGH14). **Electric dipole moment** is a T-violating effect for an elementary particle. Indeed, an elementary particle has only its momentum and spin to build it from. At rest only the spin remains, so

$$\vec{d}_e = d_e \frac{\vec{s}}{|s|} \quad \Longrightarrow \quad H_{int} = -\vec{E}\vec{d}_e = -\vec{E}\frac{\vec{s}}{|s|}d_e. \qquad (P10.69)$$

This interaction is T-odd. Indeed, Under T transformation the electric field does not change, while the angular momentum changes sign. For a spin-$\frac{1}{2}$ particle the matrix element for the dipole moment reads

$$\langle p'|J_\mu^{dem}|p\rangle = i\bar{u}_{p'}\sigma^{\mu\nu}q_\nu\gamma^5 u_p, \quad q^\mu = (p' - p)^\mu, \qquad (P10.70)$$

which correspondes to

$$\mathcal{H}_{int} = id_e\partial_\mu(\bar{\psi}\sigma^{\mu\nu}\gamma^5\psi)A_\nu = -i\frac{d_e}{2}\bar{\psi}\sigma^{\mu\nu}\gamma^5\psi F_{\mu\nu}. \qquad (P10.71)$$

Since $F^{0i} = -E^i$ and in the standard representation

$$\gamma^0\sigma^{0i}\gamma^5 = i\begin{pmatrix}\sigma^i & \\ & -\sigma^i\end{pmatrix} \quad \Longrightarrow$$

$$H_{int} = \int d^3x \mathcal{H}_{int} = -d_e E^i \int d^3x \psi^\dagger \begin{pmatrix}\sigma^i & \\ & -\sigma^i\end{pmatrix}\psi$$

$$= |at\ rest| = -d_e E^i \int d^3x \omega^\dagger \sigma^i \omega = -d_e E^i \langle \sigma^i\rangle. \qquad (P10.72)$$

Then the matrix element for the dipole moment of the neutron in the second order of the perturbation theory reads

$$d_e\bar{u}_{p'}\sigma^{\mu\nu}q_\nu\gamma^5 u_p = \sum_I \frac{\langle n_{p'}|J_{em}^\mu|I\rangle\langle I|\mathcal{L}^{Todd}|n_p\rangle}{E_n - E_I}. \qquad (P10.73)$$

To estimate this contribution one can transform back $\bar{\theta}$ into the quark mass corrections (see problem 9.15 in Chapter 9)

$$\bar{\theta} = \arg\det\tilde{M} = \arg\det(M + i\eta \times 1)$$

$$= \arg[(m_u + i\eta)(m_d + i\eta)(m_s + i\eta)]$$

$$\simeq \arg[m_u m_d m_s e^{i\eta\frac{m_u m_d + m_u m_s + m_d m_s}{m_u m_d m_s}}]$$

$$= \eta\frac{m_u m_d + m_u m_s + m_d m_s}{m_u m_d m_s}. \qquad (P10.74)$$

Here one writes the violating term as a mass correction proportional to the identity matrix $M + i\eta \times 1$ since otherwise its contribution to the effective chiral Lagrangian will be unbounded in energy. Indeed, suppose one had the contribution $\sim \tau^3$, then in the chiral Lagrangian one would have a term linear in field π, i.e. the potential would be unbounded from below

$$\sim tr[U^\dagger(M + i\eta\tau^3) + U(M - i\eta\tau^3)]$$

$$= \ldots + \frac{i\eta}{F} tr[-i\vec{\pi}\vec{\tau}\,\tau^3 - i\vec{\pi}\vec{\tau}\,\tau^3] = \ldots + \frac{4\eta}{F}\pi^3. \qquad (P10.75)$$

The mass term in the Lagrangian finally becomes

$$-\mathcal{L}_m = \bar{\psi}_L \tilde{M}\psi_R + \bar{\psi}_R \tilde{M}^\dagger\psi_L$$

$$= m_u\bar{u}u + m_d\bar{d}d + m_s\bar{s}s + i\eta(\bar{u}\gamma^5 u + \bar{d}\gamma^5 d + \bar{s}\gamma^5 s)$$

$$= m_u\bar{u}u + m_d\bar{d}d + m_s\bar{s}s$$

$$+ i\bar{\theta}\frac{m_u m_d m_s}{m_u m_d + m_u m_s + m_d m_s}(\bar{u}\gamma^5 u + \bar{d}\gamma^5 d + \bar{s}\gamma^5 s). \qquad (P10.76)$$

Then

$$\langle n_{p'}|J^{em}|I\rangle \simeq \mu_n, \quad E_n - E_I \simeq \Lambda_{QCD},$$

$$\langle I|\mathcal{L}^{T_{odd}}|n_p\rangle \simeq \bar{\theta}\frac{m_u m_d m_s}{m_u m_d + m_u m_s + m_d m_s} \quad \Longrightarrow \qquad (P10.77)$$

$$d_n \simeq \bar{\theta}\frac{m_u m_d m_s}{m_u m_d + m_u m_s + m_d m_s}\frac{\mu_n}{\Lambda_{QCD}} = \bar{\theta}e\frac{\hbar c}{m_p c^2}\frac{3\times 5}{8\times 300}$$

$$= \bar{\theta}e\frac{200\times 10^{-13}}{1000}\frac{3\times 5}{8\times 300}cm \simeq 10^{-15}\bar{\theta}e\,cm, \qquad (P10.78)$$

with the experimental value $d_e \lesssim 10^{-25}e\,cm$ (B$^+$06). As a result, one has $\bar{\theta}_e < 10^{-10}$.

10.14 Calculate $\langle\frac{\alpha_s}{\pi}\varepsilon^{\mu\nu g\alpha\beta}F^a_{\mu\nu}F^a_{\alpha\beta}\rangle$

a. in the theory with 2 light quarks;

b. in the theory with 3 light quarks.

a. SOLUTION. We follow (SVZ80). Since $\theta \ll 1$, one may write

$$\langle \frac{\alpha_s}{\pi} \varepsilon^{\mu\nu g \alpha\beta} F_{\mu\nu}^a(0) F_{\alpha\beta}^a(0) \rangle = K\theta,$$

$$\theta K = \int \mathcal{D}A \mathcal{D}\psi \mathcal{D}\bar{\psi} \frac{\alpha_s}{\pi} \varepsilon^{\mu\nu g \alpha\beta} F_{\mu\nu}^a(0) F_{\alpha\beta}^a(0)$$

$$\times e^{iS_{QCD} + \frac{i\theta\alpha_s}{16\pi} \int \varepsilon_{\mu\nu\alpha\beta} F^{a\mu\nu} F^{a\alpha\beta} d^4 x}$$

$$\simeq \int \mathcal{D}A \mathcal{D}\psi \mathcal{D}\bar{\psi} \frac{\alpha_s}{\pi} \varepsilon^{\mu\nu g \alpha\beta} F_{\mu\nu}^a(0) F_{\alpha\beta}^a(0)$$

$$\times \frac{i\theta\alpha_s}{16\pi} \int \varepsilon_{\mu\nu\alpha\beta} F_x^{a\mu\nu} F_x^{a\alpha\beta} d^4 x e^{iS_{QCD}}, \qquad (P10.79)$$

$$K = \frac{i}{16} \langle \int dx e^{iqx} \frac{\alpha_s}{\pi} \varepsilon^{\mu\nu g \alpha\beta} F_{\mu\nu}^a(0) F_{\alpha\beta}^a(0)$$

$$\times \frac{\alpha_s}{\pi} \varepsilon_{\mu\nu\alpha\beta} F_x^{a\mu\nu} F_x^{a\alpha\beta} \rangle|_{q=0}. \qquad (P10.80)$$

Working with 2 light quarks, one has

$$\partial_\mu J_A^\mu = \partial_\mu (\bar{Q} \gamma^\mu \gamma^5 Q) = 2i\bar{Q} \gamma^5 MQ - \frac{\alpha_s}{4\pi} \varepsilon^{\alpha\beta\mu\nu} F_{\alpha\beta}^a F_{\mu\nu}^a, \quad (P10.81)$$

and introduces

$$\Pi_{\mu\nu}(q) = i \int dx e^{iqx} \langle J_{A\mu}(x) J_{A\nu}(0) \rangle, \qquad (P10.82)$$

$$\tilde{\Pi}_\mu(q) = i \int dx e^{iqx} \langle J_{A\mu}(x) \frac{\alpha_s}{4\pi} \varepsilon^{\rho\sigma\alpha\beta} F_{\rho\sigma}^a(0) F_{\alpha\beta}^a(0) \rangle. \qquad (P10.83)$$

Here one assumes that there re no massless physical states coupled to $J_{A\mu}$, i.e. the pion has a finite nonzero mass and the quarks are massive. Then

$$q^\mu q^\nu \Pi_{\mu\nu}(q)|_{q\to 0} \to 0, \quad q^\mu \tilde{\Pi}_\mu(q)|_{q\to 0} \to 0. \qquad (P10.84)$$

Next, integrating by parts,

$$iq^\mu \tilde{\Pi}_\mu(q)|_{q\to 0} = -i \int dx e^{iqx} \langle \partial^\mu J_{A\mu}(x) \frac{\alpha_s}{4\pi} \varepsilon^{\rho\sigma\alpha\beta} F_{\rho\sigma}^a(0) F_{\alpha\beta}^a(0) \rangle|_{q\to 0}$$

$$= K - i \int dx \langle (2i\bar{Q}_x \gamma^5 MQ_x) \frac{\alpha_s}{4\pi} \varepsilon^{\rho\sigma\alpha\beta} F_{\rho\sigma}^a(0) F_{\alpha\beta}^a(0) \rangle$$

$$= 0. \qquad (P10.85)$$

$$q^\mu q^\nu \Pi_{\mu\nu}(q)|_{q\to 0} = i \int dx e^{iqx} q^\mu q^\nu \langle J_{A\mu}(x) J_{A\nu}(0)\rangle|_{q\to 0}$$

$$= -\int dx e^{iqx} q^\nu \langle \partial^\mu J_{A\mu}(x) J_{A\nu}(0)\rangle|_{q\to 0}$$

$$= -\int dx e^{iqx} q^\nu \langle (2i\bar{Q}_x \gamma^5 M Q_x$$

$$-\frac{\alpha_s}{4\pi} \varepsilon^{\alpha\beta\mu\nu} F^a_{x\alpha\beta} F^a_{x\mu\nu}) J_{A\nu}(0)\rangle|_{q\to 0}$$

$$= -\int dx (-i\partial^\nu e^{iqx}) \langle (2i\bar{Q}_0 \gamma^5 M Q_0$$

$$-\frac{\alpha_s}{4\pi} \varepsilon^{\alpha\beta\mu\nu} F^a_{0\alpha\beta} F^a_{0\mu\nu}) J_{A\nu}(-x)\rangle|_{q\to 0}$$

$$= i\int dx e^{iqx} \langle (2i\bar{Q}_x \gamma^5 M Q_x$$

$$-\frac{\alpha_s}{4\pi} \varepsilon^{\alpha\beta\mu\nu} F^a_{x\alpha\beta} F^a_{x\mu\nu}) \partial^\nu J_{A\nu}(0)\rangle|_{q\to 0} \qquad \text{(P10.86)}$$

$$= K + i\int dx \langle (2i\bar{Q}_0 \gamma^5 M Q_0)(2i\bar{Q}_x \gamma^5 M Q_x)\rangle$$

$$-2i\int dx \langle (2i\bar{Q}_0 \gamma^5 M Q_0) \frac{\alpha_s}{4\pi} \varepsilon^{\alpha\beta\mu\nu} F^a_{x\alpha\beta} F^a_{x\mu\nu})\rangle$$

$$+4\int dx \langle \bar{Q}_0 M Q_0\rangle = 0 \qquad \text{(P10.87)}$$

Here the last term is a contact term, which was obtained for this Swinger-Dyson equation in problem 9.9 in Chapter 9. Therefore, one can find K

$$K = 4\langle \bar{Q}_0 M Q_0\rangle + i\int dx \langle (2i\bar{Q}_x \gamma^5 M Q_x)(2i\bar{Q}_0 \gamma^5 M Q_0)\rangle. \qquad \text{(P10.88)}$$

Here

$$4\langle \bar{Q}_0 M Q_0\rangle = -4(m_u + m_d)BF^2 = -4m_\pi^2 F^2. \qquad \text{(P10.89)}$$

In the first matrix element one considers the dominant contribution of the pion in the intermediate state

$$i\int dx \langle (2i\bar{Q}_x \gamma^5 M Q_x)(2i\bar{Q}_0 \gamma^5 M Q_0)\rangle e^{iqx}|_{q\to 0}$$

$$= i\int dx \frac{dp}{(2\pi)^4} \frac{\langle (2i\bar{Q}_x \gamma^5 M Q_x)|\pi_p\rangle i\langle \pi_p|(2i\bar{Q}_0 \gamma^5 M Q_0)\rangle}{p^2 - m_\pi^2} e^{iqx}|_{q\to 0}$$

$$= i\int \frac{dx dp}{(2\pi)^4} \frac{\langle (2i\bar{Q}_0 \gamma^5 M Q_0) e^{-ipx}|\pi_p\rangle i\langle \pi_p|(2i\bar{Q}_0 \gamma^5 M Q_0)\rangle}{p^2 - m_\pi^2} e^{iqx}|_{q\to 0}$$

$$= \frac{|\langle (2i\bar{Q}_0 \gamma^5 M Q_0)|\pi_0\rangle|^2}{m_\pi^2} \qquad \text{(P10.90)}$$

$$= \frac{4}{m_\pi^2} \left| \frac{2BF^2}{4} \langle 0|tr\{(\frac{m_u + m_d}{2} + i\frac{m_u - m_d}{2}\tau^3)e^{-i\frac{\vec{\pi}\vec{\tau}}{F}}\right.$$

$$+ (\frac{m_u + m_d}{2} - i\frac{m_u - m_d}{2}\tau^3)e^{i\frac{\vec{\pi}\vec{\tau}}{F}}\}|\pi_0\rangle|^2$$

$$= \frac{4}{m_\pi^2} \left| \frac{2BF^2}{4} \langle 0|tr\{(\frac{m_u + m_d}{2} + i\frac{m_u - m_d}{2}\tau^3)e^{-i\frac{\vec{\pi}\vec{\tau}}{F}}\right.$$

$$+ (\frac{m_u + m_d}{2} - i\frac{m_u - m_d}{2}\tau^3)e^{i\frac{\vec{\pi}\vec{\tau}}{F}}\}|\pi_0\rangle|^2 \qquad \text{(P10.91)}$$

$$= \frac{4}{m_\pi^2} |BF^2 \frac{m_u - m_d}{2} tr\langle 0|\tau^3 \frac{\vec{\pi}\vec{\tau}}{F}|\pi_0\rangle|^2$$

$$= \frac{4B^2 F^2 (m_u - m_d)^2}{m_\pi^2} = 4m_\pi^2 F^2 \frac{(m_u - m_d)^2}{(m_u + m_d)^2}. \qquad \text{(P10.92)}$$

Finally, one has for K

$$K = 4m_\pi^2 F^2 \left(\frac{(m_u - m_d)^2}{(m_u + m_d)^2} - 1 \right) = -4m_\pi^2 F^2 \frac{4m_u m_d}{(m_u + m_d)^2}, \qquad \text{(P10.93)}$$

and

$$\langle \frac{\alpha_s}{\pi} \varepsilon^{\mu\nu g a\beta} F_{\mu\nu}^a(0) F_{\alpha\beta}^a(0) \rangle = K\theta. \qquad \text{(P10.94)}$$

Note, that we use the convention $F = 93$ MeV.

b. ANSWER.

$$K = \frac{-12F^2(m_\pi^2 + m_\eta^2)m_u m_d m_s}{(m_u + m_d + m_s)(m_u m_d + m_d m_s + m_s m_u)}. \qquad \text{(P10.95)}$$

10.15 **Kim–Shifman–Vainshtein–Zakharov axion** (SVZ80; Kim79; Pec08). Introduce a new heavy quark Ψ and a new complex scalar field ϕ with the lagrangian $\mathcal{L}_{tot} = \mathcal{L}_{QCD} + \mathcal{L}$,

$$\mathcal{L} = \bar{\Psi} i\hat{D}\Psi - h(\phi\bar{\Psi}_L\Psi_R + \phi^\dagger\bar{\Psi}_L\Psi_R) + |\partial_\mu\phi|^2 + m^2|\phi|^2 - \lambda|\phi|^4, \quad \text{(P10.96)}$$

with the large v.e.v. $\phi_0 = \frac{m}{\sqrt{2\lambda}}$.

a. Show that the **axion** field a :

$$\phi = |\phi| e^{\frac{ia}{\phi_0\sqrt{2}}} \qquad \text{(P10.97)}$$

makes the effective $\theta = 0$.

b. Calculate its mass.

c. Calculate the axion coupling constant to photons.

Chiral perturbation theory is comprehensibly introduced and thoroughly discussed in (MS16; DGH14; Eck95). Detailed discussion of the quark masses one can find in (GL82). An overview of the CP-problem is given in (Pec08).

FURTHER READING

[B+06] C. A. Baker et al. An improved experimental limit on the electric dipole moment of the neutron. *Phys. Rev. Lett.*, 97:131801, 2006.

[Cre78] R. J. Crewther. Effects of topological charge in gauge theories. *Acta Phys. Austriaca Suppl.*, 19:47–153, 1978.

[DGH14] J. F. Donoghue, E. Golowich, and Barry R. Holstein. *Dynamics of the standard model*, volume 2, chapter III, IV, VI, IX-4,5. CUP, 2014.

[Eck95] G. Ecker. Chiral perturbation theory. *Prog. Part. Nucl. Phys.*, 35:1–80, 1995.

[GL82] J. Gasser and H. Leutwyler. Quark masses. *Phys. Rept.*, 87:77–169, 1982.

[GLS91] J. Gasser, H. Leutwyler, and M. E. Sainio. Sigma term update. *Phys. Lett. B*, 253:252–259, 1991.

[Kim79] Jihn E. Kim. Weak interaction singlet and strong CP invariance. *Phys. Rev. Lett.*, 43:103, 1979.

[Kra90] A. Krause. Baryon matrix elements of the vector current in chiral perturbation theory. *Helv. Phys. Acta*, 63:3–70, 1990.

[MS16] Samirnath Mallik and Sourav Sarkar. *Hadrons at finite temperature*, chapter 2 - Spontaneous Symmetry Breaking, 3 - Chiral Perturbation Theory. Cambridge University Press, Cambridge, 2016.

[Pec08] R. D. Peccei. The strong CP problem and axions. *Lect. Notes Phys.*, 741:3–17, 2008.

[PS95] Michael E. Peskin and Daniel V. Schroeder. *An introduction to quantum field theory*, chapter 19.3. Addison-Wesley, Reading, USA, 1995.

[SVZ80] Mikhail A. Shifman, A. I. Vainshtein, and Valentin I. Zakharov. Can confinement ensure natural CP invariance of strong interactions? *Nucl. Phys. B*, 166:493–506, 1980.

[Wei77] Steven Weinberg. The problem of mass. *Trans. New York Acad. Sci.*, 38:185–201, 1977.

CHAPTER 11

Collinear Factorization: Deep Inelastic Scattering

T HIS chapter introduces the kinematics of deep inelastic scattering, parton distribution functions and Dokshitzer–Gribov–Lipatov–Altarelli–Parisi evolution equations. Then it gives the idea of the collinear factorization and discusses how one can estimate the cross-sections of QCD elementary processes at pp collides via the parton-parton luminosity.

1. **Deep Inelastic Scattering (DIS) kinematics.** Consider electron–proton scattering with the following kinematical variables

$$Q^2 = -q^2, \quad W^2 = (P+q)^2, \quad x = \frac{Q^2}{2(Pq)}, \quad y = \frac{(Pq)}{(Pl)}, \qquad (11.1)$$

$$\nu = \frac{(Pq)}{m_p} = q^0 \text{ in proton rest frame.} \qquad (11.2)$$

Here x is referred to as the **Bjorken variable,** and ν is called the **inelasticity.** We are interested in the **Bjorken limit,** i.e.

$$Q^2 \to \infty, \quad x = fixed \quad \Longrightarrow \quad \nu = \frac{Q^2}{2xm_p} \to \infty. \qquad (11.3)$$

In this limit

$$s = (P+l)^2 = 2(Pl), \quad I = (Pl) = \frac{s}{2}. \qquad (11.4)$$

In the Bjorken limit, we consider large Q^2, which corresponds to the word "deep", and all possible final states of the proton remnants, which justifies the word "inelastic". Since the proton and photon produce real particles

$$0 < W^2 \simeq 2(Pq) - Q^2 \quad \Longrightarrow \quad 0 < x < 1. \qquad (11.5)$$

DOI: 10.1201/9781003272403-11

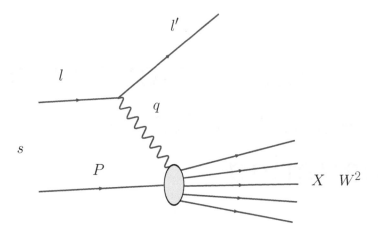

Figure 11.1 DIS kinematics.

The matrix element and the cross-section have the form

$$iM = (-ie)^2 \bar{u}_{l'} \gamma^\mu u_l \frac{-i}{q^2} \langle X | J_\mu(0)| P \rangle, \tag{11.6}$$

$$d\sigma = (2\pi)^4 \delta(l + P - l' - P') \frac{|M|^2}{4I} d\rho_X \frac{d^3 l'}{2E'(2\pi)^3}, \tag{11.7}$$

$$d\rho_X = \prod_k \frac{d^3 k}{2E_k (2\pi)^3}, \tag{11.8}$$

$$d\sigma = \frac{e^4}{Q^4} L^{\mu\nu} \frac{4\pi m_p W_{\mu\nu}}{2s} \frac{d^3 l'}{2E'(2\pi)^3}. \tag{11.9}$$

Derivation of (11.6) is discussed in problem 11.2. Here

$$L^{\mu\nu} = \frac{1}{2} tr(\hat{l}' \gamma^\mu \hat{l} \gamma^\nu) = 2(l^\mu l'^\nu + l'^\mu l^\nu - (ll')g^{\mu\nu}), \tag{11.10}$$

$$W_{\mu\nu} = -\left(g_{\mu\nu} - \frac{q_\mu q_\nu}{q^2} \right) W_1 + \left(P_\mu - \frac{q_\mu(qP)}{q^2} \right) \left(P_\nu - \frac{q_\nu(qP)}{q^2} \right) \frac{W_2}{m_p^2}. \tag{11.11}$$

We parametrized the **hadronic tensor** via the **gauge invariant** structures $W_{1,2}$. They are called **proton structure functions** and are often presented in the equivalent form

$$F_1 = m_p W_1, \quad F_2 = W_2 \nu = W_2 \frac{Q^2}{2x m_p}, \tag{11.12}$$

$$m_p W_{\mu\nu} = -\left(g_{\mu\nu} - \frac{q_\mu q_\nu}{q^2} \right) F_1 + \left(P_\mu - \frac{q_\mu(qP)}{q^2} \right) \left(P_\nu - \frac{q_\nu(qP)}{q^2} \right) \frac{F_2}{(Pq)}. \tag{11.13}$$

Rewriting the cross-section in terms of invariant variables x and y and the structure functions F one gets (see problem 11.3)

$$\frac{d\sigma}{dxdy} = \frac{4\pi\alpha^2}{s}\left[\frac{F_1}{x} + \frac{F_2}{x^2y^2}(1 - y - xy\frac{m_p^2}{s})\right] \qquad (11.14)$$

$$\simeq \frac{4\pi\alpha^2}{s}\left[\frac{F_1}{xy^2}\{(1-y)^2 + 1\} + \frac{F_L}{x^2y^2}(1-y)\right], \quad F_L = F_2 - 2xF_1. \qquad (11.15)$$

Since x and y can be reconstructed from the scattered electron, the structure functions are measurable.

2. **Parton model.** Consider DIS on a single quark with the momentum $p = \xi P$, $0 < \xi < 1$ shown in Figure 11.2. We have

$$W_{q\mu\nu} = \frac{1}{4\pi m_p\xi}\frac{1}{2}\sum_{spin\,q}\langle q_p|J_\nu(0)|X\rangle\langle X|J_\mu(0)|q_p\rangle(2\pi)^4\delta(q+p-P')d\rho_X$$

$$= \frac{1}{4\pi m_p\xi}\frac{1}{2}\int\frac{d^4p_j\delta(p_j^2)}{(2\pi)^3}\sum_{spin\,q}\langle q_p|J_\nu(0)|p_j\rangle\langle p_j|J_\mu(0)|q_p\rangle$$

$$\times (2\pi)^4\delta^{(4)}(q+p-p_j)$$

$$= \frac{e_q^2}{4\pi m_p\xi}\frac{1}{2}\int\frac{d^4p_j\delta(p_j^2)}{(2\pi)^3}tr(\hat{p}\gamma_\nu\hat{p}_j\gamma_\mu)(2\pi)^4\delta^{(4)}(q+p-p_j)$$

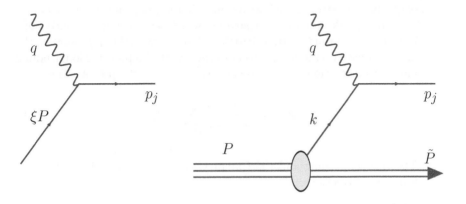

Figure 11.2 Left: DIS on a free quark. Right: DIS on a single quark within the proton.

$$= \frac{e_q^2}{2m_p\xi}\delta((q+p)^2)\frac{1}{2}tr(\hat{p}\gamma_\nu(\hat{q}+\hat{p})\gamma_\mu)$$

$$= e_q^2\frac{\delta(2pq - Q^2)}{\xi m_p}(p^\nu(p+q)^\mu + p^\mu(p+q)^\nu - p(p+q)g^{\mu\nu})$$

$$= e_q^2\frac{\delta(\xi - x)}{2(Pq)xm_p}$$

$$\times \left\{2x^2\left(P^\mu - \frac{(qP)}{q^2}q^\mu\right)\left(P^\nu - \frac{(qP)}{q^2}q^\nu\right) - \frac{Q^2}{2}\left(g^{\mu\nu} - \frac{q^\mu q^\nu}{q^2}\right)\right\}.$$

$$(11.16)$$

Here the factor $\frac{1}{\xi}$ is present since the hadronic tensor is defined without $\frac{1}{2E_P}$ (P11.2) taken into the invariant I. Therefore, we identify the Bjorken x variable with the proton longitudinal momentum fraction ξ carried by the quark. We have for one quark

$$W_1 = e_q^2\frac{\delta(\xi - x)}{m_p}\frac{1}{2}, \quad W_2 = m_pe_q^2\frac{\delta(\xi - x)}{2(Pq)}2x \quad \Longrightarrow$$

$$F_1 = e_q^2\delta(\xi - x)\frac{1}{2}, \quad F_2 = e_q^2\delta(\xi - x)x. \quad (11.17)$$

As a result, we get the **Callan-Gross relation**

$$2xF_1(x) = F_2(x), \quad F_L = 0, \quad (11.18)$$

and the **Bjorken scaling**, i.e. independence of DIS structure functions of Q^2. In the parton model, we assume that the proton consists of noninteracting quarks and gluons (partons) which scatter independently. One introduces the **parton distribution functions (PDF)** $f_{i/P}(x)$ as the number density to find a parton of type i with the proton's longitudinal momentum fraction in the interval $(x, x + dx)$ in the proton. Then

$$F_1(x) = \frac{F_2(x)}{2x} = \frac{1}{2}\sum_{i=q,\bar{q}} e_i^2 f_{i/P}(x). \quad (11.19)$$

The Bjorken scaling and the Callan–Gross relation are violated due to radiative corrections.

3. **PDF.** In the tree approximation, one can single out the parton interacting with the photon to produce a jet with the momentum p_j so that $|X\rangle = |\tilde{X}, p_j\rangle$. If this parton is a quark, as is shown in Figure 11.2, we

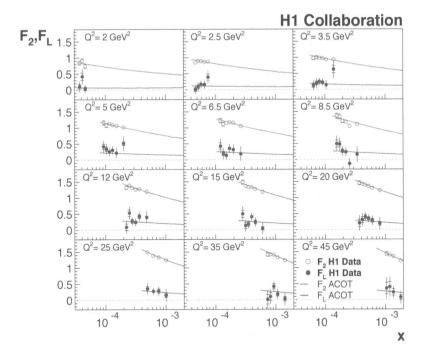

Figure 11.3 F_2 and F_L structure functions from (A$^+$11). One can see that $F_2 \gg F_L$ in support of the Callan–Gross relation (11.18).

have

$$m_p W_{\mu\nu} = \frac{1}{4\pi} \frac{1}{2} \int \sum_{spin\,p} \langle P | J_\nu(0) | X \rangle \langle X | J_\mu(0) | P \rangle$$

$$\times (2\pi)^4 \delta(q + P - P') d\rho_X$$

$$= \frac{1}{4\pi} \int \frac{d^3 p_j}{2E_j(2\pi)^3} \frac{1}{2} \sum_{spin\,p} \langle P \left| J_\nu(0) a^\dagger_{p_j s_j} \right| \tilde{X} \rangle \langle \tilde{X} \left| a_{p_j s_j} J_\mu(0) \right| P \rangle$$

$$\times (2\pi)^4 \delta(q + P - \tilde{P} - p_j) d\rho_{\tilde{X}}$$

$$= \frac{e_q^2}{4\pi} \int \frac{d^4 p_j \delta(p_j^2)}{(2\pi)^3} \int dz e^{iz(q + P - \tilde{P} - p_j)}$$

$$\times \frac{1}{2} \sum_{spin\,p,p_j} \langle P | \bar{\psi}(0) | \tilde{X} \rangle \gamma_\nu u_{p_j} \bar{u}_{p_j} \gamma_\mu \langle \tilde{X} | \psi(0) | P \rangle d\rho_{\tilde{X}} \qquad (11.20)$$

$$= \frac{e_q^2}{4\pi} \int \frac{d^4 p_j \delta(p_j^2)}{(2\pi)^3} \int dz e^{iz(q - p_j)}$$

$$\times \frac{1}{2} \sum_{spin\,p} \langle P | \bar{\psi}(z) \gamma_\nu \hat{p}_j \gamma_\mu \psi(0) | P \rangle. \qquad (11.21)$$

Then working in the frame, where

$$P_\mu = P^+ n_1, \quad P^+ \gg m_p, \quad q_\mu = -xP^+ n_{1\mu} + \frac{Q^2}{2xP^+} n_{2\mu}, \quad (11.22)$$

$$n_1 n_2 = 1, \quad P^2 = n_i^2 = 0, \quad (11.23)$$

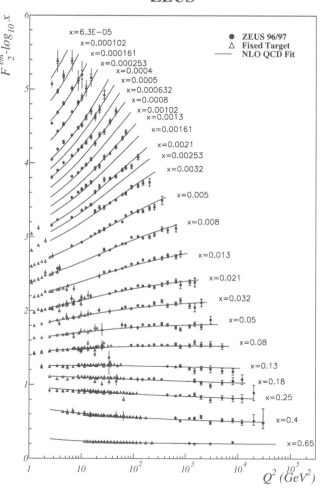

Figure 11.4 F_2 structure function from (C$^+$01). One can see the Bjorken scaling (i.e. independence of Q) for moderate x and its violation for small x. Reprinted by permission from Springer Nature The European Physical Journal C - Particles and Fields, Vol. 21, p. 443, Measurement of the neutral current cross-section and **F$_2$** structure function for deep inelastic **ep** scattering at HERA, S. Chekanov et al, copyright 2001.

one uses (P11.11) to find

$$F_1 = \frac{e_q^2}{4\pi} \int \theta(p_j^-) \frac{d^2 p_j dp_j^-}{2p_j^- (2\pi)^3}$$

$$\times \int dz e^{iz(q-p_j)} \frac{1}{2} \sum_{spin\,p} \langle P|\bar{\psi}(z) \{ \gamma^- p_j^+ + \gamma^+ p_j^- \} \psi(0)|P\rangle. \quad (11.24)$$

We also introduce the momentum of the initial parton

$$k = p_j - q = \xi P^+ n_1 + \frac{\vec{k}^{\,2}}{2\xi P^+} n_2 + k_\perp \simeq \xi P^+ n_1. \quad (11.25)$$

In the matrix element in (11.24), we neglect the contribution $\sim \hat{P} \sim \gamma^-$ in the collinear limit, i.e. assuming that the partons have negligible transverse momenta w.r.t. the longitudinal one $k_\perp \ll P^+$. Then

$$\delta(p_j^2) = \delta((k+q)^2) = \delta(2kq - Q^2) = \frac{\delta(\xi - \frac{Q^2}{2Pq})}{2Pq} \quad \Longrightarrow \quad \xi = x, \quad (11.26)$$

$$F_1 = \frac{e_q^2}{4\pi} \int_{-q^- \to -\infty}^{+\infty} dk^- \int \frac{d^2 k_\perp}{2(2\pi)^3} \int dz e^{-iz^+ k^- - iz^- k^+ + iz\vec{k}}$$

$$\times \frac{1}{2} \sum_{spin\,p} \langle P|\bar{\psi}(z)\gamma^+ \psi(0)|P\rangle$$

$$= \frac{e_q^2}{2} \int \frac{d^2 k_\perp}{(2\pi)^3} f_{q/P}(x, \vec{k}), \quad (11.27)$$

$$f_{q/P}(x, \vec{k}) = \frac{1}{2} \int dz^- d^2 z_\perp e^{-iz^- x P^+ + iz\vec{k}}$$

$$\times \frac{1}{2} \sum_{spin\,p} \langle P|\bar{\psi}(0^+, z^-, \vec{z})\gamma^+ \psi(0)|P\rangle. \quad (11.28)$$

Integrating over the transverse momenta of the parton $\vec{k}$, we get

$$F_1(x, Q^2) = \frac{e_q^2}{2} \int^{Q^2} \frac{d^2 k_\perp}{(2\pi)^3} f_{q/P}(x, \vec{k}) = \frac{e_q^2}{2} f_{q/P}(x, Q^2), \quad (11.29)$$

$$f_{q/P}(x, Q^2) = \frac{1}{2} \int \frac{dz^-}{2\pi} e^{-iz^- x P^+} \frac{1}{2} \sum_{spin\,p} \langle P|\bar{\psi}(0^+, z^-, 0)\gamma^+ \psi(0)|P\rangle. \quad (11.30)$$

For a free quark we reproduce (11.17) :

$$f_{q/q}(x, Q^2) = \frac{1}{2} \int \frac{dz^-}{2\pi} e^{-iz^- x P^+ + i\xi P^+ z^-} \frac{1}{2\xi} \sum_{spin\,p} \bar{u}_{\xi P} \gamma^+ u_{\xi P}$$

$$= \frac{\delta(x - \xi)}{2P^+} \frac{1}{2\xi} tr(\gamma^+ \xi \hat{P}) = \delta(x - \xi). \quad (11.31)$$

Expressions (11.28) and (11.30) are not gauge invariant since the quark fields are in different points. As is shown in (BJY03), (see problem 11.6) interaction of the produced quark with the proton remnants adds a Wilson line to (11.28) and (11.30) and restores gauge invariance. Finally, we have the definition of the **quark PDF** $f_{q/P}(x, Q^2) \equiv q(x, Q^2)$

$$f_{q/P}(x, Q^2) = \frac{1}{2} \int \frac{dz^-}{2\pi} e^{-iz^- xP^+}$$
$$\times \frac{1}{2} \sum_{spin\, p} \langle P | \bar{\psi}(0^+, z^-, 0) P e^{ig \int_0^{z^-} dr^- A^+(0^+, r^-, 0)} \gamma^+ \psi(0) | P \rangle.$$

$$(11.32)$$

This object is gauge invariant and contains both quarks and gluons. However, in the light cone gauge

$$A^+ = An_2 = 0, \qquad (11.33)$$

the Wilson line reduces to 1. In this gauge one may rewrite the quark field in terms of creation and annihilation operators to get

$$f_{q/P}(x, Q^2) = \frac{1}{2x} \int^{Q^2} \frac{dp_\perp}{(2\pi)^2} \frac{\frac{1}{2} \sum_{spin\, p} \langle P | \sum_s b_s^\dagger(xP^+, p_\perp) b_s(xP^+, p_\perp) | P \rangle}{\langle P | P \rangle}.$$

$$(11.34)$$

Therefore, in this gauge $f_{q/P}(x, Q^2)dx$ has a probabilistic interpretation as the average number of quarks with the proton longitudinal momentum fraction in the interval $[x, x + dx]$. Along the same lines one defines the **gluon PDF** $f_{g/P}(x, Q^2) \equiv g(x, Q^2)$ as

$$f_{g/P}(x, Q^2) = \int \frac{dz^-}{2\pi xP^+} e^{-iz^- xP^+}$$
$$\times \frac{1}{2} \sum_{spin\, p} \langle P | F^a(0^+, z^-, 0)^{+\nu} (P e^{ig \int_0^{z^-} dr^- A^+(0^+, r^-, 0)})^{ab} F^b(0)_\nu^+ | P \rangle,$$

$$(11.35)$$

where the Wilson line is in the adjoint representation and $F^{a\mu\nu}$ is the gluon field tensor.

4. The **quark number sum rule** states

$$\int_0^1 dx [f_{q/P}(x, M^2) - f_{\bar{q}/P}(x, M^2)] = const. \qquad (11.36)$$

Indeed thanks to (P11.49), we have

$$\int_0^1 dx [f_{q/P}(x, M^2) - f_{\bar{q}/P}(x, M^2)]$$

$$= \int_0^1 dx [f_{q/P}(x, M^2) + f_{q/P}(-x, M^2)] = \int_{-1}^1 dx \, f_{q/P}(x, M^2)$$

$$= \int_{-1}^1 dx \frac{1}{2} \int \frac{dz^-}{2\pi} e^{-ixP^+z^-}$$

$$\times \frac{1}{2} \sum_{spin \, p} \langle P | \bar{\psi} \left(0^+, z^-, 0_\perp\right) P e^{ig \int_0^{z^-} dr^- A^+(0^+, r^-, 0)} \gamma^+ \psi(0) | P \rangle$$

$$= \int_{-\infty}^{+\infty} \frac{d(P^+x)}{2P^+} \int \frac{dz^-}{2\pi} e^{-ixP^+z^-}$$

$$\times \frac{1}{2} \sum_{spin \, p} \langle P | \bar{\psi} \left(0^+, z^-, 0_\perp\right) P e^{ig \int_0^{z^-} dr^- A^+(0^+, r^-, 0)} \gamma^+ \psi(0) | P \rangle$$

$$= \int \frac{dz^- \delta(z^-)}{2P^+}$$

$$\times \frac{1}{2} \sum_{spin \, p} \langle P | \bar{\psi} \left(0^+, z^-, 0_\perp\right) P e^{ig \int_0^{z^-} dr^- A^+(0^+, r^-, 0)} \gamma^+ \psi(0) | P \rangle$$

$$= \frac{1}{2P^+} \frac{1}{2} \sum_{spin \, p} \langle P | \bar{\psi}(0) \gamma^+ \psi(0) | P \rangle = \frac{1}{2P^+} \frac{1}{2} \sum_{spin \, p} \langle P | j_B^+(0) | P \rangle$$

$$= \frac{1}{2P^+} \frac{1}{2} \sum_{spin \, p} \langle P | j_B^+(z) | P \rangle$$

$$= \frac{\frac{1}{2} \sum_{spin \, p} \langle P | \int d^2z dz^- j_B^+(z) | P \rangle}{2P^+ \int d^2z dz^-} = \frac{\frac{1}{2} \sum_{spin \, p} \langle P | \int d^2z dz^- j_B^+(z) | P \rangle}{\frac{1}{2} \sum_{spin \, p} \langle P | P \rangle}$$

$$= \frac{\frac{1}{2} \sum_{spin \, p} \langle P | \int \frac{d^2k dk^+}{(2\pi)^3} \left\{ b_{k,\lambda}^\dagger b_{k,\lambda} - d_{k,\lambda}^\dagger d_{k,\lambda} + const \right\} | P \rangle}{\frac{1}{2} \sum_{spin \, p} \langle P | P \rangle}. \qquad (11.37)$$

Here we used the fact that $f = 0$ for $|x| > 1$ and reduced the integral to the matrix element of the conserved $^+$ component of the baryon number current (10.35). The quarks which define the flavor content of the proton are referred to as **valence quarks**, while the quarks and antiquarks produced by gluons are called **sea quarks**. They are schematically depicted in Figure 11.5. We have

$$f_{u/P} \equiv u = u_{val} + u_{sea}, \quad f_{\bar{u}/P} = \bar{u}_{sea} = u_{sea} \quad \Longrightarrow \quad u_{val} = u - \bar{u} \qquad (11.38)$$

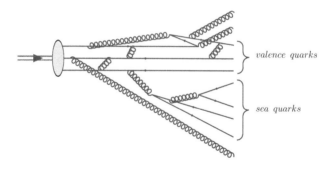

Figure 11.5 Sea and valence quarks inside the proton.

and the same formula for the d quarks. Therefore, the sum rule states that the number of the valence quarks is fixed:

$$\sum_q \int_0^1 dx [f_{q/P}(x, M^2) - f_{\bar{q}/P}(x, M^2)] = 3, \tag{11.39}$$

$$\int_0^1 dx [f_{u/P}(x, M^2) - f_{\bar{u}/P}(x, M^2)] = 2, \tag{11.40}$$

$$\int_0^1 dx [f_{d/P}(x, M^2) - f_{\bar{d}/P}(x, M^2)] = 1, \tag{11.41}$$

$$\int_0^1 dx [f_{q/P}(x, M^2) - f_{\bar{q}/P}(x, M^2)] = 0, \quad q \neq u, d. \tag{11.42}$$

5. Consider **one loop corrections to** quark **PDF**. They come from the one loop corrections to **the QCD string operator**

$$\bar{\psi}(0^+, z^-, 0) P e^{ig \int_0^{z^-} dr^- A^+(0^+, r^-, 0)} \gamma^+ \psi(0) = \bar{\Phi}(z^-) \gamma^+ \Phi(0), \tag{11.43}$$

$$\Phi(z^-) = P e^{ig \int_{z^-}^{+\infty} dr^- A^+(0^+, r^-, 0)} \psi(0^+, z^-, 0). \tag{11.44}$$

This is a gauge invariant operator and one can calculate corrections in any gauge. We will calculate them in the light cone gauge. In this gauge the Wilson line is 1 identically. Therefore, the diagrams 2, 4, and 5 in Figure 11.6 vanish. We start from the PDF $f_{q/P}(x, M_1^2)$ defined at the transverse momentum scale M_1^2, i.e. it includes all quarks in the proton with the transverse momentum squared up to M_1^2. In the loops, we want to integrate over the partons with transverse momenta in the strip $M_1^2 < \vec{l}^{\,2} < M_2^2$. We thus include them into the definition of the PDF and get $f_{q/P}(x, M_2^2)$. The contribution of diagram 1 in Figure 11.6

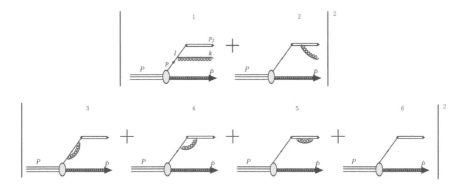

Figure 11.6 Quark contribution to the evolution of quark PDF. The double line stands for the Wilson line. In the light cone gauge diagrams 2, 4, 5 vanish.

reads

$$f_{q/P}(x, M_2^2)_1 = \frac{1}{2} \int \frac{dz^-}{2\pi} e^{-iz^- xP^+} \frac{1}{2} \sum_{spin\, p} \langle P|\bar{\Phi}(z^-)\gamma^+\Phi(0)|P\rangle_1$$

$$= \frac{1}{2} \int \frac{dz^-}{2\pi} e^{-iz^- xP^+} \frac{1}{2} \sum_{spin\, p, \lambda} \langle P|\bar{\Phi}(z^-)a_{k\lambda}^{\dagger a}|\tilde{P}\rangle_1 \gamma^+$$

$$\times \langle \tilde{P}|a_{k\lambda}^a \Phi(0)|\tilde{P}\rangle_1|_{M_1^2} d\tilde{\rho} \frac{d^4 k \delta(k^2)}{(2\pi)^3}|_{M_1^2 < \bar{k}^2 < M_2^2}. \qquad (11.45)$$

The real correction to the matrix element has the form

$$\langle P|\bar{\Phi}(z^-)a_{k\lambda}^{\dagger}|\tilde{P}\rangle_1 = ig\langle P| \int d^4x \bar{\psi}_x \hat{A}_x \psi_x \bar{\psi}(0^+, z^-, 0)a_{k\lambda}^{\dagger a}|\tilde{P}\rangle$$

$$= ig \int d^4x e^{-ikx+i(P-\tilde{P})x} \langle P|\bar{\psi}_0 \hat{\varepsilon}_{k\lambda} G(x - z^-)|\tilde{P}\rangle t^a$$

$$= ig \int \frac{d^4l}{(2\pi)^4} \int d^4x e^{-ikx+i(P-\tilde{P})x - il(x-z^-)} \langle P|\bar{\psi}_0|\tilde{P}\rangle \hat{\varepsilon}_{k\lambda} \frac{i\hat{l}}{l^2} t^a$$

$$= ig \int d^4l \delta(P - \tilde{P} - k - l) e^{il^+ z^-} \langle P|\bar{\psi}_0|\tilde{P}\rangle \hat{\varepsilon}_{k\lambda} \frac{i\hat{l}}{l^2} t^a$$

$$= ig e^{i(P-\tilde{P}-k)^+ z^-} \langle P|\bar{\psi}_0|\tilde{P}\rangle \hat{\varepsilon}_{k\lambda} \frac{i(\hat{p}-\hat{k})}{(p-k)^2} t^a$$

$$= -g e^{-ik^+ z^-} \langle P|\bar{\psi}_{z^-}|\tilde{P}\rangle \hat{\varepsilon}_{k\lambda} \frac{(\hat{p}-\hat{k})}{(p-k)^2} t^a, \qquad (11.46)$$

where we neglect the transverse components of the initial parton since they are below M_1

$$p = P - \tilde{P} = yP^+ n_1, \qquad (11.47)$$

and denote the other vectors as

$$k = (y - x)P^+ n_1 + \frac{\vec{k}^2}{2(y-x)P^+} n_2 + k_\perp, \quad k^2 = 0, \qquad (11.48)$$

$$l = p - k = xP^+ n_1 - \frac{\vec{k}^2}{2(y-x)P^+} n_2 - k_\perp. \qquad (11.49)$$

As a result,

$$f_{q/P}(x, M_2^2)_1 = g^2 \frac{1}{2} \int \frac{dz^-}{2\pi} e^{-iz^- xP^+} \frac{1}{2} \sum_{spin\,p,\lambda} e^{-ik^+ z^-} \langle P|\bar{\psi}_{z^-}|\tilde{P}\rangle$$

$$\times \hat{\varepsilon}_{k\lambda} \frac{(\hat{p} - \hat{k})}{(p-k)^2} \gamma^+ t^a t^a \frac{(\hat{p} - \hat{k})}{(p-k)^2} \hat{\varepsilon}_{k\lambda}^* \langle \tilde{P}|\psi_0|P\rangle|_{M_1^2} d\tilde{\rho} \frac{d^4 k \delta(k^2)}{(2\pi)^3}|_{M_1^2 < \vec{k}^2 < M_2^2}. \tag{11.50}$$

One can decompose the intermediate Dirac matrix into the 16 basis Γ $(1, i\gamma^5, \gamma^\mu, i\gamma^5\gamma^\mu, \sigma^{\mu\nu})$ matrices

$$\hat{\varepsilon}_{k\lambda} \frac{(\hat{p} - \hat{k})}{(p-k)^2} \gamma^+ \frac{(\hat{p} - \hat{k})}{(p-k)^2} \hat{\varepsilon}_{k\lambda}^* = \sum a_i \Gamma^i, \tag{11.51}$$

$$a_i = \frac{1}{4} tr[\hat{\varepsilon}_{k\lambda} \frac{(\hat{p} - \hat{k})}{(p-k)^2} \gamma^+ \frac{(\hat{p} - \hat{k})}{(p-k)^2} \hat{\varepsilon}_{k\lambda}^* \Gamma_i]. \tag{11.52}$$

We take only the leading contribution to the convolution of the hadronic matrix elements coming from γ^+. Using

$$-\sum_\lambda \varepsilon_{k\lambda}^{\mu*} \varepsilon_{k\lambda}^\nu = d^{\mu\nu}(k) = g^{\mu\nu} - \frac{k^\mu n_2^\nu + k^\nu n_2^\mu}{k^+}$$

$$= g_\perp^{\mu\nu} - \frac{k_\perp^\mu n_2^\nu + k_\perp^\nu n_2^\mu}{(y-x)P^+} - \frac{\vec{k}^2}{(y-x)^2 P^{+2}} n_2^\mu n_2^\nu, \tag{11.53}$$

we get

$$\sum_\lambda a_+ = \sum_\lambda a^- = \sum_\lambda \frac{1}{4} tr[\hat{\varepsilon}_{k\lambda} \frac{(\hat{p} - \hat{k})}{(p-k)^2} \gamma^+ \frac{(\hat{p} - \hat{k})}{(p-k)^2} \hat{\varepsilon}_{k\lambda}^* \gamma^-] = 2 \frac{y^2 + x^2}{\vec{k}^2 y^2}. \tag{11.54}$$

Hence

$$f_{q/P}(x, M_2^2)_1 = C_F g^2 \frac{1}{2} \int \frac{dz^-}{2\pi} e^{-iz^- yP^+}$$

$$\times \frac{1}{2} \sum_{spin\, p,\lambda} \langle P|\bar{\psi}_{z^-}|\tilde{P}\rangle \gamma^+ \langle \tilde{P}|\psi_0|P\rangle|_{M_1^2} d\tilde{\rho}$$

$$\times 2 \frac{y^2 + x^2}{\vec{k}^2 y^2} \frac{d^4 k \delta(k^2)}{(2\pi)^3}|_{M_1^2 < \vec{k}^2 < M_2^2}$$

$$= \left| P_1(z) = C_F \frac{1 + z^2}{1 - z} \right|$$

$$= \frac{\alpha_s}{2\pi} \int_x^1 \frac{dy}{y} f_{q/P}(y, M_1^2)_0 P_1(\frac{x}{y}) \frac{d\vec{k}^2}{\vec{k}^2}$$

$$= \frac{\alpha_s}{2\pi} \int_x^1 \frac{dy}{y} f_{q/P}(y, M_1^2)_0 P_1(\frac{x}{y}) \ln \frac{M_2^2}{M_1^2}. \tag{11.55}$$

This result is valid while $\alpha_s \ln \frac{M_2^2}{M_1^2} < 1$ and one can take the limit $M_2 \to M_1$ to get the **evolution equation**

$$\frac{df_{q/P}(x, M^2)}{d\ln M^2} = \frac{\alpha_s(M^2)}{2\pi} \int_x^1 \frac{dy}{y} f_{q/P}(y, M^2) P_1(\frac{x}{y}). \tag{11.56}$$

Here we took the argument of the coupling constant to be M^2 with the $O(\alpha_s)$ accuracy. This expression is singular as $y \to x$ since $P_1(z)$ has a pole at $z = 1$. In this limit the emitted gluon is soft and this singularity must cancel the soft singularity in the virtual correction from diagram 3 in Figure 11.6. However, one can find the virtual correction without calculating diagram 3. Indeed, this contribution will not change the longitudinal momentum of the quark and must be proportional to $f_{q/P}(x, M^2)$. Therefore

$$P_2(\frac{x}{y}) \sim \delta(1 - \frac{x}{y}) \implies$$

$$P_{qq}(z) = P_1(z) + P_2(z) = C_F \left[\frac{1 + z^2}{1 - z} + A\delta(1 - z) \right], \tag{11.57}$$

where A is a divergent constant which must cancel the soft singularity in $P_1(z)$ after the integration w.r.t. z. One can perform this cancellation via the $+$ **prescription**:

$$\int_0^1 dz \frac{f(z)}{(1 - z)_+} = \int_0^1 dz \frac{f(z) - f(1)}{1 - z}. \tag{11.58}$$

Indeed,

$$P_{qq}(z) = C_F \left[\frac{1 + z^2}{1 - z} + A\delta(1 - z) \right] = C_F \left[\frac{1 + z^2}{(1 - z)_+} + \tilde{A}\delta(1 - z) \right], \tag{11.59}$$

where $\tilde{A}$ is now finite. One can find it from the **conservation of the number of quarks** (11.36) since all diagrams in Figure 11.6 do not change this number. Hence

$$\int_0^1 dx f_{q/q}(x, M^2) = 1 \quad \Longrightarrow \quad \int_0^1 dx \int_x^1 \frac{dy}{y} f_{q/q}(y, M^2) P_{qq}(\frac{x}{y}) = 0.$$
(11.60)

In the leading order (11.31), $f_{q/q}(x, M^2) = \delta(x - \xi)$. Therefore

$$0 = \int_0^1 dx \int_x^1 \frac{dy}{y} \delta(y - \xi) P_{qq}(\frac{x}{y}) = \int_0^\xi \frac{dx}{\xi} P_{qq}(\frac{x}{\xi})$$

$$= \int_0^1 dz P_{qq}(z) = C_F \left[-\frac{3}{2} + \tilde{A} \right].$$
(11.61)

Finally,

$$P_{qq}(z) = C_F \left[\frac{1 + z^2}{(1 - z)_+} + \frac{3}{2} \delta(1 - z) \right].$$
(11.62)

Similar calculation gives the other **splitting functions**

$$P_{qg}(z) = \frac{1}{2} \left[z^2 + (1 - z)^2 \right], \quad P_{gq}(z) = C_F \frac{1 + (1 - z)^2}{z},$$
(11.63)

$$P_{gg}(z) = 2N_c \left[\frac{1 - z}{z} + \frac{z}{(1 - z)_+} + z(1 - z) \right] + \frac{\beta_0}{2} \delta(1 - z).$$
(11.64)

6. **Dokshitzer–Gribov–Lipatov–Altarelli–Parisi (DGLAP)** equations are the $2n_f + 1$ evolution equations for the PDFs. They have the form

$$\frac{df_{q/P}(x, M^2)}{d \ln M^2} = \frac{\alpha_s(M^2)}{2\pi} \int_x^1 \frac{dy}{y}$$
$$\times \left[f_{q/P}(y, M^2) P_{qq}(\frac{x}{y}) + f_{g/P}(y, M^2) P_{qg}(\frac{x}{y}) \right],$$
(11.65)

$$\frac{df_{g/P}(x, M^2)}{d \ln M^2} = \frac{\alpha_s(M^2)}{2\pi} \int_x^1 \frac{dy}{y}$$
$$\times \left[\sum_{q,\bar{q}} f_{q/P}(y, M^2) P_{gq}(\frac{x}{y}) + f_{g/P}(y, M^2) P_{gg}(\frac{x}{y}) \right].$$
(11.66)

Here the first equation is valid for all $2n_f$ active quark and antiquark flavors. The flavor is active if the hard scale $M^2 > m_q^2$. Introducing 1 **singlet** distribution function

$$\Sigma = \sum_{q,\bar{q}, \text{ all active flavors}} \left[f_{q/P} + f_{\bar{q}/P} \right],$$
(11.67)

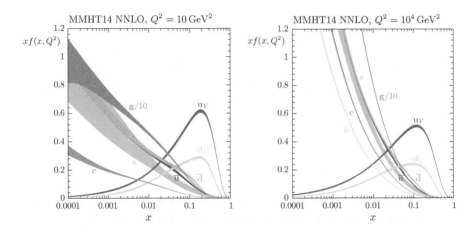

Figure 11.7 PDFs for sea and valence quarks and gluons from (B$^+$15).

$n_f - 1$ **flavor nonsinglet** distribution functions

$$f_{q/P}^{NSi} = f_{q/P}^i + f_{\bar{q}/P}^i - f_{q/P}^1 - f_{\bar{q}/P}^1, \quad i = 2, ...n_f, \qquad (11.68)$$

and n_f **valence** distribution functions

$$f_{q/P}^v = f_{q/P} - f_{\bar{q}/P}, \qquad (11.69)$$

one rewrites the DGLAP equations as

$$\frac{df_{q/P}^v(x, M^2)}{d\ln M^2} = \frac{\alpha_s(M^2)}{2\pi} \int_x^1 \frac{dy}{y} P_{qq}\left(\frac{x}{y}\right) f_{q/P}^v(y, M^2), \qquad (11.70)$$

$$\frac{df_{q/P}^{NSi}(x, M^2)}{d\ln M^2} = \frac{\alpha_s(M^2)}{2\pi} \int_x^1 \frac{dy}{y} P_{qq}\left(\frac{x}{y}\right) f_{q/P}^{NSi}(y, M^2), \qquad (11.71)$$

$$\frac{d}{d\ln M^2}\begin{pmatrix} \Sigma \\ g \end{pmatrix}(x, M^2) = \frac{\alpha_s(M^2)}{2\pi} \int_x^1 \frac{dy}{y} \begin{pmatrix} P_{qq}\left(\frac{x}{y}\right) & 2n_f P_{qg}\left(\frac{x}{y}\right) \\ P_{gq}\left(\frac{x}{y}\right) & P_{gg}\left(\frac{x}{y}\right) \end{pmatrix}\begin{pmatrix} \Sigma \\ g \end{pmatrix}. \qquad (11.72)$$

Taking the **Mellin transform** of these equations

$$f^{(n)} = \int_0^1 \frac{dx}{x} x^n f(x), \quad f(x) = \frac{1}{2\pi i}\int_{c-i\infty}^{c+i\infty} dn z^{-n} f^{(n)}, \qquad (11.73)$$

we get the equations for the **Mellin moments** of the PDFs $f^{(n)}$. Here the contour for the integration in the inverse Mellin transform lies on the right of all singularities of $f^{(n)}$.

$$
\begin{aligned}
\frac{df^{v(n)}}{d\ln M^2} &= \frac{\alpha_s(M^2)}{2\pi} \int_0^1 \frac{dx}{x} x^n \int_x^1 \frac{dy}{y} P_{qq}(\frac{x}{y}) f_{q/P}^v(y, M^2) \\
&= \frac{\alpha_s(M^2)}{2\pi} \int_0^1 \frac{dy}{y} \int_0^y \frac{dx}{x} x^n P_{qq}(\frac{x}{y}) f_{q/P}^v(y, M^2) = |x = yz| \\
&= \frac{\alpha_s(M^2)}{2\pi} \int_0^1 \frac{dy}{y} y^n f_{q/P}^v(y, M^2) \int_0^1 \frac{dz}{z} z^n P_{qq}(z) \\
&= \frac{\alpha_s(M^2)}{2\pi} f^{v(n)} P_{qq}^{(n)}.
\end{aligned}
\tag{11.74}
$$

Therefore for the moments we get the equations, which are the Callan-Symanzik equations with the **anomalous dimensions** $\frac{\alpha_s(M^2)}{2\pi} P_{qq}^{(n)}$:

$$
\frac{df^{v(n)}}{d\ln M^2} = \frac{\partial f^{v(n)}}{\partial \ln M^2} + \frac{d\alpha_s}{d\ln M^2} \frac{\partial f^{v(n)}}{\partial \alpha_s} = \frac{\alpha_s(M^2)}{2\pi} f^{v(n)} P_{qq}^{(n)} \quad \Longrightarrow
\tag{11.75}
$$

$$
\left(\frac{\partial}{\partial \ln M^2} + \beta(\alpha_s) \frac{\partial}{\partial \alpha_s} - \frac{\alpha_s(M^2)}{2\pi} P_{qq}^{(n)} \right) f^{v(n)} = 0.
\tag{11.76}
$$

With the $O(\alpha_s)$ accuracy (6.33)

$$
\beta(\alpha_s) = \frac{d\alpha_s}{d\ln M^2} = -\beta_0 \alpha_s^2 \quad \Longrightarrow
$$

$$
f^{v(n)}(M_2^2) = f^{v(n)}(M_1^2) \left(\frac{\alpha_s(M_2^2)}{\alpha_s(M_1^2)} \right)^{-\frac{P_{qq}^{(n)}}{2\pi\beta_0}}.
\tag{11.77}
$$

For the singlet and gluon moments, we have

$$
\frac{d}{d\ln M^2} \begin{pmatrix} \Sigma^{(n)} \\ g^{(n)} \end{pmatrix} = \frac{\alpha_s(M^2)}{2\pi} \begin{pmatrix} P_{qq}^{(n)} & 2n_f P_{qg}^{(n)} \\ P_{gq}^{(n)} & P_{gg}^{(n)} \end{pmatrix} \begin{pmatrix} \Sigma^{(n)} \\ g^{(n)} \end{pmatrix},
\tag{11.78}
$$

$$
\begin{pmatrix} P_{qq}^{(n)} & 2n_f P_{qg}^{(n)} \\ P_{gq}^{(n)} & P_{gg}^{(n)} \end{pmatrix} = Q \begin{pmatrix} \lambda_1 & 0 \\ 0 & \lambda_2 \end{pmatrix} Q^{-1} \quad \Longrightarrow
\tag{11.79}
$$

$$
\begin{pmatrix} \Sigma^{(n)} \\ g^{(n)} \end{pmatrix}(M_2^2) = Q \begin{pmatrix} \left(\frac{\alpha_s(M_2^2)}{\alpha_s(M_1^2)} \right)^{-\frac{\lambda_1}{2\pi\beta_0}} & 0 \\ 0 & \left(\frac{\alpha_s(M_2^2)}{\alpha_s(M_1^2)} \right)^{-\frac{\lambda_2}{2\pi\beta_0}} \end{pmatrix} Q^{-1}
$$

$$
\times \begin{pmatrix} \Sigma^{(n)} \\ g^{(n)} \end{pmatrix}(M_1^2).
\tag{11.80}
$$

Taking the PDFs at some low energy scale like $M_1 \sim GeV$ from experimental data or models one can find them at higher energies taking inverse Mellin transform of (11.77), (11.80).

7. **Collinear factorization.** In the **leading order (LO)** or in parton model, the proton structure function F_1 reads

$$F_1(x) = \frac{1}{2} \sum_{i=q,\bar{q}} e_i^2 f_i(x) = \sum_{i=q,\bar{q}} \int_0^1 \frac{dy}{y} f_{i/P}(y) \frac{e_i^2}{2} \delta(1 - \frac{x}{y}). \quad (11.81)$$

In the **next to leading order (NLO)**, the structure functions and PDFs get the dependence on the transverse momentum scale:

$$F_1(x, Q^2) = \sum_{i=q,\bar{q}} \int_0^1 \frac{dy}{y} f_{i/P}(y, M^2) \frac{e_i^2}{2}$$

$$\times \left[\delta(1 - \frac{x}{y}) + \frac{\alpha_s(M^2)}{2\pi} \left\{ P_{qq}(\frac{x}{y}) \ln \frac{Q^2}{M^2} + \tilde{C}_q(\frac{x}{y}) \right\} \right]$$

$$+ \int_0^1 \frac{dy}{y} f_{g/P}(y, M^2) \frac{\alpha_s(M^2)}{2\pi} \left\{ P_{qg}(\frac{x}{y}) \ln \frac{Q^2}{M^2} + \tilde{C}_g(\frac{x}{y}) \right\}$$

$$= \sum_{i=q,\bar{q},g} \int_0^1 \frac{dy}{y} f_{i/P}(y, M^2) C_i(\frac{x}{y}, Q^2, M^2). \quad (11.82)$$

Here the corrections without logarithms are $\tilde{C}$. The terms with logarithms in the above expression have the same structure as in the evolution of PDFs since they come from the same diagrams. In fact, the contribution of transverse momenta below M^2 is absorbed into the PDFs and the momenta from M^2 to Q^2 are in the **coefficient functions** $C_i(\frac{x}{y}, Q^2, M^2)$. In NLO the DGLAP equations give

$$f_{q/P}(x, Q^2) = f_{q/P}(x, M^2)$$

$$+ \frac{\alpha_s(M^2)}{2\pi} \int_x^1 \frac{dy}{y} \left[f_{q/P}(y, M^2) P_{qq}(\frac{x}{y}) + f_{g/P}(y, M^2) P_{qg}(\frac{x}{y}) \right] \ln \frac{Q^2}{M^2}. \quad (11.83)$$

Therefore, one can absorb the logarithms into the PDFs:

$$F_1(x, Q^2) = \sum_{i=q,\bar{q}} \int_0^1 \frac{dy}{y} f_{i/P}(y, Q^2) \frac{e_i^2}{2} \left[\delta(1 - \frac{x}{y}) + \frac{\alpha_s(Q^2)}{2\pi} \tilde{C}_q(\frac{x}{y}) \right]$$

$$(11.84)$$

$$+ \int_0^1 \frac{dy}{y} f_{g/P}(y, Q^2) \frac{\alpha_s(Q^2)}{2\pi} \tilde{C}_g(\frac{x}{y}) \quad (11.85)$$

$$= \sum_{i=q,\bar{q},g} \int_0^1 \frac{dy}{y} f_{i/P}(y, Q^2) C_i(\frac{x}{y}, Q^2, Q^2). \quad (11.86)$$

There is a proof (Col13) that such a factorized formula holds for any structure function in DIS. This factorization is called **collinear factorization**

$$F_a(x, Q^2) = \sum_{i=q,\bar{q},g} \int_0^1 \frac{dy}{y} f_{i/P}(y, \mu_F^2) C_{a,i}(\frac{x}{y}, Q^2, \mu_F^2). \qquad (11.87)$$

Here μ_F^2 is the **factorization scale** and Q^2 is the **hard scale** of the DIS. The low energy nonperturbative physics responsible for the partonic content of the proton is encoded in the PDFs as initial conditions for the DGLAP evolution at low transverse momenta $\sim GeV$. The PDFs contain the contribution of the partons with the transverse momenta from this nonperturbative scale to the factorization scale μ_F^2. **The PDFs are universal, process independent functions.** They can be measured in one process and used to calculate the cross-section of another process.

The contribution of the large transverse momenta from the factorization scale to the hard scale Q^2 is included into the **coefficient functions** C. **They are process dependent.** They contain hard momenta and can be calculated perturbatively. If $\mu_F \neq Q$ then the coeffitient functions contain logarithms of the ratio $\ln \frac{Q^2}{\mu_F^2}$. Therefore for $\alpha_s \ln \frac{Q^2}{\mu_F^2} \sim 1$, one has to calculate many terms in the perturbative expansion of C. However, for $\mu_F = Q$ the coefficient functions do not contain large logarithms and one can calculate them in the LO.

Collinear factorization is proved (Col13) for other processes besides DIS. In particular for **hard processes in** pp **collisions** $p(p_1) + p(p_2) \rightarrow H(Q^2) + X$

$$\sigma(Q^2) = \sum_{i,j} \int_{x_{\min}}^1 dx_1 dx_2 f_{i/P}(x_1, \mu_F^2) f_{j/P}(x_2, \mu_F^2) \sigma_{ij}(x_1 p_1, x_2 p_2, Q^2, \mu_F^2).$$

$$(11.88)$$

Here σ_{ij} is the hard cross-section for the production of the system H with the hard momentum $\sim Q^2$ in the collision of type i parton from the first proton and type j parton from the second parton.

8. **Differential parton-parton luminosity.** Here we follow (CHS07). Differential parton-parton luminosity is defined as

$$\frac{dL_{ij}}{d\hat{s} dy} = \frac{1}{S} \frac{1}{1 + \delta_{ij}} [f_{i/P}(x_1, \mu_F^2) f_{j/P}(x_2, \mu_F^2) + (i \leftrightarrow j)], \qquad (11.89)$$

where $S = (p_1 + p_2)^2$ is the proton-proton c.m.s. energy squared, $\hat{s} = (x_1 p_1 + x_2 p_2)^2 = x_1 x_2 S$ is the partonic c.m.s. energy squared and

$$x_1 = \sqrt{\hat{s}/S} e^y, \qquad x_2 = \sqrt{\hat{s}/S} e^{-y}, \qquad dx_1 dx_2 = dy d\hat{s}/S. \qquad (11.90)$$

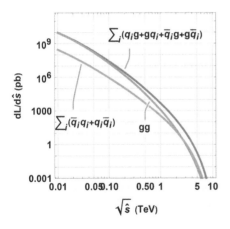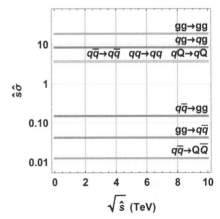

Figure 11.8 The integrated over y parton-parton luminosity at 14 TeV. We used the CT18 PDF dataset (H+21; HWNO19) and ManeParse Mathematica package (CGO17) to draw the plots.

Figure 11.9 Parton level cross-sections ($\hat{s}\hat{\sigma}$) for massless partons in the final state with the cut $p_\perp^2 > 0.01\hat{s}$ and the fixed $\alpha_s = 0.118$.

In terms of parton-parton luminosity the factorized cross-section reads

$$\sigma = \int \frac{d\hat{s}}{\hat{s}} dy \times \frac{dL_{ij}}{d\hat{s}dy} \times \hat{s}\hat{\sigma}_{ij}. \qquad (11.91)$$

The parton-parton luminosity has the dimension of the cross-section, while the other factors in this formula are dimensionless. For total cross-sections when $\hat{\sigma}$ depends only on $\hat{s}$, one can integrate w.r.t. y and get

$$\sigma = \int \frac{d\hat{s}}{\hat{s}} \times \frac{dL_{ij}}{d\hat{s}} \times \hat{s}\hat{\sigma}_{ij}(\hat{s}), \qquad \frac{dL_{ij}}{d\hat{s}} = \int dy \frac{dL_{ij}}{d\hat{s}dy}. \qquad (11.92)$$

The integrated over y parton-parton luminosity is given in Figure 11.8, while the dominant parton level cross-sections are given in Figure 11.9. These plots can be used to estimate QCD production cross-sections for a given interval $\Delta\hat{s}$ and the given p_t cut ($p_\perp^2 > 0.01\hat{s}$ in Figure 11.9):

$$\sigma \simeq \frac{\Delta\hat{s}}{\hat{s}} \times \frac{dL_{ij}}{d\hat{s}} \times \hat{s}\hat{\sigma}_{ij}(\hat{s}). \qquad (11.93)$$

For example for the $gg \to gg$ production at $\hat{s} = 1$ TeV in the interval $\Delta\hat{s} = 0.01\hat{s}$ with the p_t cut $p_\perp^2 > 0.01\hat{s}$ one has $\frac{dL_{gg}}{d\hat{s}} \simeq 10^3$pb from Figure 11.8 and $\hat{s}\hat{\sigma}_{gg}(\hat{s}) \simeq 20$ from Figure 11.9. Therefore, $\sigma \simeq 0.01 \times 10^3$ pb $\times 20 \simeq 200$ pb. More detailes are given in (CHS07) 6.2.

EXERCISES

11.1 Show that the Bjorken variable $1 > x > 0$.

SOLUTION. It follows from (11.5) and

$$Q^2 = -q^2 = -(l - l')^2 = 2ll' = 2EE'(1 - \cos\theta) > 0 \quad \Longrightarrow$$

$$x = \frac{Q^2}{2Pq} \simeq \frac{Q^2}{s} > 0. \tag{P11.1}$$

11.2 Derive the DIS matrix element (11.6).

SOLUTION. Using A and ψ for the photon and electron fields and $J_\nu(x)$ for the quark EM current, we have for the T-matrix

$$T = (-ie)^2 \langle X|a_{l'} T(\int dy \bar\psi_y \hat A_y \psi_y \int dx J_\nu(x) A_x^\nu) a_l^\dagger |\tilde P\rangle$$

$$= \frac{(-ie)^2}{\sqrt{2E2E'}} \int dy e^{-i(l-l')y} \int dx$$

$$\times \int \frac{dq}{(2\pi)^4} e^{-iq(x-y)} \frac{(-ig_{\mu\nu})}{q^2} \bar u_{l'} \gamma^\mu u_l \langle X|J^\nu(x)|\tilde P\rangle$$

$$= \frac{(-ie)^2}{\sqrt{2E2E'}} (2\pi)^4 \delta(l - l' - q) \int dx$$

$$\times \int \frac{dq}{(2\pi)^4} e^{-iqx} \frac{-i}{q^2} \bar u_{l'} \gamma^\mu u_l \langle X|e^{i\hat P x} J_\mu(0) e^{-i\hat P x}|\tilde P\rangle$$

$$= \frac{(-ie)^2}{\sqrt{2E2E'}} (2\pi)^4 \delta(l - l' - q) \int dx$$

$$\times \int \frac{dq}{(2\pi)^4} e^{-i(q+P-P')x} \frac{-i}{q^2} \bar u_{l'} \gamma^\mu u_l \langle X|J_\mu(0)|\tilde P\rangle$$

$$= \frac{(-ie)^2}{\sqrt{2E2E'}} (2\pi)^4 \delta(l - l' + P - P') \frac{-i}{q^2} \bar u_{l'} \gamma^\mu u_l \langle X|J_\mu(0)|\tilde P\rangle$$

$$= \frac{(-ie)^2}{\sqrt{2E2E'2E_P}} (2\pi)^4 \delta(l - l' + P - P') \frac{-i}{q^2} \bar u_{l'} \gamma^\mu u_l \langle X|J_\mu(0)|P\rangle. \tag{P11.2}$$

Here the proton state normalization reads

$$\langle \tilde P_p|\tilde P_k\rangle = (2\pi)^3 \delta^{(3)}(\vec p - \vec k), \quad \langle P_p|P_k\rangle = (2\pi)^3 2E_p \delta^{(3)}(\vec p - \vec k). \tag{P11.3}$$

11.3 Derive the DIS cross-section (11.14).

SOLUTION. In the proton rest frame

$$x = \frac{2l^0 l'^0 (1 - \cos\theta)}{2m_p(l^0 - l'^0)}, \quad y = \frac{l^0 - l'^0}{l^0} \quad \Longrightarrow \quad \left|\frac{\partial(x,y)}{\partial(l'^0, \cos\theta)}\right| = \frac{l'^0}{m_p(l^0 - l'^0)}, \tag{P11.4}$$

$$\frac{d^3l'}{2E'(2\pi)^3} = 2\pi\frac{l'^0\,dl'^0\,d\cos\theta}{2(2\pi)^3} = \frac{dx\,dy}{2(2\pi)^2}m_p(l^0 - l'^0)$$

$$= \frac{dx\,dy}{2(2\pi)^2}m_p y l^0 = \frac{dx\,dy}{2(2\pi)^2}\frac{s}{2}y = \frac{dx\,dy}{(4\pi)^2}sy. \tag{P11.5}$$

Using

$$Q^2 = sxy = 2ll' = -2ql = 2ql', \quad s = 2Pl = 2\frac{(Pl')}{1-y}, \tag{P11.6}$$

one gets (11.14) from (11.10–11.11) and (11.9).

11.4 Derive the DIS cross-section differential in the final electron variables in the proton rest frame.

HINT. Using

$$Q^2 = 2ll' = 2EE'(1 - \cos\theta) = 4EE'\sin^2\frac{\theta}{2}, \tag{P11.7}$$

$$\frac{d\sigma}{dE'\,d\Omega} = \frac{\alpha^2}{4E^2\sin^4\frac{\theta}{2}}\left[2W_1\sin^2\frac{\theta}{2} + W_2\cos^2\frac{\theta}{2}\right]. \tag{P11.8}$$

11.5 Show that in the frame where

$$q_\mu = -xP^+ n_{1\mu} + \frac{Q^2}{2xP^+}n_{2\mu}, \quad n_1 n_2 = 1, \quad P^2 = n_i^2 = 0, \tag{P11.9}$$

$$g_{\mu\nu\perp} = g_{\mu\nu} - n_{1\mu}n_{2\nu} - n_{1\nu}n_{2\mu}, \tag{P11.10}$$

$$F_1 = -\frac{1}{2}m_p W_{\mu\nu}g_\perp^{\mu\nu}, \quad F_L = \frac{8x^3}{Q^2}m_p W_{\mu\nu}P^\mu P^\nu. \tag{P11.11}$$

SOLUTION. Indeed, using (11.13)

$$m_p W_{\mu\nu}g_\perp^{\mu\nu} = -2F_1, \tag{P11.12}$$

$$m_p W_{\mu\nu}P^\mu P^\nu = -\frac{(Pq)^2}{Q^2}F_1 + \left(\frac{(qP)^2}{Q^2}\right)^2\frac{F_2}{(Pq)}$$

$$= \frac{(Pq)^3}{Q^4}\left(F_2 - F_1\frac{Q^2}{(Pq)}\right) = \frac{Q^2}{8x^3}F_L. \tag{P11.13}$$

11.6 Derive the Wilson line in the definition of the quark PDF.

SOLUTION. In (11.21) we singled out one quark from the hadronic matrix element

$$\langle X\,|J_\mu(0)|\,P\rangle_0 = \langle\tilde{X}\,|a_{p_j s_j}J_\mu(0)|\,P\rangle = \bar{u}_j\gamma_\mu\langle\tilde{X}\,|\psi(0)|\,P\rangle. \tag{P11.14}$$

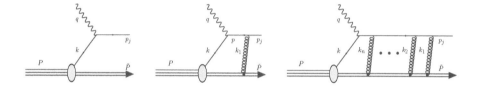

Figure 11.10 Gluon contribution to the DIS matrix element.

Consider a correction from the interaction with one gluon from the proton remnants, which we treat as the external field

$$\langle X\,|J_\mu(0)|\,P\rangle_1 = \langle \tilde{X}|a_{p_j s_j} ig \int d^4x \bar{\psi}_x \hat{A}_x \psi_x J_\mu(0)|P\rangle$$

$$= ig\bar{u}_j \int d^4x e^{ip_j x}\langle \tilde{X}|\hat{A}_x S(x)\gamma_\mu \psi(0)|P\rangle, \qquad \text{(P11.15)}$$

$$S(x) = \int \frac{d^4p}{(2\pi)^4}\frac{e^{-ipx}i\hat{p}}{p^2+i0}, \qquad A_x = \int \frac{d^4k_1}{(2\pi)^4}e^{-ik_1 x}A_{k_1}. \qquad \text{(P11.16)}$$

$$\langle X\,|J_\mu(0)|\,P\rangle_1 = ig\bar{u}_j \int d^4x \frac{d^4k_1}{(2\pi)^4}\frac{d^4p}{(2\pi)^4}e^{ip_j x - ipx - ik_1 x}$$

$$\times \langle \tilde{X}|\hat{A}_{k_1}\frac{i\hat{p}}{p^2+i0}\gamma_\mu \psi(0)|P\rangle$$

$$= ig\bar{u}_j \int \frac{d^4k_1}{(2\pi)^4}\frac{d^4p}{(2\pi)^4}(2\pi)^4\delta(p_j - k_1 - p)$$

$$\times \langle \tilde{X}|\hat{A}_{k_1}\frac{i\hat{p}}{p^2+i0}\gamma_\mu \psi(0)|P\rangle$$

$$= ig\bar{u}_j \int \frac{d^4k_1}{(2\pi)^4}\langle \tilde{X}|\hat{A}_{k_1}\frac{i(\hat{p}_j - \hat{k}_1)}{(p_j - k_1)^2 + i0}\gamma_\mu \psi(0)|P\rangle. \qquad \text{(P11.17)}$$

In frame (11.23) in the Bjorken limit $Q^2 \to \infty$

$$p_j = p_j^+ n_1 + p_j^- n_2 + p_{j\perp}, \quad p_j = q + k + k_1 \qquad \Longrightarrow$$

$$p_j^- \simeq q^- \to \infty \quad p_j^+ = \frac{\vec{p}_j^{\,2}}{2p_j^-} \to 0. \qquad \text{(P11.18)}$$

Therefore

$$\frac{i(\hat{p}_j - \hat{k}_1)}{(p_j - k_1)^2 + i0} = i\frac{(p_j - k_1)^+\gamma^- + (p_j - k_1)^-\gamma^+ + (\hat{p}_j - \hat{k}_1)_\perp}{2(p_j - k_1)^+(p_j - k_1)^- - (\vec{p}_j - \vec{k}_1)^2 + i0}$$

$$\simeq \frac{ip_j^-\gamma^+}{2(p_j - k_1)^+ p_j^- + i0} \simeq -\frac{i}{2}\frac{\gamma^+}{k_1^+ - i0}. \qquad \text{(P11.19)}$$

As a result,

$$\hat{A}_{k_1}\gamma^+\gamma_\mu\psi(0) = (A^+_{k_1}\gamma^- + A^-_{k_1}\gamma^+ + \hat{A}_{k_1\perp})\gamma^+\gamma_\mu\psi(0) \simeq 2A^+_{k_1}\gamma_\mu\psi(0),$$
(P11.20)

since

$$\gamma^+\gamma^+ = 0, \quad \gamma^\pm\gamma^\mu_\perp = -\gamma^\mu_\perp\gamma^\pm, \quad \gamma^-\gamma^+\gamma^- = 2\gamma^-, \quad \hat{P} = P^+\gamma^-,$$
(P11.21)

$$\gamma^-\gamma^+\psi(0) \sim \gamma^-\gamma^+ u_{xP} \sim \gamma^-\gamma^+\hat{P} = 2\hat{P} \quad \Longrightarrow \quad \gamma^-\gamma^+\psi(0) = 2\psi(0).$$
(P11.22)

Here we used $\gamma^\mu = \gamma^\mu_\perp$ because to get F_1 we have to convolute with $g^{\mu\nu}_\perp$, and neglected the contribution of $\hat{A}_{k_1\perp}$ since it gives a power suppressed contribution. Then

$$\int \frac{d^4k_1}{(2\pi)^4} \hat{A}_{k_1} \frac{i(\hat{p}_j - \hat{k}_1)}{(p_j - k_1)^2 + i0} \rightarrow -\frac{i}{2} \int \frac{d^4k_1}{(2\pi)^4} \frac{2A^+_{k_1}}{k_1^+ - i0}$$

$$= -i \int \frac{dk_1^+}{2\pi} \frac{A^+(k_1^+, z_1^+ = 0, \vec{z} = 0)}{k_1^+ - i0}$$

$$= -i \int dz_1^- A^+(z_1^-, z_1^+ = 0, \vec{z} = 0) \int \frac{dk_1^+}{2\pi} \frac{e^{ik_1^+ z_1^-}}{k_1^+ - i0}$$

$$= \int dz_1^- A^+(z_1^-, 0^+, \vec{0})\theta(z_1^-).$$
(P11.23)

Hence

$$\langle X | J_\mu(0) | P \rangle_1 = ig\bar{u}_j\gamma_\mu\langle\tilde{X}| \int_0^{+\infty} dz_1^- A^+(z_1^-, 0^+, \vec{0})\psi(0)|P\rangle.$$
(P11.24)

Repeating this procedure for n gluons, one gets

$$\langle X | J_\mu(0) | P \rangle_n = (ig)^n\bar{u}_j \prod_{l=1}^{n} \int \frac{d^4k_l}{(2\pi)^4} \langle\tilde{X}|\hat{A}_{k_1} \frac{i(\hat{p}_j - \hat{k}_1)}{(p_j - k_1)^2 + i0} \hat{A}_{k_2}$$

$$\times \frac{i(\hat{p}_j - \hat{k}_1 - \hat{k}_2)}{(p_j - k_1 - k_2)^2 + i0} ... \hat{A}_{k_n} \frac{i(\hat{p}_j - \hat{k}_1 ... - \hat{k}_n)}{(p_j - k_1 ... - k_n)^2 + i0}\gamma_\mu\psi(0)|P\rangle.$$
(P11.25)

In the Bjorken limit

$$\langle X | J_\mu(0) | P \rangle_1 = g^n\bar{u}_j \prod_{l=1}^{n} \int \frac{d^4k_l}{(2\pi)^4} \frac{1}{k_1^+ - i0} \frac{1}{(k_1 + k_2)^+ - i0} ...$$

$$\times \frac{1}{(k_1 + ... + k_n)^+ - i0}\langle\tilde{X}|A^+_{k_1} A^+_{k_2} ... A^+_{k_n}\gamma_\mu\psi(0)|P\rangle$$

$$= (ig)^n\bar{u}_j\gamma_\mu\theta(z_n^-)\theta(z_{n-1}^- - z_n^-)...\theta(z_1^- - z_2^-)$$

$$\times \langle\tilde{X}|A^+_{z_1} A^+_{z_2} ... A^+_{z_n}\gamma_\mu\psi(0)|P\rangle.$$
(P11.26)

Then

$$\sum_{n=0}^{+\infty} \langle X | J_\mu(0) | P \rangle_n = \bar{u}_j \gamma_\mu \langle \tilde{X} | P e^{ig \int_0^{+\infty} dz^- A^+(z^-,0^+,\vec{0})} \psi(0) | P \rangle.$$

(P11.27)

Substituting this expression into (11.20) instead of

$$\langle X | J_\mu(0) | P \rangle_1 = \bar{u}_j \gamma_\mu \langle \tilde{X} | \psi(0) | P \rangle,$$

(P11.28)

one gets the gauge invariant result for the quark PDF (11.34) instead of (11.30). More careful analysis of (BJY03) shows that one has to take into account an additional transverse gauge link.

11.7 Show that in the light cone gauge $A^+ = 0$ quark PDF counts the number of quarks in the proton with the proton's longitudinal momentum fraction x.

SOLUTION. In the normalization

$$\psi(x) = \int \frac{d^2 k}{(2\pi)^3} \int_0^{+\infty} \frac{dk^+}{2k^+} \sum_s (u_{ks} e^{-ikx} b_{ks} + v_{ks} e^{ikx} d^\dagger_{ks}),$$

$$\{b_{ks} b^\dagger_{k's'}\} = 2k^+ (2\pi)^3 \delta(k - k') \delta_{ss'},$$

(P11.29)

$$\int dx^- dx_\perp e^{ip^+ x^- + i(px)_\perp} \psi(0^+, x^-, x_\perp) = \frac{1}{2p^+} \sum_s u_{ps} b_{ps},$$

(P11.30)

$$\int dx^- dx_\perp e^{-ip^+ x^- - i(px)_\perp} \bar{\psi}(0^+, x^-, x_\perp) = \frac{1}{2p^+} \sum_s \bar{u}_{ps} b^\dagger_{ps}.$$

(P11.31)

Using

$$\bar{u}_{ps} \gamma^+ u_{ps'} = 2p^+ \delta_{ss'} \implies$$

$$\frac{1}{2p^+} \sum_s \bar{u}_{ps} b^\dagger_{ps} \gamma^+ \frac{1}{2p^+} \sum_{s'} u_{ps'} b_{ps'} = \frac{1}{2p^+} \sum_s b^\dagger_{ps} b_{ps},$$

(P11.32)

we get

$$\frac{1}{2p^+} \int \frac{dp_\perp}{(2\pi)^2} \sum_s b^\dagger_{ps} b_{ps}$$

$$= \int \frac{dp_\perp}{(2\pi)^2} \int dy^- dy_\perp e^{-ip^+ y^- - i(py)_\perp} \bar{\psi}(0^+, y^-, y_\perp)$$

$$\times \gamma^+ \int dx^- dx_\perp e^{ip^+ x^- + i(px)_\perp} \psi(0^+, x^-, x_\perp)$$

$$= \int dx^- dx_\perp \int dy^- e^{ip^+(x^- - y^-)} \bar{\psi}(0^+, y^-, x_\perp) \gamma^+ \psi(0^+, x^-, x_\perp).$$

(P11.33)

Next,

$$\langle P|\bar\psi\left(0^+,y^-,x_\perp\right)\gamma^+\psi\left(0^+,x^-,x_\perp\right)|P\rangle$$
$$=\langle P|e^{iP^+x^-+i(Px)_\perp}\bar\psi\left(0^+,y^--x^-,0_\perp\right)$$
$$\times\gamma^+\psi\left(0^+,0^-,0_\perp\right)e^{-iP^+x^--i(Px)_\perp}|P\rangle$$
$$=\langle P|\psi\left(0^+,y^--x^-,0_\perp\right)\gamma^+\bar\psi\left(0\right)|P\rangle. \qquad \text{(P11.34)}$$

Therefore

$$\frac{1}{2p^+}\int\frac{dp_\perp}{(2\pi)^2}\frac{1}{2}\sum_{spin\,p}\langle P|\sum_s b^\dagger_{ps}b_{ps}|P\rangle$$
$$=\int dx^- dx_\perp dz^- e^{-ip^+z^-}\frac{1}{2}\sum_{spin\,p}\langle P|\bar\psi\left(0^+,z^-,0_\perp\right)\gamma^+\psi\left(0\right)|P\rangle.$$
$$\text{(P11.35)}$$

Recall that the proton state normalization reads

$$\frac{1}{2}\sum_{spin\,p}\langle P|P'\rangle=(2\pi)^3 2P^+\delta(P^+-P'^+)\delta(P^+_\perp-P'_\perp)$$
$$=2P^+\int dx^- dx_\perp e^{i(P-P')^+x^-+i(P-P')_\perp x_\perp}. \qquad \text{(P11.36)}$$

Hence, substituting $p^+=xP^+$

$$f_{q/P}(x,Q^2)=\frac{1}{2}\int\frac{dz^-}{2\pi}e^{-ip^+z^-}\frac{1}{2}\sum_{spin\,p}\langle P|\bar\psi\left(0^+,z^-,0_\perp\right)\gamma^+\psi\left(0\right)|P\rangle$$
$$=\frac{1}{2x}\int^{Q^2}\frac{dp_\perp}{(2\pi)^3}\frac{\frac{1}{2}\sum_{spin\,p}\langle P|\sum_s b^\dagger_s(xP^+,p_\perp)b_s(xP^+,p_\perp)|P\rangle}{\frac{1}{2}\sum_{spin\,p}\langle P|P\rangle}.$$
$$\text{(P11.37)}$$

For the single quark state $|P\rangle=b^\dagger_{\xi P^+,\lambda}|0\rangle$

$$f_{q/q}(x,Q^2)=\delta(\xi-x). \qquad \text{(P11.38)}$$

11.8 Show that in the light cone gauge $A^+=0$ gluon PDF counts the number of gluons in the proton with the proton's longitudinal momentum fraction x.

SOLUTION. The calculation is analogous to the quark case. In the light cone gauge
$$A^+=0 \quad\Longrightarrow\quad F^{+\mu}=\partial^+A^\mu.$$

Using the normalization
$$[a_{ks}a^\dagger_{k's'}]=2k^+(2\pi)^3\delta(k-k')\delta_{ss'}, \qquad \text{(P11.39)}$$

we have

$$F^{+\mu}\left(0^+, x^-, x_\perp\right) = \int \frac{d^2k}{(2\pi)^3} \int_0^{+\infty} \frac{dk^+}{2k^+}$$
$$\times \sum_s (-ik^+ \varepsilon^\mu_{ks} e^{-ik^+ x^- - i(kx)_\perp} a_{ks} + h.c.),$$ (P11.40)

$$\int dx^- dx_\perp e^{ip^+ x^- + i(px)_\perp} F^{+\mu}\left(0^+, x^-, x_\perp\right) = \frac{-i}{2} \sum_s \varepsilon^\mu_{ps} a_{ps}.$$ (P11.41)

Using

$$\sum \varepsilon^\mu_{ps} \varepsilon^*_{ps'\mu} = -\delta_{ss'},$$ (P11.42)

we get

$$\frac{1}{4} \int \frac{dp_\perp}{(2\pi)^2} \sum_s a^\dagger_{ps} a_{ps} = \int dx^- dx_\perp \int dy^- e^{ip^+(x^- - y^-)}$$
$$\times F^{+\mu}\left(0^+, y^-, x_\perp\right) F_\mu{}^+\left(0^+, x^-, x_\perp\right).$$ (P11.43)

Next,

$$\langle P|F^{+\mu}\left(0^+, y^-, x_\perp\right) F_\mu{}^+\left(0^+, x^-, x_\perp\right)|P\rangle$$
$$= \langle P|F^{+\mu}\left(0^+, y^- - x^-, 0_\perp\right) F_\mu{}^+\left(0\right)|P\rangle.$$ (P11.44)

Therefore

$$\frac{1}{4} \int \frac{dp_\perp}{(2\pi)^2} \frac{1}{2} \sum_{spin\,p} \langle P| \sum_s a^\dagger_{ps} a_{ps}|P\rangle$$
$$= \int dx^- dx_\perp dz^- e^{-ip^+ z^-} \frac{1}{2} \sum_{spin\,p} \langle P|F^{+\mu}\left(0^+, z^-, 0_\perp\right) F_\mu{}^+\left(0\right)|P\rangle.$$
(P11.45)

Recall that the proton state normalization reads

$$\frac{1}{2} \sum_{spin\,p} \langle P|P\rangle = 2P^+ \int dx^- dx_\perp.$$ (P11.46)

Hence, substituting $p^+ = xP^+$

$$f_{g/P}(x, Q^2) = \int \frac{dz^-}{2\pi x P^+} e^{-ip^+ z^-} \frac{1}{2} \sum_{spin\,p} \langle P|F^{+\mu}\left(0^+, z^-, 0_\perp\right) F_\mu{}^+\left(0\right)|P\rangle$$
$$= \frac{1}{4\pi x} \int \frac{dp_\perp}{(2\pi)^2} \frac{\frac{1}{2} \sum_{spin\,p} \langle P| \sum_s a^\dagger_s(xP^+, p_\perp) a_s(xP^+, p_\perp)|P\rangle}{\frac{1}{2} \sum_{spin\,p} \langle P|P\rangle}.$$
(P11.47)

For a single gluon state

$$f_{g/g}(x, Q^2) = \delta(\xi - x).$$ (P11.48)

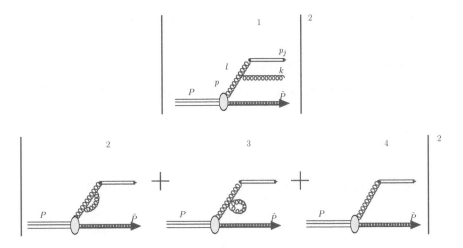

Figure 11.11 Gluon contribution to the evolution of gluon PDF in the light cone gauge. The double line stands for the Wilson line, which is 1 in the light cone gauge.

11.9 Show that

$$f_{\bar{q}/P}(x, Q^2) = -f_{q/P}(-x, Q^2).$$ (P11.49)

11.10 Derive the **momentum sum rule**

$$1 = \int_0^1 z dz (g(z) + \sum_{q,\bar{q}} f_{q/P}(z)).$$ (P11.50)

11.11 Find P_{gg}.

SOLUTION. In the light cone gauge, the Wilson line in (11.35) is 1 identically. Therefore, the evolution comes only from the contribution of the diagrams in Figure 11.11. We start from the PDF $g(x, M_1^2)$ defined at the transverse momentum scale M_1^2, i.e. it includes all gluons in the proton with the transverse momentum squared up to M_1^2. In the loops, we want to integrate over the partons with transverse momenta in the strip $M_1^2 < \vec{l}^2 < M_2^2$. We thus include them into the definition of the PDF and get $g(x, M_2^2)$. The contribution of diagram 1 reads

$$g(x, M_2^2)_1 = \int \frac{dz^-}{2\pi x P^+} e^{-iz^- x P^+} \frac{1}{2} \sum_{spin\ p,\lambda,e} \langle P | F^{+\mu}{}_e \left(0^+, z^-, 0_\perp\right) a_{k\lambda}^{\dagger a} | \tilde{P} \rangle_1$$

$$\times \langle \tilde{P} | a_{k\lambda}^a F_{\mu\perp}^{+e} (0) | \tilde{P} \rangle_1 |_{M_1^2} d\tilde{\rho} \frac{d^4 k \delta(k^2)}{(2\pi)^3} |_{M_1^2 < \vec{k}^2 < M_2^2}.$$ (P11.51)

We denote the vectors according to (11.47) and (11.49). The real correction to the matrix element has the form

$$\langle P|\partial^+ A^{\mu\perp e}\left(0^+, z^-, 0_\perp\right) a_{k\lambda}^{\dagger d}|\tilde{P}\rangle_1$$

$$= -igf^{abc}\langle P|\int d^4x (\partial_\rho A_\nu^a(x)) A^{b\rho}(x) A^{c\nu}(x)\partial^+ A^{\mu\perp e}\left(0^+, z^-, 0_\perp\right) a_{k\lambda}^{\dagger d}|\tilde{P}\rangle$$

$$= g\int d^4x e^{-ikx+i(P-P_n)x} f^{ade}\langle P|A^{a\sigma}(0)|\tilde{P}\rangle$$

$$\times \varepsilon_{k\lambda}^\rho [g_{\sigma\rho}(p_\nu + k_\nu) + (i\partial_\sigma - k_\sigma)g_{\nu\rho} - (i\partial_\rho + p_\rho)g_{\nu\sigma}]\partial^+ D^{\mu\perp\nu}(x - z^-)$$

$$= g\int \frac{d^4l}{(2\pi)^4}\int d^4x e^{-ikx+i(P-\tilde{P})x-il(x-z^-)} f^{ade}\langle P|A^{a\sigma}(0)|\tilde{P}\rangle$$

$$\times \varepsilon_{k\lambda}^\rho [g_{\sigma\rho}(p_\nu + k_\nu) + (l_\sigma - k_\sigma)g_{\nu\rho} - (l_\rho + p_\rho)g_{\nu\sigma}]\frac{-id^{\mu\perp\nu}(l)}{l^2}(-il^+)$$

$$= g\int d^4l\delta(P - \tilde{P} - k - l)e^{il^+z^-} f^{ade}\langle P|A^{a\sigma}(0)|\tilde{P}\rangle$$

$$\times \varepsilon_{k\lambda}^\rho [g_{\sigma\rho}(p_\nu + k_\nu) + (l_\sigma - k_\sigma)g_{\nu\rho} - (l_\rho + p_\rho)g_{\nu\sigma}]\frac{-id^{\mu\perp\nu}(l)}{l^2}(-il^+) \tag{P11.52}$$

$$= ge^{i(p-k)^+z^-} f^{ade}\langle P|A^{a\sigma}(0)|\tilde{P}\rangle\frac{-id^{\mu\perp\nu}(p-k)}{(p-k)^2}(-i(p-k)^+)$$

$$\times \varepsilon_{k\lambda}^\rho [g_{\sigma\rho}(p_\nu + k_\nu) + (p_\sigma - 2k_\sigma)g_{\nu\rho} - (2p_\rho - k_\rho)g_{\nu\sigma}] \tag{P11.53}$$

$$= ge^{-ik^+z^-}\langle P|\partial^+ A^{a\sigma}(z^-)|\tilde{P}\rangle f^{ade}\frac{id^{\mu\perp\nu}(p-k)}{(p-k)^2}\frac{(p-k)^+}{p^+}$$

$$\times \varepsilon_{k\lambda}^\rho [g_{\sigma\rho}(p_\nu + k_\nu) + (p_\sigma - 2k_\sigma)g_{\nu\rho} - (2p_\rho - k_\rho)g_{\nu\sigma}] \tag{P11.54}$$

$$= ge^{-ik^+z^-}\langle P|\partial^+ A^{a\sigma}(z^-)|\tilde{P}\rangle f^{ade}\varepsilon_{k\lambda}^\rho [2g_{\sigma\rho\perp}p_\nu - 2k_{\sigma\perp}g_{\nu\rho} - 2p_\rho g_{\nu\sigma\perp}]$$

$$\times \frac{id^{\mu\perp\nu}(p-k)}{(p-k)^2}\frac{(p-k)^+}{p^+} \tag{P11.55}$$

$$= -2ige^{-ik^+z^-}\langle P|\partial^+ A^{a\sigma}(z^-)|\tilde{P}\rangle f^{ade}\varepsilon_{k\lambda}^\rho [g_{\sigma\rho\perp}p_\nu - k_{\sigma\perp}g_{\nu\rho} - p_\rho g_{\nu\sigma\perp}]$$

$$\times d^{\mu\perp\nu}(p-k)\frac{(y-x)x}{y^2\vec{k}^2}. \tag{P11.56}$$

We have

$$d^{\mu\nu}(p-k)d^{\mu\nu'}(p-k)g_{\mu\mu'\perp} = d^{\nu\nu'}(p-k) + \frac{(p-k)^2}{[(p-k)^+]^2}n_2^\nu n_2^{\nu'}$$

$$= d^{\nu\nu'}(p-k) - \frac{y\vec{k}^2 n_2^\nu n_2^{\nu'}}{x^2(y-x)[P^+]^2}. \tag{P11.57}$$

As a result,

$$g(x, M_2^2)_1 = \int \frac{dz^-}{2\pi x P^+} e^{-iz^- x P^+}$$

$$\times \frac{1}{2} \sum_{spin\, \mathbf{p}, \lambda} e^{-ik^+ z^-} \langle P|\partial^+ A^{a\sigma}(z^-)|\tilde{P}\rangle\langle\tilde{P}|\partial^+ A^{a'\sigma'}(0)|P\rangle|_{M_1^2}$$

$$\times \varepsilon_{k\lambda}^{\rho} \varepsilon_{k\lambda}^{\rho'}[g_{\sigma\rho\perp}p_\nu - k_{\sigma\perp}g_{\nu\rho} - p_\rho g_{\nu\sigma\perp}]$$

$$\times [g_{\sigma'\rho'\perp}p_{\nu'} - k_{\sigma'\perp}g_{\nu'\rho'} - p_{\rho'}g_{\nu'\sigma'\perp}]$$

$$\times 4g^2 f^{ade} f^{a'de}[d^{\nu'\nu}(p-k) - \frac{y\vec{k}^2 n_2^\nu n_2^{\nu'}}{x^2(y-x)[P^+]^2}]$$

$$\times \frac{(y-x)^2 x^2}{y^4 \vec{k}^4} d\tilde{\rho} \frac{dk^+ d\vec{k}}{2k^+(2\pi)^3}|_{M_1^2 < \vec{k}^2 < M_2^2}. \tag{P11.58}$$

The color structures read

$$-\frac{y\vec{k}^2 n_2^\nu n_2^{\nu'}}{x^2(y-x)[P^+]^2} d(k)^{\rho\rho'}[g_{\sigma\rho\perp}p_\nu - k_{\sigma\perp}g_{\nu\rho} - p_\rho g_{\nu\sigma\perp}]$$

$$\times [g_{\sigma'\rho'\perp}p_{\nu'} - k_{\sigma'\perp}g_{\nu'\rho'} - p_{\rho'}g_{\nu'\sigma'\perp}]$$

$$= -\frac{y\vec{k}^2 n_2^\nu n_2^{\nu'}}{x^2(y-x)[P^+]^2} d(k)_{\sigma\perp}^{\rho'} yP^+[g_{\sigma'\rho'\perp}yP^+ - n_{2\rho'}k_{\sigma'\perp}]$$

$$= -\frac{y^3\vec{k}^2 g_{\sigma'_\perp \sigma_\perp}}{x^2(y-x)}, \tag{P11.59}$$

$$d(k)^{\rho\rho'}[g_{\rho\sigma_\perp}p_\nu - k_{\sigma\perp}g_{\nu\rho} - p_\rho g_{\nu\sigma_\perp}]$$
$$\times d^{\nu'\nu}(p-k)[g_{\rho'\sigma'_\perp}p_{\nu'} - k_{\sigma'\perp}g_{\nu'\rho'} - p_{\rho'}g_{\nu'\sigma'_\perp}]$$
$$=[d(k)_{\sigma_\perp}^{\rho'}p_\nu - d(k)_\nu^{\rho'}k_{\sigma\perp} - d(k)^{\rho\rho'}p_\rho g_{\nu\sigma\perp}]$$
$$\times [d^{\nu'\nu}(p-k)g_{\rho'\sigma'_\perp}p_{\nu'} - d_{\rho'}^\nu(p-k)k_{\sigma'\perp} - d_{\sigma'\perp}^\nu(p-k)p_{\rho'}]$$
$$=g_{\sigma'_\perp \sigma_\perp}p_\nu d^{\nu'\nu}(p-k)p_{\nu'} - d(k)_{\sigma_\perp}^{\rho'}p_\nu d_{\rho'}^\nu(p-k)k_{\sigma'\perp}$$
$$- d(k)_{\sigma_\perp}^{\rho'}p_\nu d_{\sigma'\perp}^\nu(p-k)p_{\rho'} - d(k)_{\nu\sigma'_\perp}k_{\sigma\perp}d^{\nu'\nu}(p-k)p_{\nu'}$$
$$+ d(k)_\nu^{\rho'}d_{\rho'}^\nu(p-k)k_{\sigma\perp}k_{\sigma'\perp} + d(k)_\nu^{\rho'}k_{\sigma\perp}d_{\sigma'\perp}^\nu(p-k)p_{\rho'}$$
$$- d(k)_{\sigma'_\perp}^\rho p_\rho d_{\sigma\perp}^\nu(p-k)p_{\nu'} + d(k)^{\rho\rho'}p_\rho d_{\rho'\sigma\perp}(p-k)k_{\sigma'\perp}$$
$$+ d(k)^{\rho\rho'}p_\rho p_{\rho'}g_{\sigma'_\perp \sigma_\perp} \tag{P11.60}$$

$$= g_{\sigma'_\perp \sigma_\perp} \vec{k}^{\,2} [\frac{y^2}{x(y-x)} - \frac{y^2}{(y-x)^2}]$$

$$+ 2[1 - \frac{y}{(y-x)} - \frac{y}{x} + \frac{y^2}{(y-x)x}] k_{\sigma'_\perp} k_{\sigma_\perp} \tag{P11.61}$$

$$= g_{\sigma'_\perp \sigma_\perp} \vec{k}^{\,2} [\frac{y^2}{x(y-x)} - \frac{y^2}{(y-x)^2} - 1] \tag{P11.62}$$

$$= \frac{y^2}{x(y-x)} g_{\sigma'_\perp \sigma_\perp} \vec{k}^{\,2} [1 - \frac{\frac{x}{y}}{1 - \frac{x}{y}} - \frac{x}{y}(1 - \frac{x}{y})]. \tag{P11.63}$$

Here we used the convolutions

$$p_\nu d^{\nu' \nu}(p-k) p_{\nu'} = 2\frac{y}{x}(pk) = \frac{y^2}{x(y-x)} \vec{k}^{\,2}, \tag{P11.64}$$

$$d(k)^{\rho'}_\nu d^\nu_{\rho'}(p-k) = g^{\rho'}_\nu d^\nu_{\rho'}(p-k) = 2, \tag{P11.65}$$

$$d(k)^{\rho\rho'} p_\rho p_{\rho'} = -2\frac{y(pk)}{(y-x)} = -\frac{y^2}{(y-x)^2} \vec{k}^{\,2}, \tag{P11.66}$$

$$-d(k)^{\rho'}_{\sigma_\perp} p_\nu d^\nu_{\rho'}(p-k) = -d(k)^{\rho'}_{\sigma_\perp}(p_{\rho'} - \frac{p_{\rho'} y}{x})$$

$$= \frac{x-y}{x}\frac{y}{y-x} k_{\sigma_\perp} = -\frac{y}{x} k_{\sigma_\perp}, \tag{P11.67}$$

$$-d(k)^{\rho'}_{\sigma_\perp} p_\nu d^\nu_{\sigma'\perp}(p-k) p_{\rho'} = \frac{y}{y-x} k_{\sigma_\perp}(\frac{-y k_{\sigma'\perp}}{x}) = \frac{y^2}{(y-x)x} k_{\sigma_\perp} k_{\sigma'\perp}, \tag{P11.68}$$

$$d(k)^{\rho'}_\nu d^\nu_{\sigma'\perp}(p-k) p_{\rho'} = d^\nu_{\sigma'\perp}(p-k)(p_\nu - \frac{y k_\nu}{(y-x)}) = -\frac{-k_{\sigma'\perp} y}{x}$$

$$- \frac{y k_{\sigma'\perp}}{(y-x)} + \frac{y(y-x)}{(y-x)}\frac{-k_{\sigma'\perp}}{x} = -\frac{y}{(y-x)} k_{\sigma'\perp}. \tag{P11.69}$$

Finally,

$$g(x, M_2^2)_1 = \int \frac{dz^-}{2\pi y P^+} e^{-iz^- y P^+}$$

$$\times \frac{1}{2} \sum_{spin\, p} \langle P|\partial^+ A^{a\sigma}(z^-)|\tilde{P}\rangle \langle \tilde{P}|\partial^+ A^{a\sigma'}(0)|P\rangle|_{M_1^2}$$

$$\times \frac{y^2}{x(y-x)} g_{\sigma'_\perp \sigma_\perp} \vec{k}^{\,2} [\frac{1 - \frac{x}{y}}{\frac{x}{y}} + \frac{\frac{x}{y}}{1 - \frac{x}{y}} + \frac{x}{y}(1 - \frac{x}{y})]$$

$$\times 2 N_c \frac{\alpha_s}{2\pi} \frac{(y-x)^2 x}{y^3 \vec{k}^{\,4}} d\tilde{\rho} \frac{dy d\vec{k}^{\,2}}{y-x}|_{M_1^2 < \vec{k}^{\,2} < M_2^2}. \tag{P11.70}$$

Hence,

$$g(x, M_2^2)_1 = \frac{\alpha_s}{2\pi} \int_x^1 \frac{dy}{y} g(y, M_1^2) P_1(\frac{y}{x}) \ln \frac{M_2^2}{M_1^2}, \qquad \text{(P11.71)}$$

$$P_1(z) = 2N_c[\frac{1-z}{z} + \frac{z}{1-z} + (1-z)z]. \qquad \text{(P11.72)}$$

Therefore

$$P_{gg}(z) = 2N_c[\frac{1-z}{z} + \frac{z}{(1-z)_+} + (1-z)z] + A\delta(1-z). \qquad \text{(P11.73)}$$

The contribution of the virtual corrections to the gluon splitting function is proportional to $\delta(1-z)$. One can find them from the requirement that the sum of the longitudinal momentum fractions of all partons equals 1 (P11.50), i.e. they carry the total proton's momentum:

$$1 = \int_0^1 zdz(g(z) + \Sigma(z)). \qquad \text{(P11.74)}$$

Using singlet DGLAP equation (11.72) this condition leads to

$$0 = \int_0^1 xdx \int_x^1 \frac{dy}{y} \left\{ [P_{qq}(\frac{x}{y}) + P_{gq}(\frac{x}{y})]\Sigma(y, M^2) \right.$$
$$\left. + [2n_f P_{qg}(\frac{x}{y}) + P_{gg}(\frac{x}{y})]g(y, M^2) \right\}. \qquad \text{(P11.75)}$$

Thus for a single gluon state

$$\Sigma = 0, \quad g = \delta(y - \xi), \qquad \text{(P11.76)}$$

$$0 = \int_0^\xi \frac{xdx}{\xi} \left\{ 2n_f P_{qg}(\frac{x}{\xi}) + P_{gg}(\frac{x}{\xi}) \right\} = \xi \int_0^1 zdz \left\{ 2n_f P_{qg}(z) + P_{gg}(z) \right\}$$
$$= \xi \int_0^1 zdz \left\{ 2n_f \frac{1}{2} [z^2 + (1-z)^2] \right.$$
$$\left. + 2N_c[\frac{1-z}{z} + \frac{z}{(1-z)_+} + (1-z)z] + A\delta(1-z) \right\} \qquad \text{(P11.77)}$$

$$= \xi \left\{ \frac{n_f}{3} - \frac{11N_c}{6} + A \right\} \quad \Longrightarrow \quad A = \frac{\beta_0}{2}. \qquad \text{(P11.78)}$$

11.12 Find the **Mellin moments for the DGLAP splitting functions**.

SOLUTION.

$$P_{qq}^{(n)} = C_F \int_0^1 \frac{dz}{z} z^n \left[\frac{1+z^2}{(1-z)_+} + \frac{3}{2}\delta(1-z) \right] \tag{P11.79}$$

$$= C_F \int_0^1 dz \left[-\frac{1-z^{n-1}}{1-z} - \frac{1-z^{n+1}}{1-z} + \frac{3}{2}\delta(1-z) \right] \tag{P11.80}$$

$$= C_F \left[-\psi(n) - \psi(n+2) + 2\psi(1) + \frac{3}{2} \right], \tag{P11.81}$$

$$P_{gq}^{(n)} = C_F \left[\frac{2}{n-1} - \frac{2}{n} + \frac{1}{n+1} \right], \tag{P11.82}$$

$$P_{qg}^{(n)} = \frac{1}{2} \left[\frac{1}{n} - \frac{2}{n+1} + \frac{2}{n+2} \right], \tag{P11.83}$$

$$P_{gg}^{(n)} = 2N_c \left[\frac{1}{n-1} - \frac{1}{n} - \psi(n+1) + \psi(1) + \frac{1}{n+1} - \frac{1}{n+2} \right] + \frac{\beta_0}{2}, \tag{P11.84}$$

$$\psi(n) = \psi(n+1) - \frac{1}{n} = \int_0^1 dz \frac{1-z^{n-1}}{1-z} - \gamma = \frac{d\ln\Gamma(z)}{dz}. \tag{P11.85}$$

11.13 Estimate the proton's momentum fraction carried by gluons in the asymptotic Bjorken limit.

SOLUTION. The proton's momentum fractions carried by gluons and quarks are given by the second Mellin moments of the gluon and singlet distributions. The DGLAP equations for the second moments of the gluon and singlet distribution functions read

$$\frac{d}{d\ln Q^2} \begin{pmatrix} \Sigma^{(2)} \\ g^{(2)} \end{pmatrix} = \frac{\alpha_s(Q^2)}{2\pi} \begin{pmatrix} -\frac{4}{3}C_F & \frac{n_f}{3} \\ \frac{4}{3}C_F & -\frac{n_f}{3} \end{pmatrix} \begin{pmatrix} \Sigma^{(2)} \\ g^{(2)} \end{pmatrix}. \tag{P11.86}$$

The first eigenvalue of this system is zero. It is consistent with momentum conservation rule (P11.50)

$$\Sigma^{(2)} + g^{(2)} = 1. \tag{P11.87}$$

The second eigenvalue gives

$$(\Sigma^{(2)} - \frac{n_f}{4C_f}g^{(2)}) = -\frac{2(4C_f + n_f)}{3\beta_0 \ln\frac{Q^2}{\Lambda^2}}(\Sigma^{(2)} - \frac{n_f}{4C_f}g^{(2)}), \tag{P11.88}$$

$$(\Sigma^{(2)} - \frac{n_f}{4C_f}g^{(2)})(Q^2) = (\Sigma^{(2)} - \frac{n_f}{4C_f}g^{(2)})(Q_0^2)$$

$$\times \left[\frac{\ln\frac{Q_0^2}{\Lambda_{QCD}^2}}{\ln\frac{Q^2}{\Lambda_{QCD}^2}} \right]^{2\frac{4C_f+n_f}{3\beta_0}} \xrightarrow[Q\to\infty]{} 0. \tag{P11.89}$$

Therefore, in the asymptotic region $Q \to 0$

$$\frac{\Sigma^{(2)}}{g^{(2)}} = \frac{n_f}{4C_f}. \tag{P11.90}$$

Hence, for 5 active flavors the gluons carry $\frac{16}{16+15} \simeq 50\%$ of the proton's momentum.

11.14 Find the **gluon PDF at low** x.

SOLUTION. In the low z region

$$P_{qq}(z) \simeq C_F, \quad P_{qg}(z) \simeq \frac{1}{2}, \quad P_{gg}(z) = 2N_c \frac{1}{z} > P_{gq}(z) \simeq C_F \frac{2}{z}. \tag{P11.91}$$

Therefore, we can neglect the quark PDF compared to the gluon one and get the equation

$$\frac{dg(x, M^2)}{d\ln M^2} = \frac{\alpha_s(M^2)}{2\pi} 2N_c \int_x^1 \frac{dy}{x} g(y, M^2) \quad \Longleftrightarrow \tag{P11.92}$$

$$\frac{d^2 x g(x, M^2)}{d\ln M^2 d\ln \frac{1}{x}} = \frac{\alpha_s(M^2)N_c}{\pi} x g(y, M^2). \tag{P11.93}$$

In the moment space

$$\frac{dg^{(n)}(M^2)}{d\ln M^2} = \frac{\alpha_s(M^2)N_c}{\pi} \frac{g^{(n)}}{n-1} \quad \Longrightarrow \tag{P11.94}$$

$$g^{(n)}(M_2^2) = g^{(n)}(M_1^2) \left(\frac{\alpha_s(M_2^2)}{\alpha_s(M_1^2)}\right)^{-\frac{N_c}{\pi\beta_0}\frac{1}{n-1}}. \tag{P11.95}$$

Taking the inverse Mellin transform, we get

$$g(x, M_2^2) = \frac{1}{2\pi i} \int_{-i\infty}^{+i\infty} dn g^{(n)}(M_1^2) x^{-n} \left(\frac{\alpha_s(M_2^2)}{\alpha_s(M_1^2)}\right)^{-\frac{N_c}{\pi\beta_0}\frac{1}{n-1}} \tag{P11.96}$$

$$= \frac{1}{2\pi i} \int_{-i\infty}^{+i\infty} dn g^{(n)}(M_1^2) e^{f(n)}, \tag{P11.97}$$

$$f(n) = n \ln \frac{1}{x} - \frac{N_c}{\pi\beta_0} \frac{1}{n-1} \ln \frac{\alpha_s(M_2^2)}{\alpha_s(M_1^2)}. \tag{P11.98}$$

We assume that $M_2^2 \gg M_1^2$ and $x \ll 1$. Then one can take this integral in the saddle point approximation. The saddle points are located at

$$f'(n) = 0 \quad \Longrightarrow \quad n_\pm = 1 \pm \left[\frac{N_c}{\pi\beta_0} \frac{\ln \frac{\alpha_s(M_1^2)}{\alpha_s(M_2^2)}}{\ln \frac{1}{x}}\right]^{\frac{1}{2}}, \tag{P11.99}$$

among which only n_+ can be the uppermost point on the path from $-i\infty$ to $+i\infty$. Therefore, we take n_+ as the saddle point on our contour.

$$f''(n_+) = \frac{N_c}{\pi\beta_0}\frac{2}{(n_+ - 1)^3}\ln\frac{\alpha_s(M_1^2)}{\alpha_s(M_2^2)} = 2\left[\frac{\ln^3\frac{1}{x}}{\frac{N_c}{\pi\beta_0}\ln\frac{\alpha_s(M_1^2)}{\alpha_s(M_2^2)}}\right]^{\frac{1}{2}} > 0,$$

(P11.100)

$$f(n) = f(n_+) + \frac{1}{2}f''(n_+)(n - n_+)^2, \quad n - n_+ = i\rho,$$

(P11.101)

$$f(n_+) = n_+\ln\frac{1}{x} + \frac{N_c}{\pi\beta_0}\frac{1}{n_+ - 1}\ln\frac{\alpha_s(M_1^2)}{\alpha_s(M_2^2)}$$

(P11.102)

$$= \ln\frac{1}{x} + 2\left[\frac{N_c}{\pi\beta_0}\ln\frac{1}{x}\ln\frac{\alpha_s(M_1^2)}{\alpha_s(M_2^2)}\right]^{\frac{1}{2}},$$

(P11.103)

$$g(x, M_2^2) = \frac{e^{f(n_+)}}{2\pi}g^{(n_+)}(M_1^2)\int_{-\infty}^{+\infty}d\rho e^{-\frac{1}{2}f''(n_+)\rho^2}$$

(P11.104)

$$= \frac{e^{f(n_+)}}{2\pi}g^{(n_+)}(M_1^2)\sqrt{\frac{2\pi}{f''(n_+)}},$$

(P11.105)

$$xg(x, M_2^2) = g^{(n_+)}(M_1^2)\frac{e^{2\sqrt{\frac{N_c}{\pi\beta_0}\ln\frac{1}{x}\ln\frac{\alpha_s(M_1^2)}{\alpha_s(M_2^2)}}}}{\sqrt{4\pi}}\left[\frac{\frac{N_c}{\pi\beta_0}\ln\frac{\alpha_s(M_1^2)}{\alpha_s(M_2^2)}}{\ln^3\frac{1}{x}}\right]^{\frac{1}{4}}.$$

(P11.106)

If one neglects running of the coupling

$$\frac{dg^{(n)}(M^2)}{d\ln M^2} = \frac{\alpha_s N_c}{\pi}\frac{g^{(n)}}{n-1}$$

(P11.107)

$$\implies \quad g^{(n)}(M_2^2) = g^{(n)}(M_1^2)\left(\frac{M_2^2}{M_1^2}\right)^{\frac{\alpha_s N_c}{\pi}\frac{1}{n-1}}$$

(P11.108)

$$\implies \quad xg(x, M_2^2) = g^{(\tilde n_+)}(M_1^2)\frac{e^{2\sqrt{\frac{\alpha_s N_c}{\pi}\ln\frac{1}{x}\ln\frac{M_2^2}{M_1^2}}}}{\sqrt{4\pi}}\left[\frac{\frac{\alpha_s N_c}{\pi}\ln\frac{M_2^2}{M_1^2}}{\ln^3\frac{1}{x}}\right]^{\frac{1}{4}},$$

(P11.109)

which is a series in $\left(\frac{\alpha_s N_c}{\pi}\ln\frac{1}{x}\ln\frac{M_2^2}{M_1^2}\right)^k$ called the **double logarithm approximation**. One can observe the rise of F_2 as $x \to 0$ and $Q^2 \to \infty$ in the small x region in Figure 11.4.

11.15 Show that the Bjorken limit corresponds to the light cone limit $x^2 \to 0$ for the hadronic tensor.

SOLUTION. The hadronic tensor can be rewritten as a matrix element of current commutator

$$W_{\mu\nu} = \frac{1}{4\pi m_p} \frac{1}{2} \sum_{spin\,p} \langle P|J_\nu(0)|X\rangle\langle X|J_\mu(0)|P\rangle (2\pi)^4 \delta(q+P-P') d\rho_X$$

$$= \frac{1}{4\pi m_p} \int dx e^{ix(q+P-P')} \frac{1}{2} \sum_{spin\,p} \langle P|J_\nu(0)|X\rangle\langle X|J_\mu(0)|P\rangle d\rho_X$$

$$= \left| O(x) = e^{i\hat{P}x} O(0) e^{-i\hat{P}x} \right| = \frac{1}{4\pi m_p} \int dx e^{ix(q+P-P')}$$

$$\times \frac{1}{2} \sum_{spin\,p} \langle P|e^{-iPx} J_\nu(x) e^{iP'x}|X\rangle\langle X|J_\mu(0)|P\rangle d\rho_X$$

$$= \frac{1}{4\pi m_p} \int dx e^{ixq} \frac{1}{2} \sum_{spin\,p} \langle P|J_\nu(x)|X\rangle\langle X|J_\mu(0)|P\rangle d\rho_X$$

$$= \left| \int |X\rangle\langle X| d\rho_X = 1 \right|$$

$$= \frac{1}{4\pi m_p} \int dx e^{ixq} \frac{1}{2} \sum_{spin\,p} \langle P|J_\nu(x)J_\mu(0)|P\rangle$$

$$= \frac{1}{4\pi m_p} \int dx e^{ixq} \frac{1}{2} \sum_{spin\,p} \langle P|[J_\nu(x)J_\mu(0)]|P\rangle. \qquad \text{(P11.110)}$$

where we used the completeness of the states X. In the final equality, we added the extra term

$$\sim \int dx e^{ixq} \langle P|J_\mu(0)J_\nu(x)|P\rangle = \int d\rho_X dx e^{ixq} \langle P|J_\mu(0)|X\rangle\langle X|J_\nu(x)|P\rangle$$

$$= \int d\rho_X dx e^{ix(q+P'-P)} \langle P|J_\mu(0)|X\rangle\langle X|J_\nu(0)|P\rangle \sim \delta(q+P'-P)$$

$$\sim |\text{proton rest frame}| \sim \delta(q^0 - m_p + P'^0) = |q^0 \to \infty, P'^0 > 0| = 0 \qquad \text{(P11.111)}$$

to have a commutator. For causally independent events the commutator is 0. Therefore, the hadronic tensor $W_{\mu\nu}$ vanishes for $x^2 < 0$. In the frame where q^μ has only ^{0} and ^{3} components, we introduce the light cone variables

$$q^+ = q^0 + q^3, \quad q^- = \frac{q^0 - q^3}{2}, \quad \vec{q}_\perp = 0, \quad q^2 = 2q^+ q^-, \qquad \text{(P11.112)}$$

$$x^2 = 2x^+ x^- - \vec{x}_\perp^2 < 2x^+ x^-, \quad qx = q^+ x^- + q^- x^+. \qquad \text{(P11.113)}$$

In the Bjorken limit in the proton rest frame

$$q^0 = \frac{q^+}{2} + q^- = \nu \to \infty, \tag{P11.114}$$

$$q^3 = \frac{q^+}{2} - q^- = \sqrt{\nu^2 + Q^2} \simeq \nu + \frac{Q^2}{2\nu} = \nu + x m_p, \tag{P11.115}$$

$$q^+ \simeq 2\nu, \quad q^- \simeq -\frac{x m_p}{2}. \tag{P11.116}$$

The dominant contribution to the hadronic tensor $W_{\mu\nu}$ (P11.110) comes from the region where

$$e^{iqx} \sim 1 \implies x^- \lesssim \frac{1}{q^+} \simeq \frac{1}{\nu}, \quad x^+ \lesssim \frac{1}{|q^-|} \simeq \frac{1}{m_p x} \implies x^2 \lesssim \frac{1}{Q^2} \to 0. \tag{P11.117}$$

Therefore, the Bjorken limit corresponds to the light cone limit of the hadronic tensor.

11.16 Estimate the cross-section for inclusive Higgs boson production at LHC: $pp \to H + X$.

    a. Use the quark - Higgs interaction vertex

$$\mathcal{L}_{q\bar{q}H} = -\frac{m_Q}{v} \int dx \bar{\psi}(x)\psi(x)H(x), \quad v = 246\,GeV, \tag{P11.118}$$

to calculate the $gg \to H$ cross-section as a function of the Higgs mass $m_H = \sqrt{\hat{s}}$. Take into account the top quark loop only and assume $m_t \gg m_H$. Get

$$\hat{\sigma}(\hat{s}) = \frac{\alpha_s^2}{\pi} \frac{1}{64} \frac{1}{9} \frac{m_H^2}{v^2} \delta(\hat{s} - m_H^2). \tag{P11.119}$$

    b. Download any PDF database, e.g.
CTEQ: http://www.physics.smu.edu/scalise/cteq/#PDFs,
MMHT: https://www.hep.ucl.ac.uk/mmht/code.shtml,
LHAPDF: https://lhapdf.hepforge.org/
and learn how to work with it, i.e. check the sum rules (11.36), (P11.50), reproduce the plots in Figure 11.5.

    c. Calculate the integrated gluon-gluon luminosity $\frac{dL_{gg}}{d\hat{s}}$ (11.92). Draw a plot of $\frac{dL_{gg}}{d\hat{s}}$ vs $\sqrt{\hat{s}}$ for $S = 14$ TeV. Compare it with the green line in Figure 11.8.

    d. Draw a plot of the $\sigma_{pp \to H+X}(m_H)$ for 100 GeV $< m_H <$ 500 GeV for $S = 14$ TeV.

    e. Draw a plot of the $\sigma_{pp \to H+X}(S)$ for $m_H = 125$ GeV and 1 TeV $< \sqrt{S} <$ 30 TeV.

f. Estimate the cross-section of the background QCD processes.

g. Explain whether the Higgs boson can decay into 3 gluons, or, into 3 photons.[1]

There are different approaches to the DGLAP equations. One can expand the hadronic tensor into the series of local operators and derive DGLAP equations as the Kallan-Symanzik equations for the Green functions of these operators. This approach is discussed in detail in (PS95). One can derive these equations in the old-fashioned perturbation theory (IFL10), or in the light-cone perturbation theory (KL12). The cited books discussing these 3 approaches together with (Col13) can give the comprehensive outlook on the evolution equations. Here we used Wilson lines and followed the logic of (BJY03) and (Mar08). Further applications of the collinear factorization to the collider phenomenology the reader can find in (CHS07) and (Mal18), which can help to solve the last problem of this chapter. As a very helpful practical tool giving access to PDF datasets one can use the Mathematica package ManeParse (CGO17).

FURTHER READING

[A⁺11] F. D. Aaron et al. Measurement of the inclusive e\pmp scattering cross section at high inelasticity y and of the structure function F_L. *Eur. Phys. J. C*, 71:1579, 2011.

[B⁺15] Richard D. Ball et al. Parton distributions for the LHC Run II. *JHEP*, 04:040, 2015.

[BJY03] Andrei V. Belitsky, X. Ji, and F. Yuan. Final state interactions and gauge invariant parton distributions. *Nucl. Phys. B*, 656:165–198, 2003.

[C⁺01] S. Chekanov et al. Measurement of the neutral current cross-section and F(2) structure function for deep inelastic e + p scattering at HERA. *Eur. Phys. J. C*, 21:443–471, 2001.

[CGO17] D. B. Clark, E. Godat, and F. I. Olness. ManeParse: A mathematica reader for Parton distribution functions. *Comput. Phys. Commun.*, 216:126–137, 2017.

[CHS07] John M. Campbell, J. W. Huston, and W. J. Stirling. Hard interactions of quarks and gluons: A primer for LHC physics. *Rept. Prog. Phys.*, 70:89, 2007.

[Col13] John Collins. *Foundations of perturbative QCD*, volume 32. Cambridge University Press, 11, 2013.

---

[1] This question was asked by A. I. Milstein who passed by at the time of the exam on this course. Although problem 2.1 in Chapter 2 was compulsory, it was not an easy question.

[H⁺21] Tie-Jiun Hou et al. New CTEQ global analysis of quantum chromodynamics with high-precision data from the LHC. *Phys. Rev. D*, 103(1):014013, 2021.

[HWNO19] T. J. Hobbs, Bo-Ting Wang, Pavel M. Nadolsky, and Fredrick I. Olness. Charting the coming synergy between lattice QCD and high-energy phenomenology. *Phys. Rev. D*, 100(9):094040, 2019.

[IFL10] Boris Lazarevich Ioffe, Victor Sergeevich Fadin, and Lev Nikolaevich Lipatov. *Quantum chromodynamics: Perturbative and nonperturbative aspects*. Cambridge University Press, 2010.

[KL12] Yuri V. Kovchegov and Eugene Levin. *Quantum chromodynamics at high energy*, volume 33. Cambridge University Press, 8, 2012.

[Mal18] F. Maltoni. Basics of QCD for the LHC: $pp \to H + X$ as a case study. *CERN Yellow Rep. School Proc.*, 2:41–67, 2018.

[Mar08] Alan D. Martin. Proton structure, Partons, QCD, DGLAP and beyond. *Acta Phys. Polon. B*, 39:2025–2062, 2008.

[PS95] Michael E. Peskin and Daniel V. Schroeder. *An introduction to quantum field theory*, chapter 18. Addison-Wesley, Reading, USA, 1995.

Index

$R = \frac{\sigma_{e^+e^- \to hadrons}}{\sigma_{e^+e^- \to \mu^+\mu^-}}$
    2-loop, 181
Λ-term, 220
Λ_{QCD}, 164
    n_f-dependence, 177
α_s, 164
    2-loop, 178
β-function, 163
    2-loop in $\overline{MS}$, 178
σ-model, 239
$+$ prescription, 279

angular momentum tensor, 213
anomalous dimension, 191
    mass, 200
    mass at 2-loop, 202
area derivative, 66
area law, 144
asymptotic freedom, 164
axion, 265

baryon
    interval rule mass formula, 25
    number, 244
    magnetic moment, 29
    mass splitting due to em
        interaction, 31
Becchi-Rouet-Stora-Tyutin
        transformation, 128
Belinfante construction, 213
Belinfante tensor, 213
    in QCD, 216
Bjorken
    limit, 267
    scaling, 270
    variable, 267
Britto-Cachazo-Feng-Witten
        recursion relation, 122

Callan-Gross relation, 270
Callan-Symanzik equation, 190
    for local operators, 192
    with mass, 193
Casimir operator
    cubic, 5
    quadratic, 4
Cayley representation, 12
center of $SU(N)$ - Z_N, 2
chiral
    effective Lagrangian, 247
    effective theory, 247
    expansion, 250
    symmetry, 246
classical field, 159
coefficient functions, 283
collinear factorization, 284
color, 17
color group, 39
completeness relation, 8
confinement, 143
    Wilson criterion of, 144
contact terms, 76
Cornell potential, 144

decoupling identity, 118
dilatation current, 210
dimensional transmutation, 207
Dokshitzer - Gribov - Lipatov -
        Altarelli - Parisi equations,
    280
double logarithm approximation, 300

electric dipole moment of neutron,
    260
energy-momentum tensor, 213, 217
Euclidean time, 139
evolution operator, 69